Date Due

Fast Fourier Transforms

Studies in Advanced Mathematics

Series Editor

Steven G. Krantz
Washington University in St. Louis

Editorial Board

R. Michael Beals
Rutgers University

Dennis de Turck
University of Pennsylvania

Ronald DeVore
University of South Carolina

L. Craig Evans
University of California at Berkeley

Gerald B. Folland
University of Washington

William Helton
University of California at San Diego

Norberto Salinas
University of Kansas

Michael E. Taylor
University of North Carolina

Volumes in the Series

Real Analysis and Foundations, *Steven G. Krantz*

CR Manifolds and the Tangential Cauchy–Riemann Complex, *Albert Boggess*

Elementary Introduction to the Theory of Pseudodifferential Operators,
   Xavier Saint Raymond

Fast Fourier Transforms, *James S. Walker*

Measure Theory and Fine Properties of Functions, *L. Craig Evans and
   Ronald Gariepy*

JAMES S. WALKER
University of Wisconsin

Fast Fourier Transforms

CRC PRESS
Boca Raton   Ann Arbor   Boston   London

Library of Congress Cataloging-in-Publication Data

Walker, James S.
    Fast Fourier Transforms / James S. Walker.
        p.   cm.
    Includes bibliographical references and index.
    ISBN 0-8493-7154-6
    1. Fourier transformations.   I. Title.
QA403.W33   1991
515'.723–dc20   91-28128
                                                            CIP

This book was formatted with LaTeX by Archetype Publishing Inc., P.O. Box 6567, Champaign, IL 61821.

Direct all inquiries to CRC Press, Inc., 2000 Corporate Blvd., N.W., Boca Raton, Florida, 33431.

International Standard Book Number 0-8493-7154-6

Printed in the United States of America   3 4 5 6 7 8 9 0

Printed on acid-free paper

To my mother,

and the memory of my father.

Contents

Introduction

This book is intended to be an introduction to the *fast fourier transform* (FFT) and some of its applications. Since its creation in the mid-1960s, the field of computerized Fourier analysis, based on the FFT, has been near the heart of a revolution in scientific understanding made possible through the aid of the digital computer. No single book can describe all the features of this new science. I have tried to show how the FFT arises from classical Fourier analysis and describe how the FFT is used to do Fourier analysis.

I wrote this book for as wide an audience as I could, given the mathematical requirements needed to appreciate the basic definitions of Fourier analysis. Hopefully, the text will be helpful to students of electrical engineering, optical engineering, physics, physical chemistry, and mathematics. The mathematical prerequisites are a solid understanding of calculus and, for some of the applications, differential equations. To understand the proofs in Sections 4.9 and 5.8–5.12, a first course in advanced calculus would be ideal, although a determined reader could gain the necessary knowledge by pursuing a few references (for example, [Ba] or [Ru,2]). I have generally tried to keep the mathematical proofs to a bare minimum. Readers who wish to pursue the pure mathematical theory might begin with the following reference:

James S. Walker, *Fourier Analysis*. Oxford University Press, 1988.

This text is referred to as [Wa] in this book. Since I wrote [Wa] for the purpose of explaining the mathematical theory of classical Fourier analysis, I saw no need to reproduce its pure mathematical arguments in this book.

What makes this book unique among the vast literature on the FFT is that it comes together with computer software, *Fourier Analysis Software* (*FAS*) for doing Fourier analysis on any PC (using DOS version 2.1 and up). Using *FAS*, almost all readers will be able to immediately generate computer images of Fourier series, sine and cosine series, Fourier transforms, convolutions, and all the other aspects of Fourier analysis described in the text.

Here is a summary of the main topics covered. In Chapter 1, I discuss the principal features of Fourier series. The emphasis here is on using *FAS* to gain a firm grasp of the main aspects of Fourier series, as well as Fourier sine and cosine series. In Chapter 2, the digital (discrete) version of Fourier analysis is introduced, showing how the Fourier series can be put into a discrete form intended for computer computation. An efficient means for carrying out this computer computation is called a *Fast Fourier Transform* or FFT. Chapter 3 is an introduction to one type of FFT. Although there are a great many FFT algorithms, I discuss just one. Any readers who are familiar with Bracewell's book on the Hartley transform [Br] will recognize the debt I owe to him and his colleague at Stanford, Oscar Buneman. The work done by Buneman has significantly improved the original FFT algorithms. Suprisingly, relatively few (if any) books on the FFT discuss Buneman's methods; I have tried to give them the prominence that I think they deserve. Besides the basic FFT algorithm, I also describe the standard method for computing a real FFT and for computing discrete sine and discrete cosine transforms efficiently.

Since the purpose of the FFT is to do Fourier analysis, Chapter 4 consists of applications of Fourier series. The classic Fourier series solutions to the heat and wave equations are described with the principal emphasis being on using *FAS* to study the time evolution of these solutions. Another area where the computer allows one to do much more than the typical textbook presentations is the famous *particle in a box problem*, in one dimension, from quantum mechanics. Section 3 shows how *FAS* allows one to study the time evolution of a potential-free quantum mechanical particle. This problem, which has important applications to electron diffraction, is too often treated in a sterile fashion with too much emphasis on stationary states. Many students are left with the impression that the stationary states are the *only* states. The rest of Chapter 4 describes aspects of Fourier series that are important in signal processing, in particular, emphasizing the concept of *filtering* of Fourier series. *FAS* makes it possible for students to see with their own eyes why hanning and Hamming filters, as well as other types of filters, are used. The subtle notion of *point spread functions* (*kernels*) should be easier to learn with the aid of *FAS*, too.

In Chapter 5, I take up the other main area of Fourier analysis, *Fourier transforms*. Here *FAS* is used to aid in understanding the fundamentals of Fourier transforms, Fourier inversion, and convolution. In particular, the difficult concept of convolution integrals should be more accessible with the help of *FAS*. A student will be able to see very quickly what the convolution of two functions looks like. The examples I have treated (heat equation, Laplace's equation, potential-free Schrödinger equation) will show the practical importance of convolutions. I have made a point of showing that *FAS* can be used to implement the classic convolution solutions to these problems in a practical way. The second half of Chapter 5 (Sections 5.9–5.12) covers some very important, but sometimes difficult, material on Poisson summation and sampling theory, stressing the close connection between these two topics. It would be hard to underestimate the importance of sampling theory in modern electrical engineering and communication systems design. *FAS* should help make some of the principles of sampling theory easier to understand.

The book concludes with a chapter on Fourier optics. I think that it is important to describe in as much detail as possible one major area of application of computerized Fourier analysis. In Chapter 6, the topics of Fresnel diffraction, diffraction from circular apertures, interference, diffraction gratings, Fourier transforming properties of a lens, and imaging with a single lens are discussed. Fourier optics is an essential element of modern optics. It plays a vital role in understanding diffraction and imaging. And these two areas, through their roles in crystallography, chemical analysis, and optical and electron microscopy, underlie a great deal of what we have learned about the physical world during the last century.

Acknowledgments

I would like to thank Steve Krantz for suggesting this project to me and for all his help and encouragement. Thanks also to Guido Weiss for his assistance during my sabbatical year at Washington University and for his many writings in Fourier analysis, from which I have learned so much. Also, a thank you to all the other members of the Washington University Mathematics Department who made my stay there such a pleasant one. And thanks to the faculty of University of Wisconsin–Eau Claire for granting me a sabbatical leave to work on this project. My colleague at UWEC, Bob Langer, has my gratitude for introducing me to the world of computers.

My students have been very influential in this project, especially in developing the software, and I would like to thank them (my apologies if I have left anyone out). Thanks to Brian Hicks, Dave Larson, Eduardo Fernandez, May Vang, Lo Pao Vang, Brad Hinaus, Dave Kruger, Scott Nelson, Ty Prosa, Steve Kortenkamp, Michael Ohl, Joel Woletz, Michael Meyer, Dena Laramy, and John Schmidt for all their interest and help.

My thanks to Lori Pickert for an excellent job of typesetting, turning my raw TEX file into a polished book. Thanks to Sharon Forsyth for her excellent drawings in Chapter 6. And thanks to Wayne Yuhasz and everyone else at CRC Press for their enthusiastic support for this project.

Finally, I want to thank my wife, Dawn Manire, for her steady love and support. She has shown extraordinary patience in putting up with a new member of our household (my PC, known affectionately as "Sara").

— James S. Walker
August, 1991
Eau Claire, Wisconsin

1

Basic Aspects of Fourier Series

This chapter is a summary of the basic theory of Fourier series. Some of the deeper theorems are only quoted; their proofs can be found in [Wa]. Besides their importance in applications, Fourier series provide a foundation for understanding the FFT.

1 Definition of Fourier series

To understand the definition of Fourier series we will begin with the essential idea: representing a wave form in terms of frequency as opposed to time. We will denote time by x rather than t.

Suppose our wave form is described by $4\cos 2\pi\nu x$, which has frequency ν. Using *Euler's identity*

$$e^{i\phi} = \cos\phi + i\sin\phi \tag{1.1}$$

we can write $4\cos 2\pi\nu x$ in complex exponential form

$$4\cos 2\pi\nu x = 2e^{i2\pi\nu x} + 2e^{-i2\pi\nu x}$$

where the complex exponentials have amplitudes of 2 and frequencies of ν and $-\nu$. See Figure 1.1.

Or, suppose our wave form is described by $6\sin 2\pi\nu x$, which also has frequency ν. Using Euler's identity (1.1) again, we obtain

$$6\sin 2\pi\nu x = 3ie^{-i2\pi\nu x} - 3ie^{i2\pi\nu x}$$

where the complex exponentials have complex amplitudes of $3i$ and $-3i$ and frequencies of $-\nu$ and ν. See Figure 1.2.

These two examples show how the waves $\cos 2\pi\nu x$ and $\sin 2\pi\nu x$ can be expressed in frequency terms *and distinguished from each other* using complex

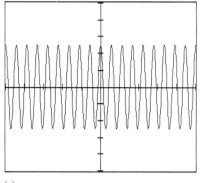
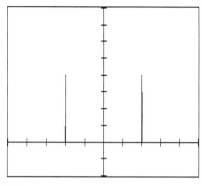

(a)

X interval: $[-1, 1]$ X increment = .2

Y interval: $[-8, 8]$ Y increment = 1.6

(b)

X interval: $[-22.5, 22.5]$ X increment = 4.5

Y interval: $[-1, 4]$ Y increment = .5

FIGURE 1.1
Frequency representation of $4\cos 2\pi\nu x$, $\nu = 9$. **(a) Graph in** x **domain (time or space). (b) Graph in frequency domain.**

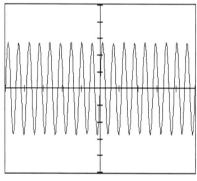
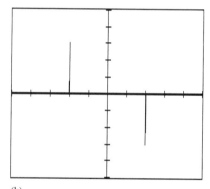

(a)

X interval: $[-1, 1]$ X increment = .2

Y interval: $[-11, 11]$ Y increment = 2.2

(b)

X interval: $[-22.5, 22.5]$ X increment = 4.5

Y interval: $[-5, 5]$ Y increment = 1

FIGURE 1.2
Frequency representation of $6\sin 2\pi\nu x$, $\nu = 9$. **(a) Graph in** x **domain (time or space). (b) Graph in frequency domain.**

exponentials. Also, Figures 1.1(b) and 1.2(b) show how, in a certain sense, the frequency representation of these waves is simpler.

The basic idea in Fourier series is to express a periodic wave as a sum of complex exponentials all of which have the same period. This is made feasible by the following property of complex exponentials having the same period.

THEOREM 1.2 ORTHOGONALITY OF COMPLEX EXPONENTIALS
For all integers m and n, the complex exponentials $\{e^{i2\pi nx/P}\}_{n=-\infty}^{\infty}$ with period P satisfy the following **orthogonality relation**

$$\frac{1}{P}\int_0^P e^{i2\pi mx/P}e^{-i2\pi nx/P}\,dx = \begin{cases} 0 & \text{for } m \neq n \\ 1 & \text{for } m = n. \end{cases}$$

PROOF We will prove the theorem for the case of $P = 2\pi$. For this case, we have

$$\frac{1}{2\pi}\int_0^{2\pi} e^{imx}e^{-inx}\,dx = \frac{1}{2\pi}\int_0^{2\pi} e^{i(m-n)x}\,dx$$

$$= \frac{1}{2\pi}\int_0^{2\pi}\cos(m-n)x\,dx$$

$$+ \frac{i}{2\pi}\int_0^{2\pi}\sin(m-n)x\,dx \qquad (1.3)$$

If $m \neq n$, then by calculus we get

$$\frac{1}{2\pi}\int_0^{2\pi}\cos(m-n)x\,dx = \frac{\sin(m-n)x}{2\pi(m-n)}\bigg|_0^{2\pi} = 0$$

and similarly $i/2\pi\int_0^{2\pi}\sin(m-n)x\,dx = 0$.

This takes care of the case when $m \neq n$. If $m = n$, then $\sin(m-n)x = \sin 0 = 0$ and $\cos(m-n)x = \cos 0 = 1$. Hence (1.3) becomes

$$\frac{1}{2\pi}\int_0^{2\pi} e^{imx}e^{-inx}\,dx = \frac{1}{2\pi}\int_0^{2\pi} 1\,dx = 1.$$

This proves the theorem. ∎

Now, suppose g is a periodic function, period P, and g is expanded in a series of complex exponentials having the same period

$$g(x) = \sum_{n=-\infty}^{\infty} c_n e^{i2\pi nx/P}. \qquad (1.4)$$

We will show that a reasonable formula for the general coefficient c_n can be obtained using the orthogonality theorem (Theorem 1.2). Multiplying both sides of (1.4) by $(1/P)e^{-i2\pi nx/P}$ (being careful to change the name of the index in the series) and integrating term by term, we get

$$\frac{1}{P}\int_0^P g(x)e^{-i2\pi nx/P}\,dx = \sum_{m=-\infty}^{\infty} c_m \frac{1}{P}\int_0^P e^{i2\pi mx/P}e^{-i2\pi nx/P}\,dx$$

$$= c_n$$

because of Theorem 1.2. Based on this result, we make the following definition.

DEFINITION 1.5   *If the function g has period P, then the **Fourier coefficients** $\{c_n\}$ for g are defined by*

$$c_n = \frac{1}{P}\int_0^P g(x)e^{-i2\pi nx/P}\,dx$$

*for all integers n. The **Fourier series** for g is defined by the right side of the following correspondence:*

$$g \sim \sum_{n=-\infty}^{\infty} c_n e^{i2\pi nx/P}\,, \qquad \left(c_n = \frac{1}{P}\int_0^P g(x)e^{-i2\pi nx/P}\,dx\right).$$

REMARK 1.6   (a) We have used the correspondence symbol $\sim$ since we have not yet discussed the conditions of validity of equation (1.4); we will do that in Section 4. (b) Note that when g is a real-valued function, then $c_{-n} = c_n^*$ where c_n^* is the *complex conjugate* of c_n. (c) The Fourier coefficients of g can also be found from the formula

$$c_n = \frac{1}{P}\int_{-\frac{1}{2}P}^{\frac{1}{2}P} g(x)e^{-i2\pi nx/P}\,dx. \qquad \blacksquare$$

2   Examples of Fourier series

In this section we will cover some basic examples of Fourier series expansions.

Example 2.1
Expand the function $g(x) = x$ in a Fourier series, period 2π, using the interval $[-\pi, \pi]$.   ▢

SOLUTION For $n = 0$, we have $c_0 = 1/2\pi \int_{-\pi}^{\pi} x \, dx = 0$. For $n \neq 0$, we split the integral for c_n into real and imaginary terms and then integrate by parts, obtaining

$$c_n = \frac{1}{2\pi} \int_{-\pi}^{\pi} x e^{-inx} \, dx$$

$$= \frac{1}{2\pi} \int_{-\pi}^{\pi} x \cos nx \, dx - \frac{i}{2\pi} \int_{-\pi}^{\pi} x \sin nx \, dx = \frac{i(-1)^n}{n}.$$

Hence,

$$g \sim \sideset{}{'}\sum_{n=-\infty}^{\infty} \frac{i(-1)^n}{n} e^{inx} \tag{2.2}$$

where the prime on the sum is used to indicate that the $n = 0$ term is omitted. ∎

Grouping $(i(-1)^n/n)e^{inx}$ and $(i(-1)^{-n}/(-n))e^{-inx}$ we obtain $(2(-1)^{n+1})/n \sin nx$, so we can rewrite (2.2) in the *real form*

$$g \sim \sum_{n=1}^{\infty} \frac{2(-1)^{n+1}}{n} \sin nx.$$

In Figure 1.3, we show the graphs of the *partial sums*

$$S_M(x) = \sum_{n=1}^{M} \frac{2(-1)^{n+1}}{n} \sin nx$$

for $M = 1, 2, 4$, and 8. The number M denotes the *number of harmonics* in the partial sum.

Notice that in Figure 1.3(d) the graph of S_8, on the interval $[-\pi, \pi]$ that we used to calculate the Fourier series, is a wavy (or wiggly) approximation to the original function $g(x) = x$. Outside the original interval, the graph of S_8 is approximating the *periodic extension* of $g(x) = x$. See Figure 1.4. ∎

Example 2.3
Expand the function $g(x) = x^2 - 2x$ in a Fourier series, period 2, using the interval $[0, 2]$. ⬚

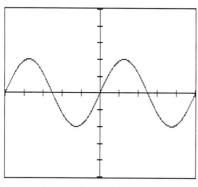

(a) partial sum, 1 harmonic
X interval: $[-2\pi, 2\pi]$ X increment $= 2\pi/5$
Y interval: $[-5, 5]$ Y increment $= 1$

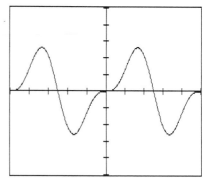

(b) partial sum, 2 harmonics
X interval: $[-2\pi, 2\pi]$ X increment $= 2\pi/5$
Y interval: $[-5, 5]$ Y increment $= 1$

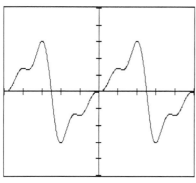

(c) partial sum, 4 harmonics
X interval: $[-2\pi, 2\pi]$ X increment $= 2\pi/5$
Y interval: $[-5, 5]$ Y increment $= 1$

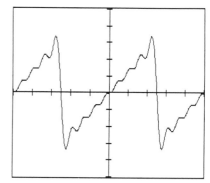

(d) partial sum, 8 harmonics
X interval: $[-2\pi, 2\pi]$ X increment $= 2\pi/5$
Y interval: $[-5, 5]$ Y increment $= 1$

FIGURE 1.3
Graphs of some partial sums of the Fourier series for $g(x) = x$, period 2π, using the interval $[-\pi, \pi]$ for computing Fourier series coefficients.

SOLUTION   For c_0 we obtain $c_0 = 1/2 \int_0^2 x^2 - 2x \, dx = -2/3$. For $n \neq 0$, we obtain

$$c_n = \frac{1}{2} \int_0^2 (x^2 - 2x) e^{-i\pi n x} \, dx$$

$$= \frac{1}{2} \int_0^2 (x^2 - 2x) \cos n\pi x \, dx - \frac{i}{2} \int_0^2 (x^2 - 2x) \sin n\pi x \, dx = \frac{2}{n^2 \pi^2}$$

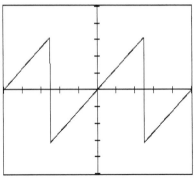

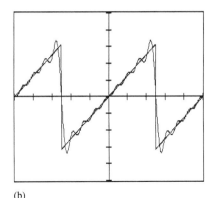

(a)

(b)

X interval: $[-2\pi, 2\pi]$ X increment = $2\pi/5$

Y interval: $[-5, 5]$ Y increment = 1

X interval: $[-2\pi, 2\pi]$ X increment = $2\pi/5$

Y interval: $[-5, 5]$ Y increment = 1

FIGURE 1.4
The periodic extension, period 2π, of the function $g(x) = x$ is shown in (a). In (b), the graph of S_8 is shown together with the graph from (a).

after integrating by parts twice. Thus,

$$g \sim -\frac{2}{3} + \sum_{n=-\infty}^{\infty}{}' \frac{2}{n^2\pi^2} e^{in\pi x}.$$

Grouping terms for n and $-n$, we obtain the real form of the Fourier series for g

$$g \sim -\frac{2}{3} + \sum_{n=1}^{\infty} \frac{4}{n^2\pi^2} \cos n\pi x.$$

In Figure 1.5, we have graphed the partial sums

$$S_M(x) = -\frac{2}{3} + \sum_{n=1}^{M} \frac{4}{n^2\pi^2} \cos n\pi x$$

for $M = 2$, 4, and 8 harmonics, and the graph of the periodic extension of g with the graph of S_8 superimposed on it. ▌

Example 2.4
Expand the function $g(x) = x^2 + x$ in a Fourier series, period 1, over the interval $[-1/2, 1/2]$. ⬜

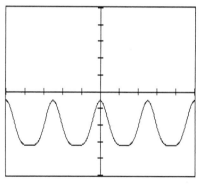

(a) partial sum, 2 harmonics
X interval: $[-4, 4]$ X increment = .8
Y interval: $[-1.5, 1.5]$ Y increment = .3

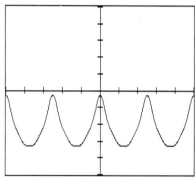

(b) partial sum, 4 harmonics
X interval: $[-4, 4]$ X increment = .8
Y interval: $[-1.5, 1.5]$ Y increment = .3

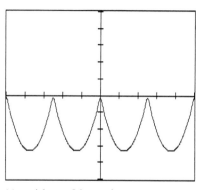

(c) partial sum, 8 harmonics
X interval: $[-4, 4]$ X increment = .8
Y interval: $[-1.5, 1.5]$ Y increment = .3

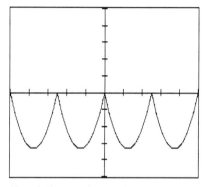

(d) periodic extension and 8 harmonics
X interval: $[-4, 4]$ X increment = .8
Y interval: $[-1.5, 1.5]$ Y increment = .3

FIGURE 1.5
Graphs illustrating Example 2.3.

SOLUTION   For c_0 we obtain $c_0 = \int_{-1/2}^{1/2} x^2 + x \, dx = 1/12$. For $n \neq 0$, we obtain

$$
c_n = \int_{-\frac{1}{2}}^{\frac{1}{2}} (x^2 + x) \cos 2\pi n x \, dx - i \int_{-\frac{1}{2}}^{\frac{1}{2}} (x^2 + x) \sin 2\pi n x \, dx
$$

$$
= \frac{(-1)^n}{2} \left[\frac{1}{n^2 \pi^2} + \frac{i}{n\pi} \right]
$$

after integrating by parts twice. Thus,

$$g \sim \frac{1}{12} + \sum_{n=-\infty}^{\infty}{}' \frac{(-1)^n}{2} \left[\frac{1}{n^2\pi^2} + \frac{i}{n\pi} \right] e^{i2\pi nx}.$$

Grouping terms for n and $-n$, we obtain the real form of the Fourier series for g,

$$g \sim \frac{1}{12} + \sum_{n=1}^{\infty} (-1)^n \left[\frac{\cos 2\pi nx}{n^2\pi^2} - \frac{\sin 2\pi nx}{n\pi} \right].$$

In Figure 1.6, we show the graphs of the partial sums

$$S_M(x) = \frac{1}{12} + \sum_{n=1}^{M} (-1)^n \left[\frac{\cos 2\pi nx}{n^2\pi^2} - \frac{\sin 2\pi nx}{n\pi} \right]$$

for $M = 3$, 12, and 36 harmonics, and the graph of the periodic extension of $x^2 + x$ superimposed on S_{36}. ∎

3 Fourier series of real functions

In each of the examples in the previous section we computed Fourier series for real-valued functions. We were able in each case to express those Fourier series in real forms, series involving real-valued sines and cosines. We will now show that this is possible in general.

Suppose that g is a *real-valued* function. Then the Fourier coefficients of g over the interval $[0, P]$ are given by

$$c_n = \frac{1}{P} \int_0^P g(x) e^{-i2\pi nx/P} \, dx. \tag{3.1}$$

If we take complex conjugates of both sides of (3.1), then by keeping in mind that $g(x)$ and x are real we can write

$$c_n^* = \frac{1}{P} \left[\int_0^P g(x) e^{-i2\pi nx/P} \, dx \right]^*$$

$$= \frac{1}{P} \int_0^P g(x)^* \left[e^{-i2\pi nx/P} \right]^* \, dx$$

$$= \frac{1}{P} \int_0^P g(x) e^{i2\pi nx/P} \, dx = c_{-n}.$$

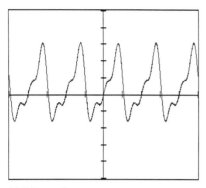

(a) 3 harmonics
X interval: $[-2.5, 2.5]$ X increment $= .5$
Y interval: $[-1, 1]$ Y increment $= .2$

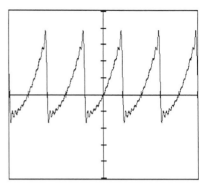

(b) 12 harmonics
X interval: $[-2.5, 2.5]$ X increment $= .5$
Y interval: $[-1, 1]$ Y increment $= .2$

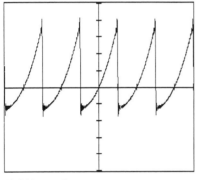

(c) 36 harmonics
X interval: $[-2.5, 2.5]$ X increment $= .5$
Y interval: $[-1, 1]$ Y increment $= .2$

(d) 36 harmonics and periodic extension
X interval: $[-2.5, 2.5]$ X increment $= .5$
Y interval: $[-1, 1]$ Y increment $= .2$

FIGURE 1.6
Graphs illustrating Example 2.4.

Thus, we have whenever g is real valued

$$c_n^* = c_{-n}. \tag{3.2}$$

Using (3.2) we can cast the Fourier series for g into a real form by grouping the nth and $-n$th terms as follows:

$$g \sim \sum_{n=-\infty}^{\infty} c_n e^{i2\pi nx/P} = c_0 + \sum_{n=1}^{\infty} \left\{ c_n e^{i2\pi nx/P} + c_{-n} e^{-i2\pi nx/P} \right\}$$

$$= c_0 + \sum_{n=1}^{\infty} \left\{ c_n e^{i2\pi nx/P} + (c_n e^{i2\pi nx/P})^* \right\} = c_0 + \sum_{n=1}^{\infty} 2Re \left\{ c_n e^{i2\pi nx/P} \right\}$$

where Re means taking the real part of a complex number.

Thus, we have

$$g \sim c_0 + \sum_{n=1}^{\infty} 2Re\left\{c_n e^{i2\pi nx/P}\right\}. \tag{3.3}$$

Now, we write c_n more explicitly as a complex number

$$c_n = \frac{1}{P}\int_0^P g(x)e^{-i2\pi nx/P}\,dx$$

$$= \frac{1}{P}\int_0^P g(x)\cos\frac{2\pi nx}{P}\,dx - \frac{i}{P}\int_0^P g(x)\sin\frac{2\pi nx}{P}\,dx$$

$$= \frac{1}{2}A_n - \frac{i}{2}B_n$$

where we *define* A_n and B_n by

$$A_n = \frac{2}{P}\int_0^P g(x)\cos\frac{2\pi nx}{P}\,dx \quad, \quad B_n = \frac{2}{P}\int_0^P g(x)\sin\frac{2\pi nx}{P}\,dx. \tag{3.4}$$

Using $c_n = (1/2)A_n - (i/2)B_n$ and $e^{i2\pi nx/P} = \cos(2\pi nx/P) + i\sin(2\pi nx/P)$ in (3.3) we obtain

$$g \sim c_0 + \sum_{n=1}^{\infty}\left\{A_n\cos\frac{2\pi nx}{P} + B_n\sin\frac{2\pi nx}{P}\right\}.$$

For notational uniformity, we define $(1/2)A_0$ to be equal to c_0. We then have the following formula:

$$g \sim \frac{1}{2}A_0 + \sum_{n=1}^{\infty}\left\{A_n\cos\frac{2\pi nx}{P} + B_n\sin\frac{2\pi nx}{P}\right\}$$

$$A_n = \frac{2}{P}\int_0^P g(x)\cos\frac{2\pi nx}{P}\,dx \quad, \qquad (n = 0, 1, 2, \ldots)$$

$$B_n = \frac{2}{P}\int_0^P g(x)\sin\frac{2\pi nx}{P}\,dx \quad, \qquad (n = 1, 2, 3, \ldots). \tag{3.5}$$

Formula (3.5) defines the real form of the Fourier series for g, using the period interval $[0, P]$. Similar formulas hold if the interval is $[-(1/2)P, (1/2)P]$. In particular, the coefficients A_n and B_n would then be given by

$$A_n = \frac{2}{P}\int_{-\frac{1}{2}P}^{\frac{1}{2}P} g(x)\cos\frac{2\pi nx}{P}\,dx \quad, \qquad (n = 0, 1, 2, \ldots)$$

$$B_n = \frac{2}{P}\int_{-\frac{1}{2}P}^{\frac{1}{2}P} g(x)\sin\frac{2\pi nx}{P}\,dx \quad, \qquad (n = 1, 2, 3, \ldots). \tag{3.6}$$

If we define the partial sum S_M having M harmonics by

$$S_M(x) = \sum_{n=-M}^{M} c_n e^{i2\pi nx/P}. \tag{3.7}$$

Then, by algebra similar to that used above, it follows that S_M has a real form given by

$$S_M(x) = \frac{1}{2}A_0 + \sum_{n=1}^{M} \left\{ A_n \cos \frac{2\pi nx}{P} + B_n \sin \frac{2\pi nx}{P} \right\}. \tag{3.8}$$

REMARK   The convention of using $1/2$ as a factor on A_0 in the constant term in the real form of Fourier series is troublesome. But, it is a common convention since it simplifies the form of (3.5). Also, it is the convention adopted in [Wa] to which we make frequent reference.   ▌

---

4   Pointwise convergence of Fourier series

In this section we will examine the validity of equating a periodic function with its Fourier series. Our approach to this problem will be to examine the limit of partial sums of the Fourier series at any given point.

To be more specific, we ask under what conditions can we have

$$\sum_{n=-\infty}^{\infty} c_n e^{i2\pi nx/P} = g(x) \tag{4.1}$$

where the left side of (4.1) is the Fourier series, period P, for g. A formal definition of (4.1) is

$$\lim_{M \to \infty} S_M(x) = g(x)$$

where S_M is the partial sum, containing M harmonics, of the Fourier series for g [see formula (3.7)].

The types of functions that we will be interested in are mostly covered by the following definition.

DEFINITION 4.2   *A function g on a finite interval $[a, b]$ is **piecewise continuous on** $[a, b]$ if the interval $[a, b]$ can be divided into a finite number of subintervals on each of which g is continuous. If g is piecewise continuous on every finite interval, then g is called **piecewise continuous**.*

Example 4.3
(a) The function

$$g(x) = \begin{cases} x & \text{for } -1 < x < 0 \\ 4 & \text{for } 0 < x < 2 \end{cases}$$

is piecewise continuous on the interval $[-1, 2]$. (b) The function g defined by $g(x) = x$ for $-\pi < x < \pi$ and having period 2π is piecewise continuous. (c) The function $g(x) = x^2$ is piecewise continuous. In fact, every continuous function is piecewise continuous. ☐

For a piecewise continuous function, the following limits always make sense and are finite:

$$g(x+) = \lim_{h \to 0+} g(x + h), \quad g(x-) = \lim_{h \to 0-} g(x + h).$$

The first limit above is called the *right-hand limit of g at* x, since $x + h$ always lies to the right of x when $h > 0$. The second limit above is called the *left-hand limit of g at* x, since $x + h$ always lies to the left of x when $h < 0$. Also, we can define some useful generalizations of derivative. Whenever the limit

$$\lim_{h \to 0+} \frac{g(x + h) - g(x+)}{h}$$

exists and is finite, then g is said to have a *right-hand derivative at* x, denoted by $g'(x+)$ (which equals the value of this limit). Similarly, whenever the limit

$$\lim_{h \to 0-} \frac{g(x + h) - g(x-)}{h}$$

exists and is finite, then g is said to have a *left-hand derivative at* x, denoted by $g'(x-)$ (which equals the value of this limit). Of course, if the left-hand and right-hand derivatives are equal, then g has a derivative, $g'(x) = g'(x+) = g'(x-)$, at x.

We can now state a pointwise convergence theorem for Fourier series. This theorem covers many of the functions dealt with in an analytic setting.

THEOREM 4.4
Let g be a piecewise continuous function, period P. At each point x where g has a right- and left-hand derivative, the Fourier series for g converges to $(1/2)[g(x+) + g(x-)]$. Thus, we can write

$$\sum_{n=-\infty}^{\infty} c_n e^{i2\pi nx/P} = \frac{1}{2}[g(x+) + g(x-)].$$

If x is a point of continuity for g, then this result simplifies to

$$\sum_{n=-\infty}^{\infty} c_n e^{i2\pi nx/P} = g(x).$$

PROOF   For a proof of this theorem, see [Wa, Chapter 2.3].   ∎

Suppose we are given a function g on the interval $[0, P]$ and we compute its Fourier series, period P. To apply Theorem 4.4 we look at the *periodic extension* g_P of our function g defined by

$$g_P(x + kP) = g(x), \qquad (k = 0, \pm 1, \pm 2, \ldots).$$

Since $g_P = g$ on $[0, P]$, g_P has the same Fourier series as g and we may apply Theorem 4.4 to the periodic extension g_P. Similar remarks apply if we are given a function on $[-(1/2)P, (1/2)P]$. Here are some examples.

Example 4.5
Let g be defined by

$$g(x) = \begin{cases} 1 & \text{for } 0 < x < 1 \\ 0 & \text{for } -1 < x < 0. \end{cases}$$

Graphs of g and its partial sums S_M for $M = 12, 36$, and 108 harmonics are shown in Figure 1.7. Fixing $x \neq 0, \pm 1$, we see $S_M(x)$ converging to $g(x)$ in these graphs. For $x = 0$, the graphs show $S_M(0)$ converging to $(1/2)[g(0+) + g(0-)] = 1/2$. While, for $x = \pm 1$, the graphs show $S_M(\pm 1)$ converging to half the sum of the right- and left-hand limits of the periodic extension of g at $x = \pm 1$, which is $1/2$ in both cases.   ☐

Example 4.6
Let $g(x) = |x - 2| + 1$ on the interval $[0, 4]$. Graphs of the function g and S_M for $M = 12, 36$, and 108 harmonics are shown in Figure 1.8. The tendency of S_M to converge to g at all points in the interval $[0, 4]$ is clearly shown in these graphs.   ☐

5   Further aspects of convergence of Fourier series

In this section we will discuss the following aspects of convergence of Fourier series: Gibbs' phenomenon, uniform convergence, and a more profound convergence theorem which applies to a wider class of functions that are encountered in measuring signals.

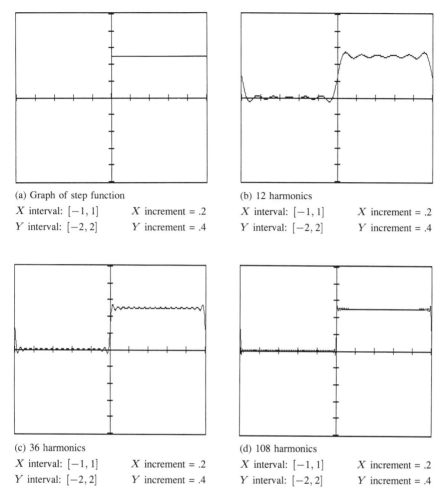

(a) Graph of step function
X interval: $[-1, 1]$ X increment = .2
Y interval: $[-2, 2]$ Y increment = .4

(b) 12 harmonics
X interval: $[-1, 1]$ X increment = .2
Y interval: $[-2, 2]$ Y increment = .4

(c) 36 harmonics
X interval: $[-1, 1]$ X increment = .2
Y interval: $[-2, 2]$ Y increment = .4

(d) 108 harmonics
X interval: $[-1, 1]$ X increment = .2
Y interval: $[-2, 2]$ Y increment = .4

FIGURE 1.7
Graphs illustrating Example 4.5.

We begin by mentioning Gibbs' phenomenon. Consider Example 4.5 again. In Figure 1.7 notice the sharp peaks in S_{36} and S_{108} near 0, the discontinuity point of g. The height of these peaks is greater than $g(0+)$ by about 9%. Figure 1.9 shows that this phenomenon, known as *Gibbs' phenomenon*, does not go away as the number of harmonics is increased. Further treatment of Gibbs' phenomenon will be given in Chapter 4. See also [Wa, Chapter 2.5].

Some functions do not exhibit Gibbs' phenomenon in their Fourier series. For example, look again at Figure 1.8. That Fourier series exhibits a property known as *uniform convergence*. Uniform convergence occurs when the partial sums

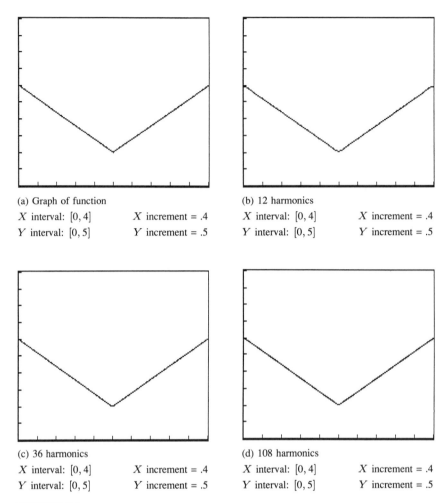

(a) Graph of function
X interval: $[0, 4]$ X increment = .4
Y interval: $[0, 5]$ Y increment = .5

(b) 12 harmonics
X interval: $[0, 4]$ X increment = .4
Y interval: $[0, 5]$ Y increment = .5

(c) 36 harmonics
X interval: $[0, 4]$ X increment = .4
Y interval: $[0, 5]$ Y increment = .5

(d) 108 harmonics
X interval: $[0, 4]$ X increment = .4
Y interval: $[0, 5]$ Y increment = .5

FIGURE 1.8
Graphs illustrating Example 4.6.

tend uniformly over some interval (or over the whole real line) to a function g. To be more precise, we have the following theorem. (The notation "sup" in this theorem means, in this case, the same thing as *maximum*.)

THEOREM 5.1
If g is a continuous function with period P and its derivative g' is piecewise continuous, then the partial sums of the Fourier series for g converge uniformly

to g over the whole real line. In particular, we have

$$\lim_{M\to\infty}\left\{\sup_{x\in\mathbf{R}}|g(x)-S_M(x)|\right\}=0.$$

REMARK   As we mentioned above, the notation "sup" in this theorem can be viewed as just shorthand for *maximum*. Hence, the theorem says that as $M\to\infty$ the maximum difference between g and S_M tends to 0.   ∎

PROOF OF THEOREM 5.1.   See [Wa, Chapter 2.4].   ∎

Theorem 5.1 applies to the periodic extension of the function g in Example 4.6. We also saw in Figure 1.8 the excellent convergence properties of the sequence of partial sums $\{S_M\}$ to g. It is interesting to note, however, that g is being thought of here as a limit of the infinitely differentiable functions S_M (that is, S_M can be differentiated as often as you please). Yet, g itself is not differentiable. In fact, $g'(2)$ does not exist. Figure 1.10 shows the Fourier series partial sums S_M near the point $(2,1)$ and how they compare to g. These graphs show the failure of the Fourier series partial sums to reproduce the sharp corner in the graph of g at the point $(2,1)$, a failure that is only eliminated in the limit as $M\to\infty$.

In an analytic setting we often have formulas for the Fourier coefficients of a specific function, such as the examples given in Section 2, for instance. When this happens there is a very useful theorem that allows us to obtain estimates on the closeness of S_M to our function.

THEOREM 5.2
If $\sum_{n=-\infty}^{\infty}|c_n|$ converges, then the Fourier series $\sum_{n=-\infty}^{\infty}c_n e^{i2\pi nx/P}$ converges uniformly to a continuous function g with period P.

PROOF   See [Wa, Chapter 1.5].   ∎

Here is an interesting result that is related to Theorem 5.2. Consider the real Fourier series

$$1+\sum_{n=1}^{\infty}\frac{\cos 2^n x}{2^n}\ .$$

Since

$$\frac{|\cos 2^n x|}{2^n}\le\frac{1}{2^n}$$

the series converges to a function g for all x-values (by the Comparison Test).

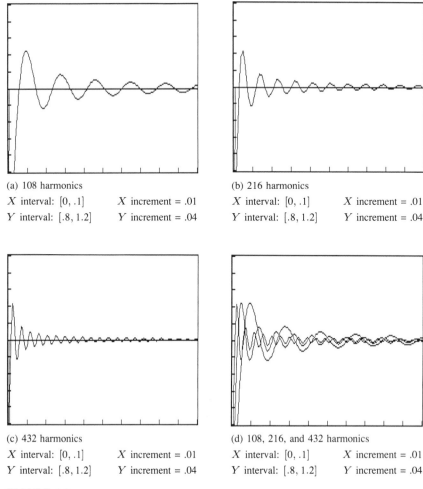

(a) 108 harmonics
X interval: $[0, .1]$ X increment = .01
Y interval: $[.8, 1.2]$ Y increment = .04

(b) 216 harmonics
X interval: $[0, .1]$ X increment = .01
Y interval: $[.8, 1.2]$ Y increment = .04

(c) 432 harmonics
X interval: $[0, .1]$ X increment = .01
Y interval: $[.8, 1.2]$ Y increment = .04

(d) 108, 216, and 432 harmonics
X interval: $[0, .1]$ X increment = .01
Y interval: $[.8, 1.2]$ Y increment = .04

FIGURE 1.9
Graphs illustrating Gibbs' phenomenon.

And

$$|g(x) - S_{2^N}(x)| = \left| \sum_{n > N}^{\infty} \frac{\cos 2^n x}{2^n} \right|$$

$$\leq \sum_{n > N}^{\infty} \frac{|\cos 2^n x|}{2^n} .$$

The inequality holding because of possible cancellation of positive and negative values. Since $|\cos 2^n x| \leq 1$ and $\sum_{n > N}^{\infty} 1/2^n = 1/2^N$ we get $|g(x) - S_{2^N}(x)| \leq$

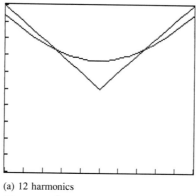

(a) 12 harmonics

X interval: $[1.9, 2.1]$	X increment = .01
Y interval: $[.9, 1.1]$	Y increment = .02

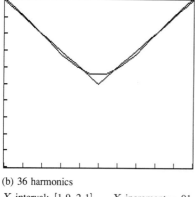

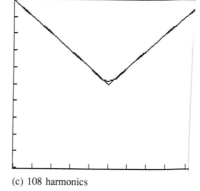

(b) 36 harmonics

X interval: $[1.9, 2.1]$	X increment = .01
Y interval: $[.9, 1.1]$	Y increment = .02

(c) 108 harmonics

X interval: $[1.9, 2.1]$	X increment =
Y interval: $[.9, 1.1]$	Y increment =

FIGURE 1.10
Graphs of $g(x) = |x - 2| + 1$ and S_M for $M = 12, 36,$ and 108, near the point $(2$

$1/2^N$. Hence

$$\sup_{x \in \mathbf{R}} |g(x) - S_{2^N}(x)| \leq \frac{1}{2^N}.$$

For example, if $N = 9$, then g and $S_{512}(x) = 1 + \sum_{n=1}^{9} (\cos 2^n x)/2^n$ di
by no more than $\pm 1/512$ over all of $\mathbf{R}$. In Figure 1.11, we have shown
graph of S_{512}. The function g, to which S_M converges as $M \to \infty$, is ca
the *Weierstrass nowhere differentiable function*. It is a continuous function
all x-values, but it has no derivative at any x-value.

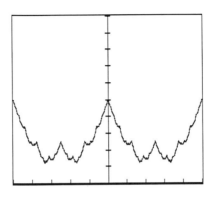

X interval: $[-\pi, \pi]$ X increment = $\pi/5$
Y interval: $[0, 4]$ Y increment = .4

FIGURE 1.11
Ten-term approximation to Weierstrass nowhere differentiable function.

We close this section by describing a theorem that is broad enough to cover most data measured from periodic signals.

DEFINITION 5.3   *A function g is said to be **Lipschitz from the right at** x if for some positive constants A, α, and δ, we have*

$$|g(x + h) - g(x+)| \le Ah^{\alpha}, \quad \text{for } 0 < h < \delta.$$

*Similarly, g is said to be **Lipschitz from the left at** x if for some positive constants B, β, and ϵ, we have*

$$|g(x - h) - g(x-)| \le Bh^{\beta}, \quad \text{for } 0 < h < \epsilon.$$

Figure 1.12 shows the geometric interpretation of the inequalities in this definition.

The following theorem covers most of the function data (signals) that arise from measurements.

THEOREM 5.4
Let g be a piecewise continuous function with period P. At each point x where g is Lipschitz from the left and right

$$\sum_{n=-\infty}^{\infty} c_n e^{i2\pi nx/P} = \frac{1}{2}[g(x+) + g(x-)].$$

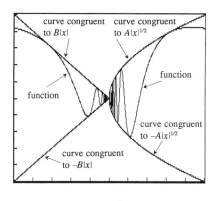

X interval: $[0, 2]$ X increment $= .2$
Y interval: $[0, 2]$ Y increment $= .2$

FIGURE 1.12
Illustration of a function, Lipschitz from the left and right with $\alpha = 1/2$ and $\beta = 1$.

If x is a point of continuity for g, then this result simplifies to

$$\sum_{n=-\infty}^{\infty} c_n e^{i2\pi nx/P} = g(x).$$

PROOF The techniques needed to prove this theorem are discussed in [Wa, Chapter 2.3]. ∎

6 Fourier sine series and cosine series

Fourier sine series and cosine series are special forms of Fourier series for functions possessing either odd or even symmetry.

DEFINITION 6.1 *A function g is **odd** on the interval $(-L, L)$ if $g(-x) = -g(x)$ for each x-value. A function g is called **even** on the interval $(-L, L)$ if $g(-x) = g(x)$ for each x-value.*

It is easy to see that the product of two odd functions, or the product of two even functions, is an even function, while the product of an odd and an even function is an odd function. We leave as an exercise for the reader the proofs of the following facts: Whenever g is odd on the interval $(-L, L)$, then

$$\int_{-L}^{L} g(x)\, dx = 0 \tag{6.2a}$$

and, whenever g is even on the interval $(-L, L)$, then

$$\int_{-L}^{L} g(x)\,dx = 2 \int_{0}^{L} g(x)\,dx. \qquad (6.2b)$$

Now, suppose we compute the Fourier series for an odd function g over the interval $(-L, L)$. Here the period of our complex exponentials will be $2L$ and we get for c_n

$$c_n = \frac{1}{2L} \int_{-L}^{L} g(x) e^{-i2\pi nx/(2L)}\,dx$$

$$= \frac{1}{2L} \int_{-L}^{L} g(x) \cos \frac{n\pi x}{L}\,dx - \frac{i}{2L} \int_{-L}^{L} g(x) \sin \frac{n\pi x}{L}\,dx.$$

Since g is odd, so is $g(x)\cos(n\pi x/L)$, and we have

$$\int_{-L}^{L} g(x) \cos \frac{n\pi x}{L}\,dx = 0.$$

Also, $g(x)\sin(n\pi x/L)$ is even, so

$$\int_{-L}^{L} g(x) \sin \frac{n\pi x}{L}\,dx = 2 \int_{0}^{L} g(x) \sin \frac{n\pi x}{L}\,dx.$$

Returning to our calculation of c_n, these last two results allow us to simplify c_n to the following form:

$$c_n = \frac{-i}{L} \int_{0}^{L} g(x) \sin \frac{n\pi x}{L}\,dx. \qquad (6.3)$$

From (6.3) we obtain, using the oddness of the sine function,

$$c_{-n} = -c_n, \quad \text{(and } c_0 = 0\text{)}. \qquad (6.4)$$

If we now examine the Fourier series for g, then (6.4) allows us to express this series as a series of sines

$$g \sim \sum_{n=-\infty}^{\infty} c_n e^{i2\pi nx/(2L)} = \sum_{n=1}^{\infty} c_n e^{i\pi nx/L} + \sum_{n=1}^{\infty} c_{-n} e^{-i\pi nx/L}$$

$$= \sum_{n=1}^{\infty} c_n \left\{ e^{i\pi nx/L} - e^{-i\pi nx/L} \right\}$$

$$= \sum_{n=1}^{\infty} (2ic_n) \sin \frac{n\pi x}{L} \; .$$

If we *define* B_n by $2ic_n$, then we can write

$$g \sim \sum_{n=1}^{\infty} B_n \sin \frac{n\pi x}{L}, \qquad \left(B_n = \frac{2}{L} \int_0^L g(x) \sin \frac{n\pi x}{L}\, dx \right). \qquad (6.5)$$

Based on our results we now make a formal definition of Fourier sine series.

DEFINITION 6.6 *If g is a function defined on the interval $(0, L)$, then the* **Fourier sine series** *for g is defined in (6.5).*

Let's look at an example.

Example 6.7
Expand $g(x) = 1$ on the interval $(0, 1)$ in a Fourier sine series.   □

SOLUTION   Using (6.5) we have

$$B_n = \frac{2}{1} \int_0^1 1 \sin n\pi x\, dx = \frac{-2}{n\pi} \cos n\pi x \Big|_0^1 = \frac{2}{n\pi}[1 - (-1)^n]$$

$$= \begin{cases} \frac{4}{n\pi} & \text{for } n \text{ odd} \\ 0 & \text{for } n \text{ even.} \end{cases}$$

Thus, *on the interval* $(0, 1)$

$$1 \sim \sum_{k=0}^{\infty} \frac{4 \sin (2k+1)\pi x}{(2k+1)\pi}. \quad \blacksquare$$

Suppose we have an even function g on the interval $(-L, L)$. The Fourier series for g can be cast into a series of cosines. In fact, we get for the nth Fourier series coefficient c_n

$$c_n = \frac{1}{2L} \int_{-L}^{L} g(x) \cos \frac{n\pi x}{L}\, dx - \frac{i}{2L} \int_{-L}^{L} g(x) \sin \frac{n\pi x}{L}\, dx$$

$$= \frac{1}{L} \int_0^L g(x) \cos \frac{n\pi x}{L}\, dx. \qquad (6.8)$$

From (6.8) we get for each n,

$$c_{-n} = c_n \qquad (6.9)$$

and then

$$g \sim \sum_{n=-\infty}^{\infty} c_n e^{i2\pi n x/(2L)} = c_0 + \sum_{n=1}^{\infty} c_n \left[e^{i\pi n x/L} + e^{-i\pi n x/L} \right]$$

$$= c_0 + \sum_{n=1}^{\infty} 2c_n \cos \frac{n\pi x}{L} . \tag{6.10}$$

Therefore, if we *define* A_n by $2c_n$ for $n = 0, 1, 2, \ldots$, then

$$g \sim \frac{1}{2} A_0 + \sum_{n=1}^{\infty} A_n \cos \frac{n\pi x}{L}, \qquad \left(A_n = \frac{2}{L} \int_0^L g(x) \cos \frac{n\pi x}{L} \, dx \right). \tag{6.11}$$

REMARK   Note that the constant term in (6.11) is $(1/2)A_0$. Although this may be troublesome to remember, it follows from the form of the series in (6.10). ∎

Based on our results we can make a formal definition of Fourier cosine series.

DEFINITION 6.12   *If g is a function defined on the interval $(0, L)$, then the* **Fourier cosine series** *for g is defined in (6.11).*

Here's an example.

Example 6.13
Expand $\sin x$ on the interval $(0, (1/2)\pi)$ in a Fourier cosine series.   ⬚

SOLUTION   Using (6.11) we get for A_n

$$A_n = \frac{4}{\pi} \int_0^{\frac{1}{2}\pi} \sin x \cos 2nx \, dx$$

$$= \frac{2}{\pi} \int_0^{\frac{1}{2}\pi} \sin (2n+1)x - \sin (2n-1)x \, dx = \frac{-4}{\pi} \frac{1}{4n^2 - 1} .$$

Thus, *on the interval* $(0, (1/2)\pi)$

$$\sin x \sim \frac{2}{\pi} + \sum_{n=1}^{\infty} \frac{-4}{\pi} \frac{\cos 2nx}{4n^2 - 1} . \qquad ∎$$

7   Convergence of Fourier sine and cosine series

The pointwise convergence of Fourier sine and cosine series can be deduced from the pointwise convergence of Fourier series. The key to this is to properly extend a function from its original interval of $(0, L)$.

Let's begin with Fourier sine series. Here the proper extension to make is an *odd, periodic extension.* Given a function g on the interval $(0, L)$ one makes an *odd extension* to $(-L, L)$. See Figure 1.13. This odd extension, if we also call it g, would have to satisfy $g(-x) = -g(x)$. Then the function can be extended to a periodic function, period $2L$. See Figure 1.14. Our discussion in Section 6 shows that the Fourier series for this odd, periodic extension is just the Fourier sine series for our original function on $(0, L)$. Therefore, the convergence theorems described in Section 5 can be applied. In particular, we have the following convergence theorem.

THEOREM 7.1   POINTWISE CONVERGENCE OF FOURIER SINE SERIES
Let g be a piecewise continuous function on $[0, L]$. At the points 0 and L the sine series for g equals 0. For $0 < x < L$ we have

$$\sum_{n=1}^{\infty} B_n \sin \frac{n\pi x}{L} = \frac{1}{2}[g(x+) + g(x-)]$$

provided x is a point where g is Lipschitz from the right and left (or where g has right- and left-hand derivatives).

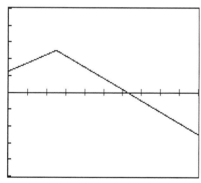

(a) Function on $(0, 2)$ (b) Odd extension to $(-2, 2)$

X interval: $[0, 2]$ X increment $= .2$ X interval: $[-2, 2]$ X increment $= .4$

Y interval: $[-4, 4]$ Y increment $= .8$ Y interval: $[-4, 4]$ Y increment $= .8$

FIGURE 1.13
Odd extension of a function.

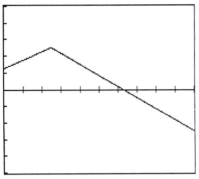

(a) Function on $(0, 2)$ (b) Odd, periodic extension
X interval: $[0, 2]$ X increment = .2 X interval: $[-5, 5]$ X increment = 1
Y interval: $[-4, 4]$ Y increment = .8 Y interval: $[-4, 4]$ Y increment = .8

FIGURE 1.14
Odd, periodic extension of a function.

Of course, when x is a point of continuity for g, then

$$\sum_{n=1}^{\infty} B_n \sin \frac{n\pi x}{L} = g(x)$$

provided g is Lipschitz from the left and the right at x (or g has right- and left-hand derivatives at x).

The rigorous definition of convergence in Theorem 7.1 is that

$$\lim_{M \to \infty} S_M(x) = \frac{1}{2}[g(x+) + g(x-)]$$

where

$$S_M(x) = \sum_{n=1}^{M} B_n \sin \frac{n\pi x}{L} .$$

The function S_M is called the *(sine series) partial sum containing M harmonics.*

Example 7.2
Let $g(x) = 1$ on the interval $(0, 1)$. In Figure 1.15 we have graphed S_M for $M = 10$, 20, and 40 harmonics. Notice the Gibbs' phenomenon occurring near $x = 0$ and $x = 1$. This is a consequence of the discontinuity at these points of the odd, periodic extension of g.   ☐

For Fourier cosine series the proper extension to make is an *even, periodic extension.* Given a function g on the interval $(0, L)$ one first makes an even extension to $(-L, L)$. See Figure 1.16. This even extension, if we also call it g,

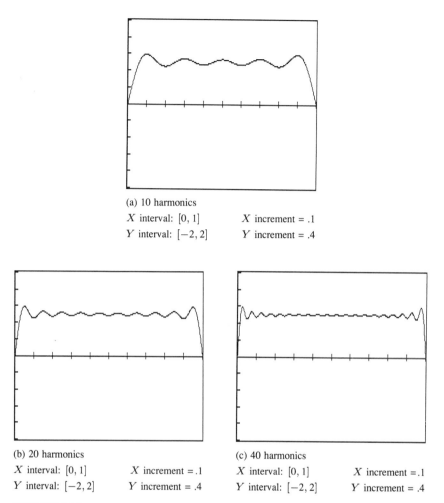

(a) 10 harmonics
X interval: $[0, 1]$ X increment $= .1$
Y interval: $[-2, 2]$ Y increment $= .4$

(b) 20 harmonics
X interval: $[0, 1]$ X increment $= .1$
Y interval: $[-2, 2]$ Y increment $= .4$

(c) 40 harmonics
X interval: $[0, 1]$ X increment $= .1$
Y interval: $[-2, 2]$ Y increment $= .4$

FIGURE 1.15
Graphs of Fourier sine series partial sums for the function in Example 7.2.

would have to satisfy $g(-x) = g(x)$. Then one extends this even extension to a periodic function, period $2L$. See Figure 1.17. Our discussion in Section 6 shows that the Fourier series for this even, periodic extension is just the Fourier cosine series for our original function on $(0, L)$. Therefore, the convergence theorems from Section 5 lead to the following convergence theorem for Fourier cosine series.

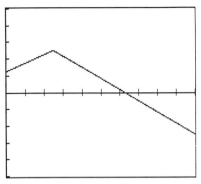
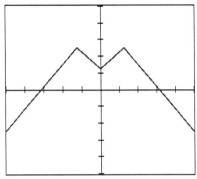

(a) Function on $(0, 2)$

| X interval: $[0, 2]$ | X increment = .2 |
| Y interval: $[-4, 4]$ | Y increment = .8 |

(b) Even extension to $(-2, 2)$

| X interval: $[-2, 2]$ | X increment = .4 |
| Y interval: $[-4, 4]$ | Y increment = .8 |

FIGURE 1.16
Even extension of a function.

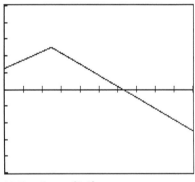
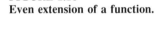

(a) Function on $(0, 2)$

| X interval: $[0, 2]$ | X increment = .2 |
| Y interval: $[-4, 4]$ | Y increment = .8 |

(b) Even, periodic extension

| X interval: $[-5, 5]$ | X increment = 1 |
| Y interval: $[-4, 4]$ | Y increment = .8 |

FIGURE 1.17
Even, periodic extension of a function.

THEOREM 7.3   POINTWISE CONVERGENCE OF FOURIER COSINE SERIES

Let g be a piecewise continuous function on $[0, L]$. At the point 0 the Fourier cosine series converges to $g(0+)$ provided g is Lipschitz from the right at 0 (or has a right-hand derivative at 0). At the point L the Fourier cosine series converges to $g(L-)$ provided g is Lipschitz from the left at L (or has a left-hand

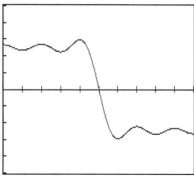

(a) 10 harmonics
X interval: $[0, 2]$ X increment = .2
Y interval: $[-2, 2]$ Y increment = .4

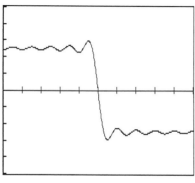

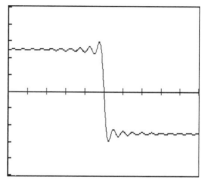

(b) 20 harmonics
X interval: $[0, 2]$ X increment = .2
Y interval: $[-2, 2]$ Y increment = .4

(c) 40 harmonics
X interval: $[0, 2]$ X increment = .2
Y interval: $[-2, 2]$ Y increment = .4

FIGURE 1.18
Graphs of Fourier cosine series partial sums for the function in Example 7.4.

derivative at L). For $0 < x < L$, we have

$$\frac{1}{2}A_0 + \sum_{n=1}^{\infty} A_n \cos \frac{n\pi x}{L} = \frac{1}{2}[g(x+) + g(x-)]$$

provided g is Lipschitz from the left and right at x (or has left- and right-hand derivatives at x).

We remind the reader again that in Theorem 7.3, $(1/2)[g(x+) + g(x-)]$ equals $g(x)$ when x is a point of continuity for g. The rigorous definition of convergence is

$$\lim_{M \to \infty} S_M(x) = \frac{1}{2}[g(x+) + g(x-)]$$

where

$$S_M(x) = \frac{1}{2}A_0 + \sum_{n=1}^{M} A_n \cos \frac{n\pi x}{L}.$$

The function S_M is called the *(cosine series) partial sum containing M harmonics.*

Example 7.4

Let g be defined on the interval $(0, 2)$ by

$$g(x) = \begin{cases} 1 & \text{for } 0 < x < 1 \\ -1 & \text{for } 1 < x < 2. \end{cases}$$

In Figure 1.18 we have graphed S_M for $M = 10$, 20, and 40 harmonics. Notice the Gibbs' phenomenon occurring near the jump discontinuity of g at $x = 1$. ▯

References

For further discussion of the theory of Fourier series, consult the following: [Ch-B], [Da], [Ka], [To], [Wa], [Zy]. A good history of Fourier analysis can be found in [Da-H].

Exercises

Section 1

1.1 Using power series expansions, justify Euler's identity (1.1).

1.2 Draw figures similar to Figures 1.1 and 1.2 for the following functions.
  (a) $3\cos 10\pi x - 6\cos 20\pi x$
  (b) $7\sin 8\pi x + 3\sin 16\pi x - \sin 32\pi x$
  (c) $6\cos 4\pi x + 3\sin 12\pi x$

1.3 Using *FAS*, have your computer display the graph of the Fourier transform of $4\cos 2\pi 9x$ using 2048 points over each of the intervals $[-2, 2]$, $[-5, 5]$, and $[-10, 10]$. (*Note: Press* **y** *(= pos. exponent) when you are asked for type of transform.*)

   Notice the similarity of your transform's real part to Figure 1.1(b). In particular, by changing the x-interval to $[-45, 45]$, check that the frequency 9 is located correctly on the horizontal axis.

1.4 Repeat Exercise 1.3, but use $6 \sin 2\pi 9x$ and compare your transform's imaginary part to Figure 1.2(b).

1.5 Repeat Exercise 1.3, but use the functions given in 1.2(a)–(c). Check that the frequencies involved for each function are accurately determined.

1.6 Prove the identity $c_{-n} = c_n^*$ described in Remark 1.6(b).

1.7 Prove the identity described in Remark 1.6(c), *which amounts to proving the identity*

$$\frac{1}{P}\int_{-\frac{1}{2}P}^{\frac{1}{2}P} h(x)\,dx = \frac{1}{P}\int_0^P h(x)\,dx$$

for a periodic function h of period P.

1.8 Explain why, for a function h having period P, the following equation is true for *all* constants c and d:

$$\int_c^{c+P} h(x)\,dx = \int_d^{d+P} h(x)\,dx.$$

Note: This shows that the integral of a function of period P is always the same over all intervals of length P.

Section 2

1.9 Expand the following functions in Fourier series using the periods specified and the given intervals.
   (a) $\cos x$, period π, interval $[-(1/2)\pi, (1/2)\pi]$
   (b) x^2, period 3, interval $[0,3]$
   (c) $\exp(x)$, period 2, interval $[-1,1]$. [*Note:* $\exp(x)$ is the same as e^x.]
   (d) $g(x) = \begin{cases} -1 & \text{for } -\frac{1}{2} < x < 0 \\ 1 & \text{for } 0 < x < \frac{1}{2} \end{cases}$, period 1, interval $[-\frac{1}{2}, \frac{1}{2}]$

   (e) $h(x) = \begin{cases} 1 & \text{for } 0 < x < 1 \\ 0 & \text{for } 1 < x < 2 \end{cases}$, period 3, interval $[0,2]$

1.10 Obtain the real forms of the Fourier series of the functions in Exercise 1.9.

1.11 Using *FAS*, graph the partial sums S_M for the Fourier series of the functions in Exercise 1.9. Use 1024 points and $M = 1, 4, 16$, and 64 harmonics.

Section 3

1.12 Using either formula (3.5) or (3.6), calculate the real form of the Fourier series for the following functions over the given intervals.
   (a) $f(x) = x^2 - 3x$, over $[-\pi, \pi]$
   (b) $f(x) = \cos x$, over $[-\frac{1}{2}\pi, \frac{1}{2}\pi]$
   (c) $f(x) = |x|$, over $[-1, 1]$
   (d) $f(x) = x^2 + 3$, over $[0, 4]$
   (e) $f(x) = \sin x$, over $[0, \frac{1}{2}\pi]$
   (f) $f(x) = 3x$, over $[0, 4]$

1.13 Using *FAS*, graph the sum of the first four nonzero terms of each Fourier series that you found in Exercise 1.12. Also, graph each function as a check of your calculations.

Section 4

1.14 Using *FAS*, draw graphs of partial sums of the Fourier series, using 10, 30, and 90 harmonics, for the following functions.
   (a) $\exp(x)$, $0 \le x \le 1$
   (b) $x^2 + x$, $-2 \le x \le 2$

$$(c) \ g(x) = \begin{cases} 0 & \text{for } 0 < x < 1 \\ 1 & \text{for } 1 < x < 2 \\ 2 & \text{for } 2 < x < 3 \end{cases}$$

(d) $g(x) = |x|$, $-1 \le x \le 1$

(e) $\sin x$, $-\frac{1}{2}\pi \le x \le \frac{1}{2}\pi$

(f) $\exp(-x)$, $-1 \le x \le 1$

1.15 Discuss the convergence of the Fourier series of the functions in Exercise 1.14. In particular, draw graphs of the functions to which the Fourier series converge over the whole real line.

1.16 For the function given in Exercise 1.14(c), find out how may harmonics are needed in the partial sum of the Fourier series in order to approximate $g(x)$ to within ± 0.005 error on the interval $[1.4, 1.6]$. (*Note*: After initially graphing g change the x-interval to $[1.4, 1.6]$ and the y-interval to $[0.995, 1.005]$.) Repeat this exercise for the intervals $[0.1, 0.3]$ and $[0.3, 0.5]$.

1.17 Same problem as 1.16, but use the function in Exercise 1.14)(d) and x-interval $[-0.1, 0.1]$. Repeat this exercise for the intervals $[0.1, 0.3]$ and $[0.3, 0.5]$.

Section 5

1.18 By changing x- and y-intervals, in order to magnify appropriate regions, display Gibbs' phenomenon for the functions in Exercise 1.14.

1.19 A more quantitative treatment of Gibbs' phenomenon (see [Wa, Chapter 2.5]) shows that as $M \to \infty$ the maxima (minima) of the Fourier series partial sum S_M near a jump discontinuity tend to overshoot (undershoot) by about 9% of the magnitude of the jump. Check this using the graphs displayed in solving Exercise 1.18.

1.20 If g is a piecewise continuous function with period P and its derivative g' is piecewise continuous, then it is known that on any closed interval where g is continuous, the Fourier series for g converges uniformly to g. (For a proof, see [Wa, Chapter 2.5].) Check this result for the function

$$g(x) = \begin{cases} -1 & \text{for } 0 < x < 1 \\ 0 & \text{for } 1 < x < 2 \\ 1 & \text{for } 2 < x < 3 \end{cases}$$

using the intervals $[0.1, 0.9]$, $[1.1, 1.9]$, and $[2.2, 2.8]$, and using 5, 10, 20, and 40 harmonics for the partial sums. Use 1024 points.

1.21 The function $g(x) = |x|$ has a uniformly convergent Fourier series, using the interval $[-1, 1]$. Show this using Theorem 5.1. Also, compute the Fourier series for g and estimate how many harmonics are needed in the Fourier series partial sum S_M in order to insure that $\sup_{x \in [-1,1]} |g(x) - S_M(x)| \le 0.01$. Using *FAS*, confirm this estimate on the region: $-0.01 \le x \le 0.01$ and $-0.01 \le y \le 0.02$. Why is $S_M(x)$ farthest from $g(x)$ at $x = 0$?

1.22 Which of the following functions has a uniformly convergent Fourier series? You are not required to compute the Fourier series, but you might check your results using *FAS*.

(a) $g(x) = x^2 - x$, interval $[0, 1]$

(b) $g(x) = \exp(x)$, interval $[-1, 1]$

(c) $g(x) = x^3$, interval $[-2, 2]$

(d) $g(x) = x^4$, interval $[-1, 1]$

Section 6

1.23 Expand the following functions in Fourier sine and cosine series over the given intervals.

(a) $\cos x$,   on interval $(0, 3)$

(b) x,   on interval $(0, 1)$

(c) $x^2 - x$,   on interval $(0, 2)$

(d) $g(x) = \begin{cases} 1 & \text{for } 0 < x < 1 \\ -1 & \text{for } 1 < x < 2, \end{cases}$ on interval $(0, 2)$

(e) $g(x) = \begin{cases} 0 & \text{for } 0 < x < 1 \\ 1 & \text{for } 1 < x < 2 \\ 0 & \text{for } 2 < x < 3, \end{cases}$ on interval $(0, 3)$

1.24 *Orthogonality of Sines.* Prove the following, for all positive integers m and n,

$$\frac{2}{L} \int_0^L \sin \frac{m\pi x}{L} \sin \frac{n\pi x}{L} \, dx = \begin{cases} 0 & \text{for } m \neq n \\ 1 & \text{for } m = n. \end{cases}$$

1.25 Use the result of Exercise 1.24 to derive (6.5) for a function g defined on the interval $(0, L)$, using a similar argument to the one used in Section 1 to derive Fourier series.

1.26 *Orthogonality of Cosines.* Prove the following, for all positive integers m and n,

$$\int_0^L \cos \frac{n\pi x}{L} \, dx = 0$$

$$\frac{2}{L} \int_0^L \cos \frac{m\pi x}{L} \cos \frac{n\pi x}{L} \, dx = \begin{cases} 0 & \text{for } m \neq n \\ 1 & \text{for } m = n. \end{cases}$$

1.27 Use the result of Exercise 1.26 to derive (6.11) for a function g defined on the interval $(0, L)$, using a similar argument to the one used in Section 1 to derive Fourier series.

1.28 Using *FAS* and the choice *Sine S*, graph $\sum_{n=1}^{M} B_n \sin(n\pi x/L)$, the partial sum containing M harmonics, for the functions in Exercise 1.23. Use $M = 10, 20,$ and 40 harmonics and 1024 points.

1.29 Using *FAS* and the choice *Cos S*, graph $(1/2)A_0 + \sum_{n=1}^{M} A_n \cos(n\pi x/L)$, the partial sum containing M harmonics, for the functions in Exercise 1.23. Use $M = 10, 20,$ and 40 harmonics and 1024 points.

Section 7

1.30 Using *FAS*, graph partial sums of *Fourier series* for odd extensions of the functions in Exercise 1.23 for $M = 10, 20,$ and 40 harmonics. Compare your results to the Fourier sine series partial sums found in Exercise 1.28.

1.31 Using *FAS*, graph partial sums of *Fourier series* for even extensions of the functions in Exercise 1.23 for $M = 10, 20,$ and 40 harmonics. Compare your results to the Fourier cosine series partial sums found in Exercise 1.29.

2

The Discrete Fourier Transform (DFT)

In this chapter we describe the digital (discrete) version of Fourier analysis. We show how Fourier series can be discretized, resulting in the DFT. We also describe discretization of the Fourier sine and Fourier cosine series.

1 Derivation of the DFT

This section will derive the Discrete Fourier Transform by approximating Fourier coefficients.

Consider the kth Fourier series coefficient c_k for a function g, using period P complex exponentials

$$c_k = \frac{1}{P} \int_0^P g(x) e^{-i2\pi kx/P} \, dx. \tag{1.1}$$

We shall approximate the integral in (1.1) by a *left-endpoint, uniform Riemann sum*

$$c_k \approx \frac{1}{P} \sum_{j=0}^{N-1} g(x_j) e^{-i2\pi kx_j/P} \frac{P}{N} \tag{1.2}$$

where $x_j = j(P/N)$ for $j = 0, 1, \ldots, N-1$. For this choice of points $\{x_j\}$ we have

$$c_k \approx \frac{1}{N} \sum_{j=0}^{N-1} g\left(j\frac{P}{N}\right) e^{-i2\pi jk/N}. \tag{1.3}$$

Formula (1.3) motivates the following definition of the Discrete Fourier Transform.

DEFINITION 1.4 *Given N complex numbers*

$$\{h_j\}_{j=0}^{N-1}$$

their N-point Discrete Fourier Transform (DFT) is denoted by $\{H_k\}$ where H_k is defined by

$$H_k = \sum_{j=0}^{N-1} h_j e^{-i2\pi jk/N}$$

for all integers $k = 0, \pm1, \pm2, \ldots$.

REMARK 1.5 (a) We see that formula (1.3) describes the kth Fourier coefficient of the function g as approximately $(1/N)G_k$ where $\{G_k\}$ is the N-point DFT of $\{g(j(P/N))\}_{j=0}^{N-1}$. (b) It is a useful notational convention to use a small letter for a given finite sequence of numbers and a capital version of that letter for the sequence's DFT. (c) In Definition 1.4 the integer j takes values from 0 to $N - 1$, but the integer k takes values from all the integers. This apparent asymmetry will be removed in the next section when we discuss the *inverse* DFT. ∎

Example 1.6
Let $h_j = e^{-jc}$ for $j = 0, 1, \ldots, N-1$ and c a constant. Show that the N-point DFT of $\{h_j\}$ is given by

$$H_k = \frac{1 - e^{-cN}}{1 - e^{-c-i2\pi k/N}} \cdot \tag{1.7}$$

SOLUTION By the definition of an N-point DFT, we have

$$H_k = \sum_{j=0}^{N-1} e^{-jc} e^{-i2\pi jk/N}. \tag{1.8}$$

Hence, because $e^x e^z = e^{x+z}$, we can write

$$H_k = \sum_{j=0}^{N-1} e^{j(-c-i2\pi k/N)}. \tag{1.9}$$

To simplify (1.9), we make use of the formula for the sum of a finite geometric series

$$1 + r + r^2 + \ldots + r^{N-1} = \frac{1 - r^N}{1 - r} \cdot \tag{1.10}$$

Putting $r = e^{-c-i2\pi k/N}$ in (1.10) and substituting into (1.9), we get

$$H_k = \frac{1 - e^{N(-c-i2\pi k/N)}}{1 - e^{-c-i2\pi k/N}} . \tag{1.11}$$

The right side of (1.11) can be further simplified by making use of the identities

$$e^{N(-c-i2\pi k/N)} = e^{-cN}e^{-i2\pi k}$$

$$= e^{-cN}. \tag{1.12}$$

The last equality in (1.12) holding because $e^{-i2\pi k} = 1$. Using (1.12) we rewrite (1.11) obtaining

$$H_k = \frac{1 - e^{-cN}}{1 - e^{-c-i2\pi k/N}}$$

which shows that (1.7) is true. ∎

REMARK 1.13   In *FAS*, the method used for approximating Fourier coefficients is the one described in Exercise 2.5. To make our description simpler in the text, however, we stick to left-endpoint sums. ∎

---

2   Basic properties of the DFT

In this section we shall discuss some basic properties of the DFT. These properties are *linearity*, *periodicity*, and *inversion*. This last property will allow us to define the *inverse DFT* and remove the asymmetry between the original sequence (of length N) versus the transformed sequence (of infinite length) mentioned in the previous section.

From now on we will use the letter W to stand for either $e^{-i2\pi/N}$ or $e^{i2\pi/N}$. Which of the two exponentials is meant should be clear from the context. In either case we have

$$W^N = 1 \tag{2.1}$$

which is the only special property of W that we will need in this section.

THEOREM 2.2
Suppose that the sequence $\{h_j\}_{j=0}^{N-1}$ has N-point DFT $\{H_k\}$ and $\{g_j\}_{j=0}^{N-1}$ has N-point DFT $\{G_k\}$, then the following properties hold:

(a) **Linearity.** *For all complex constants a and b, the sequence $\{ah_j + bg_j\}_{j=0}^{N-1}$ has N-point DFT $\{aH_k + bG_k\}$.*

(b) **Periodicity.** *For all integers k we have $H_{k+N} = H_k$.*

(c) **Inversion.** For $j = 0, 1, \ldots, N - 1$,

$$h_j = \frac{1}{N} \sum_{k=0}^{N-1} H_k e^{i2\pi jk/N}.$$

PROOF (a) is left to the reader as an exercise. To prove (b) and (c) we put W equal to $e^{-i2\pi/N}$ and make use of (2.1). The DFT $\{H_k\}$ is defined by

$$H_k = \sum_{j=0}^{N-1} h_j W^{jk}.$$

Hence

$$H_{k+N} = \sum_{j=0}^{N-1} h_j W^{j(k+N)} = \sum_{j=0}^{N-1} h_j W^{jk}(W^N)^j$$

$$= \sum_{j=0}^{N-1} h_j W^{jk}$$

which proves (b). To prove (c) we note that

$$W^{-1} = e^{i2\pi/N} \tag{2.3}$$

and then, by raising both sides of this last equation to the power jk, we get

$$e^{i2\pi jk/N} = W^{-jk}.$$

It then follows that

$$\frac{1}{N} \sum_{k=0}^{N-1} H_k e^{i2\pi jk/N} = \frac{1}{N} \sum_{k=0}^{N-1} H_k W^{-jk}$$

$$= \frac{1}{N} \sum_{k=0}^{N-1} \left[\sum_{m=0}^{N-1} h_m W^{mk} \right] W^{-jk}$$

$$= \frac{1}{N} \sum_{k=0}^{N-1} \left[\sum_{m=0}^{N-1} h_m W^{mk} W^{-jk} \right].$$

Now, since $W^{mk} W^{-jk} = W^{(m-j)k}$, we obtain upon switching the order of sums

$$\frac{1}{N} \sum_{k=0}^{N-1} H_k e^{i2\pi jk/N} = \frac{1}{N} \sum_{m=0}^{N-1} h_m \left[\sum_{k=0}^{N-1} W^{(m-j)k} \right]. \tag{2.4}$$

For *fixed* j, if $m \neq j$ then putting r equal to W^{m-j} in (1.10) yields

$$\sum_{k=0}^{N-1} W^{(m-j)k} = \frac{1 - (W^{m-j})^N}{1 - W^{m-j}} = \frac{0}{1 - W^{m-j}}$$

$$= 0$$

since $W^{m-j} \neq 1$. If, however, $m = j$, then $W^{(m-j)k} = W^0 = 1$ and (2.4) becomes

$$\frac{1}{N} \sum_{k=0}^{N-1} H_k e^{i2\pi jk/N} = \frac{1}{N} h_j \sum_{k=0}^{N-1} 1 = h_j$$

and (c) is proved. ∎

A consequence of inversion is that *no two distinct sequences can have the same DFT*. Based on Theorem 2.2(c) we make the following definition.

DEFINITION 2.5   *If $\{G_k\}_{k=0}^{N-1}$ is a sequence of N complex numbers, then the N-point **inverse DFT** is defined by*

$$g_j = \sum_{k=0}^{N-1} G_k e^{i2\pi jk/N}$$

for all integers $j = 0, \pm 1, \pm 2, \ldots$.

We can see from putting W equal to $e^{i2\pi/N}$ that (using essentially the same proof as above for Theorem 2.2) the inverse DFT has the properties of linearity, periodicity, and inversion. For the inverse DFT the inversion formula is

$$G_k = \frac{1}{N} \sum_{j=0}^{N-1} g_j e^{-i2\pi jk/N}.$$

Periodicity for the inverse DFT shows that $\{g_j\}_{j=0}^{N-1}$ is a finite subsequence of a larger periodic sequence $\{g_j\}$ defined for all integers j by the inverse DFT of $\{G_k\}_{k=0}^{N-1}$.

There are two other important properties of the DFT that are worth noting. First, by periodicity, it follows that every DFT $\{H_k\}$ satisfies

$$H_{N-k} = H_{-k}. \tag{2.6}$$

And, there is the property known as *Parseval's equality*

$$\sum_{j=0}^{N-1} |h_j|^2 = \frac{1}{N} \sum_{k=0}^{N-1} |H_k|^2 \tag{2.7}$$

where $\{H_k\}$ is the N-point DFT of $\{h_j\}$. We leave the proof of (2.7) to the reader as an exercise.

---

3 Relation of the DFT to Fourier coefficients

This section will explain further the approximation involved in replacing Fourier coefficients by the DFT. The approximation described in (1.3) must be interpreted carefully. Since the right side of (1.3) is $1/N$ times the DFT of $\{g(j(P/N))\}_{j=0}^{N-1}$ it has period N in the variable k. The left side of (1.3) is c_k, the kth Fourier coefficient of g, and will typically *not* have period N.

To make sure that (1.3) is valid we will make the assumption that $|k| \le (1/8)N$. To explain the origin of this *rule of thumb*, let's consider a specific example.

Example 3.1
Let $g(x) = e^{-x}$ on the interval $[0, 10]$. Compare the two sides of equation (1.3).
$\square$

SOLUTION For c_k we have

$$c_k = \frac{1}{10} \int_0^{10} e^{-x} e^{-i2\pi kx/10} \, dx$$

$$= \frac{1}{10} \int_0^{10} e^{(-1-i2\pi k/10)x} \, dx$$

$$= \frac{1 - e^{-10}e^{-i2\pi k}}{10 + i2\pi k} \, .$$

Thus, we have found c_k,

$$c_k = \frac{1 - e^{-10}}{10 + i2\pi k} \, . \tag{3.2}$$

On the other hand, for $(1/N)G_k$ we have

$$\frac{1}{N}G_k = \frac{1}{N} \sum_{j=0}^{N-1} g\left(j\frac{P}{N}\right) e^{-i2\pi jk/N}$$

$$= \frac{1}{N} \sum_{j=0}^{N-1} e^{-j(10/N)} e^{-i2\pi jk/N} \, .$$

Applying (1.7), putting c equal to $10/N$, we have

$$\frac{1}{N}G_k = \frac{1}{N}\frac{1 - e^{-10}}{1 - e^{-(10+i2\pi k)/N}} \ . \tag{3.3}$$

In order to compare (3.2) and (3.3), we use the approximation

$$e^{-x} \approx 1 - x \tag{3.4}$$

which is valid for $|x| \ll 1$. For $|k|/N \ll 1$ we will have, using $(10+i2\pi k)/N$ in place of x in (3.4),

$$e^{-(10+i2\pi k)/N} \approx 1 - (10 + i2\pi k)/N. \tag{3.5}$$

Using (3.5) in (3.3) we get

$$\frac{1}{N}G_k \approx \frac{1}{N}\frac{1 - e^{-10}}{(10 + i2\pi k)/N} = \frac{1 - e^{-10}}{10 + i2\pi k} = c_k. \tag{3.6}$$

Hence,

$$\frac{1}{N}G_k \approx c_k \tag{3.7}$$

provided $|k| \ll N$. Our rule of thumb that $|k| \le (1/8)N$ is not in contradiction with $|k| \ll N$.

There is another consideration involved in demanding that $|k| \le (1/8)N$. In Figure 2.1 we have graphs of part of $e^{-x}e^{-i2\pi kx/10}$ obtained by connecting the sampled values (at $x_j = j(10/N)$) by line segments. These graphs are of the *imaginary part*

$$Im\left[e^{-x}e^{-i2\pi kx/10}\right] = -e^{-x}\sin\frac{2\pi kx}{10} \tag{3.8}$$

but similar graphs could be drawn for the *real part*

$$Re\left[e^{-x}e^{-i2\pi kx/10}\right] = -e^{-x}\cos\frac{2\pi kx}{10} \ . \tag{3.9}$$

The graphs reveal (for the case of $N = 128$) that when $k = (1/2)N$ the sampled version of the function in (3.8) is *indistinguishable* from the sampled version of the constant function 0. In this case, the 0-function is called an *alias* of the function in (3.8) for $k = (1/2)N$.

The origin of our *rule of thumb* for taking $|k| \le (1/8)N$ is based on the *qualitative* observation that, for such values of k, the sampled versions of our functions generally have no aliases with other samplings at lower frequencies. In general, it is difficult to give precise estimates on the size of k that will insure a given degree of error in approximation (1.3). In any case, to pursue this matter further would obscure our attempt at introducing the fundamentals of DFTs.   ∎

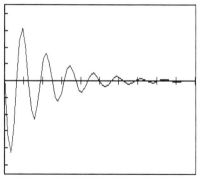

(a) $k = 16$
X interval: $[0, 5]$ X increment = .5
Y interval: $[-1.1, .9]$ Y increment = .2

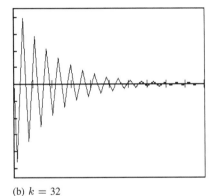

(b) $k = 32$
X interval: $[0, 5]$ X increment = .5
Y interval: $[-1.1, .9]$ Y increment = .2

(c) $k = 64$
X interval: $[0, 5]$ X increment = .5
Y interval: $[-1.1, .9]$ Y increment = .2

FIGURE 2.1
Graphs of $Im(e^{-x}e^{-i2\pi kx/10})$ **for** $N = 128$ **points and** $k = 16, 32,$ **and** 64.

4   Relation of the DFT to sampled Fourier series

An important application of the DFT is to calculations with sampled Fourier series. We will begin our treatment of this topic in this section and expand on it in Chapter 4 when we discuss filtering of Fourier series.

Using the approximation of Fourier coefficients by the DFT described in (1.3), we can write the Mth partial sum S_M of the Fourier series for a function g as

$$S_M(x) \approx \sum_{k=-M}^{M} \frac{1}{N} G_k e^{i2\pi kx/P} \tag{4.1}$$

where $\{G_k\}$ is the N-point DFT of the sequence of samples $\{g(j(P/N))\}_{j=0}^{N-1}$ of g.

In (4.1) we assume that $M \leq (1/8)N$ and P is the period for the Fourier series expansion of g.

If we substitute for x the same sampling points $\{j(P/N)\}_{j=0}^{N-1}$, which we used to get G_k, then we obtain from (4.1)

$$S_M\left(j\frac{P}{N}\right) \approx \frac{1}{N} \sum_{k=-M}^{M} G_k e^{i2\pi jk/N}. \tag{4.2}$$

Formula (4.2) can be expressed as an N-point (inverse) DFT with weight $W = e^{i2\pi/N}$ if we make the following algebraic manipulations. First, since $W^N = 1$, we have $W^{N-k} = W^{-k}$. And, from (2.6), we have $G_{N-k} = G_{-k}$ (using $\{G_k\}$ in place of $\{H_k\}$). Thus, (4.2) can be rewritten as

$$
\begin{aligned}
S_M\left(j\frac{P}{N}\right) &\approx \frac{1}{N} \sum_{k=0}^{M} G_k e^{i2\pi jk/N} + \frac{1}{N} \sum_{k=1}^{M} G_{-k} e^{-i2\pi jk/N} \\
&= \frac{1}{N} \sum_{k=0}^{M} G_k e^{i2\pi jk/N} + \frac{1}{N} \sum_{k=1}^{M} G_{N-k} e^{i2\pi j(N-k)/N} \\
&= \frac{1}{N} \sum_{k=0}^{M} G_k e^{i2\pi jk/N} + \frac{1}{N} \sum_{k=N-M}^{N-1} G_k e^{i2\pi jk/N}.
\end{aligned}
\tag{4.3}
$$

To see that the last sums can be rewritten as a single DFT sum, we define $\{H_k\}_{k=0}^{N-1}$ by

$$H_k = \begin{cases} G_k & \text{for } k = 0, 1, \ldots, M \\ 0 & \text{for } k = M+1, \ldots, N-M-1 \\ G_k & \text{for } k = N-M, \ldots, N-1. \end{cases} \tag{4.4}$$

We can interpret $\{H_k\}$ as a lopping off to 0 of $\{G_k\}$ in an interval about the index value $(1/2)N$. Based on formula (4.4) we can write

$$S_M\left(j\frac{P}{N}\right) \approx \frac{1}{N} \sum_{k=0}^{N-1} H_k e^{i2\pi jk/N} \tag{4.5}$$

which expresses the sampled Fourier series partial sum values $\{S_M(j(P/N))\}_{j=0}^{N-1}$ as the N-point DFT (weight $W = e^{i2\pi/N}$) of $\{H_k\}$, multiplied by $1/N$. When graphing S_M the software uses the method of *linear interpolation*; in other words, it connects the sampled values of S_M with straight line segments.

To summarize our results, we compute an approximation to a Fourier series partial sum S_M for a function g as follows:

Step 1. Sample the given function g over the interval $[0, P]$, using N evenly spaced sample points, obtaining $\{g(j(P/N))\}_{j=0}^{N-1}$.

Step 2. Compute the N-point DFT $\{G_k\}$ of the sequence $\{g(j(P/N))\}_{j=0}^{N-1}$ using the weight $W = e^{-i2\pi/N}$.

Step 3. Lop off to 0 the DFT $\{G_k\}$ for $M + 1 \leq k \leq N - M - 1$ where $M \leq (1/8)N$, obtaining a new sequence $\{H_k\}$.

Step 4. Compute the N-point DFT of $\{H_k\}$, using weight $W = e^{i2\pi/N}$, and multiply each number of this DFT by $1/N$. The result is the approximation to the sampled Fourier series partial sum $\{S_M(j(P/N))\}_{j=0}^{N-1}$. *Note:* By periodicity, $S_M(P) = S_M(0)$.

Step 5. Connect the points of the sequence $\{S_M(j(P/N))\}_{j=0}^{N}$ by straight line segments, yielding an approximation to $S_M(x)$ over the interval $0 \leq x \leq P$.

---

5 Discrete sine and cosine transforms

Discrete sine and cosine transforms can be derived from the formulas for sine and cosine series described in Chapter 1, Section 6. The kth sine coefficient B_k of a real-valued function g over the interval $[0, L]$ is defined by

$$B_k = \frac{2}{L} \int_0^L g(x) \sin \frac{k\pi x}{L} \, dx \tag{5.1}$$

while the kth cosine coefficient A_k is defined by

$$A_k = \frac{2}{L} \int_0^L g(x) \cos \frac{k\pi x}{L} \, dx. \tag{5.2}$$

To derive the discrete sine transform we approximate the integral in (5.1) using a uniform left-endpoint Riemann sum

$$B_k \approx \frac{2}{L} \sum_{j=0}^{N-1} g(j\frac{L}{N}) \sin \frac{k\pi j L/N}{L} \frac{L}{N}. \tag{5.3}$$

Therefore,

$$B_k \approx \frac{2}{N} \sum_{j=0}^{N-1} g\left(j\frac{L}{N}\right) \sin \frac{\pi j k}{N}.$$

Or, since the $j = 0$ term above is just 0, we have

$$B_k \approx \frac{2}{N} \sum_{j=1}^{N-1} g\left(j\frac{L}{N}\right) \sin \frac{\pi j k}{N} . \tag{5.4}$$

and, similarly,

$$A_k \approx \frac{2}{N} \sum_{j=0}^{N-1} g\left(j\frac{L}{N}\right) \cos \frac{\pi j k}{N} . \tag{5.5}$$

Based on formulas (5.4) and (5.5), we make the following definition.

DEFINITION 5.6 *For a real sequence* $\{h_j\}_{j=1}^{N-1}$ *the* **Discrete Sine Transform** *(DST),* $\{H_k^S\}$, *is defined by*

$$H_k^S = \sum_{j=1}^{N-1} h_j \sin \frac{\pi j k}{N} .$$

For a real sequence $\{h_j\}_{j=0}^{N-1}$ *the* **Discrete Cosine Transform** *(DCT) is defined by*

$$H_k^C = \sum_{j=0}^{N-1} h_j \cos \frac{\pi j k}{N} .$$

From formula (5.4), we see that the kth Fourier sine coefficient B_k is approximated by $2/N$ times the DST of the sequence $\{g(j(L/N))\}_{j=1}^{N-1}$. And from formula (5.5), we see that the kth Fourier cosine coefficient A_k is approximated by $2/N$ times the DCT of the sequence $\{g(j(L/N))\}_{j=0}^{N-1}$.

REMARK 5.7 Just as we discussed in the previous sections for the DFT and Fourier series coefficients, the approximations (5.4) and (5.5) are only valid for $k \ll N$. *We shall assume from now on that* $k \le (1/4)N$, for reasons similar to the ones given in Section 3. ∎

Just as for Fourier series, a Fourier sine series and a Fourier cosine series can be approximated using a DST and a DCT, respectively. Let's begin with a Fourier sine series. Suppose that a function g is expanded in a Fourier sine series over $[0, L]$:

$$g \sim \sum_{k=1}^{\infty} B_k \sin \frac{k\pi x}{L}, \qquad \left(B_k = \frac{2}{L} \int_0^L g(x) \sin \frac{k\pi x}{L}\, dx\right) . \tag{5.8}$$

If $S_M(x)$ is used to denote the M-harmonic partial sum of the Fourier sine series in (5.8), then we have

$$S_M(x) = \sum_{k=1}^{M} B_k \sin \frac{k\pi x}{L}, \qquad \left(B_k = \frac{2}{L} \int_0^L g(x) \sin \frac{k\pi x}{L} \, dx \right). \tag{5.9}$$

Assuming that $M \leq (1/4)N$ (see Remark 5.7), we approximate each B_k, using $2/N$ times the DST of $\{g(j(L/N))\}$. Thus,

$$S_M(x) \approx \frac{2}{N} \sum_{k=1}^{M} G_k^S \sin \frac{k\pi x}{L} \tag{5.10}$$

where $\{G_k^S\}$ is the DST of $\{g(j(L/N))\}$. Replacing x by the sample values $j(L/N)$, we have

$$S_M\left(j\frac{L}{N} \right) \approx \frac{2}{N} \sum_{k=1}^{M} G_k^S \sin \frac{\pi j k}{N}, \qquad (j = 1, \ldots, N-1). \tag{5.11}$$

Defining the sequence $\{H_k\}$ by

$$H_k = \begin{cases} G_k & \text{for } k = 1, \ldots, M \\ 0 & \text{for } k = M+1, \ldots, N-1 \end{cases} \tag{5.12}$$

we see that

$$S_M\left(j\frac{L}{N} \right) \approx \frac{2}{N} \sum_{k=0}^{N-1} H_k \sin \frac{\pi j k}{N}, \qquad (j = 1, \ldots, N-1). \tag{5.13}$$

Formula (5.13) shows that the sampled values of the M-harmonic Fourier sine series partial sum, $\{S_M(j(L/N))\}$, are approximated by multiplying each element of the DST of the sequence $\{H_k\}$, defined in (5.12), by $2/N$. If these sample values are connected by line segments [including $S_M(0) = S_M(L) = 0$], then $S_M(x)$ can be approximated for $0 \leq x \leq L$. This, of course, is exactly what *FAS* does when it graphs a sine series.

Similarly, we can approximate Fourier cosine series using a DCT. Suppose that the function g is expanded in a Fourier cosine series over the interval $[0, L]$:

$$g \sim \frac{1}{2} A_0 + \sum_{k=1}^{\infty} A_k \cos \frac{k\pi x}{L}, \qquad \left(A_k = \frac{2}{L} \int_0^L g(x) \cos \frac{k\pi x}{L} \, dx \right). \tag{5.14}$$

If $S_M(x)$ is used to denote the M-harmonic partial sum of the Fourier cosine series in (5.14), then

$$S_M(x) = \frac{1}{2} A_0 + \sum_{k=1}^{M} A_k \cos \frac{k\pi x}{L}. \tag{5.15}$$

Assuming that $M \leq (1/4)N$ (see Remark 5.7), we approximate each A_k, using $2/N$ times the DCT of $\{g(j(L/N))\}$. Thus,

$$S_M(x) \approx \frac{2}{N} \left[\frac{1}{2} G_0^C + \sum_{k=1}^{M} G_k^C \cos \frac{k\pi x}{L} \right] \qquad (5.16)$$

where $\{G_k^C\}$ is the DCT of $\{g(j(L/N))\}$. Replacing x by the sample values $j(L/N)$, we have

$$S_M\left(j\frac{L}{N}\right) \approx \frac{2}{N} \left[\frac{1}{2} G_0^C + \sum_{k=1}^{M} G_k^C \cos \frac{\pi j k}{N} \right]. \qquad (5.17)$$

Defining the sequence $\{H_k\}$ by

$$H_k = \begin{cases} \frac{1}{2} G_0^C & \text{for } k = 0 \\ G_k^C & \text{for } k = 1, \ldots, M \\ 0 & \text{for } k = M+1, \ldots, N-1 \end{cases} \qquad (5.18)$$

we see that

$$S_M\left(j\frac{L}{N}\right) \approx \frac{2}{N} \sum_{k=0}^{M} H_k \cos \frac{\pi j k}{N}, \qquad (j = 0, 1, \ldots, N-1). \qquad (5.19)$$

Formula (5.19) shows that the sampled values of the M-harmonic Fourier cosine series partial sums $\{S_M(j(L/N))\}$ are approximated by multiplying each element of the DCT of the sequence $\{H_k\}$, defined in (5.18), by $2/N$. If these sample values are connected by line segments [including $S_M(L) \approx 2/N \sum_{k=0}^{N-1} H_k(-1)^k$ which must be calculated separately], then $S_M(x)$ can be approximated for $0 \leq x \leq L$. This, of course, is just what *FAS* does when it graphs a cosine series.

In the next chapter, the Fast Sine Transform and the Fast Cosine Transform will be discussed. Also, the inversion theorems for DSTs and DCTs will be explained.

References

For more information on DFTs, see [Br,2], [Ra-R], [Bri], and [Op-S].

Exercises

Section 1

2.1 Compute the N-point DFT of the *constant sequence* $\{1\}_{j=0}^{N-1}$.

2.2 Compute the 16-point DFT of the sequence

$$1\ 1\ 1\ 1\ 0\ 0\ 0\ 0\ 0\ 0\ 0\ 0\ 1\ 1\ 1\ 1\ .$$

2.3 Compute the 16-point DFT of the sequence

$$1\ 1\ 1\ 1\ 0\ 0\ 0\ 0\ 0\ 0\ 0\ 0\ 0\ 1\ 1\ 1\ .$$

2.4 Suppose that a *right-endpoint, uniform Riemann sum* (using points $x_j = j(P/N)$ for $j = 1,\ 2,\ \dots,\ N$) is used to approximate the integral for c_k in (1.1). Show that c_k is approximated by an N-point DFT, using the sequence

$$g(P),\ g\left(\frac{P}{N}\right),\ g\left(2\frac{P}{N}\right),\dots,\ g\left((N-1)\frac{P}{N}\right).$$

2.5 Show that by averaging approximations with left- and right-endpoint sums, the Fourier coefficient c_k in (1.1) can be approximated by an N-point DFT, using the sequence

$$\frac{1}{2}[g(0) + g(P)],\ g\left(\frac{P}{N}\right),\ g\left(2\frac{P}{N}\right),\dots,\ g\left((N-1)\frac{P}{N}\right).$$

Section 2

2.6 Prove Parseval's equality (2.7), and prove the linearity property (a) in Theorem 2.2.

2.7 Suppose that a sequence $\{h_j\}_{j=0}^{N-1}$ is modified as follows: N zeroes are appended to the sequence, to obtain a new sequence $\{g_j\}_{j=0}^{2N-1}$, where $g_j = h_j$ for $j = 0,\ 1,\ \dots,\ N-1$, and $g_j = 0$ for $j = N,\ \dots,\ 2N-1$. Show that for the $2N$-point DFT $\{G_k\}$ of $\{g_j\}$,

$$G_{2k} = H_k, \qquad k = 0,\ 1,\dots,\ N-1$$

where $\{H_k\}$ is the N-point DFT of $\{h_j\}$.

Section 3

2.8 Using *FAS*, graph $\sin(k\pi x)$ over the interval $[-1, 1]$, using 1024 points, for $k = 32$, 64, 128, 256, 512, and 1024. When does aliasing begin (in a noticeable way) and how does this compare with the rule of thumb $|k| \leq 128$?

2.9 Repeat Exercise 2.8 for $\cos(k\pi x)$.

2.10 Repeat Exercise 2.8 and 2.9 for $N = 2048$ points and $k = 64,\ 128,\ 256,\ 512,$ 1024, and 2048.

Section 4

2.11 For the function

$$f(x) = \begin{cases} 1 & \text{for } 0 < x < \pi \\ 0 & \text{for } -\pi < x < 0 \end{cases}$$

graph the 7-harmonic partial sum of its Fourier series over $[-\pi, \pi]$, using 1024 points, in two ways. First, using *FAS*. Second, by graphing explicitly the eight terms of the series. Compare these two graphs over the interval $[-0.1\pi, 0]$. You should observe a slight discrepancy between these two graphs. [The DFT method, actually the FFT method, trades a small amount of inaccuracy for a *massive* increase in speed of calculation.]

2.12 Repeat Exercise 2.11, but now use 4096 points. The discrepancy between the two graphs should now be less.

Section 5

2.13 Compute a 7-harmonic Fourier sine series for the function

$$f(x) = \begin{cases} 1 & \text{for } 0 < x < 2 \\ 0 & \text{for } 2 < x < 3 \end{cases}$$

using 1024 points. Do this computation in two ways. First, using *FAS*. Second, by graphing the explicitly entered function consisting of the 7 terms of the Fourier sine series. Compare these two graphs over the interval $[1.9, 2.1]$. You should observe a slight discrepancy. (The Fast Sine Transform method exchanges a slight inaccuracy for a *massive* increase in speed.)

2.14 Repeat Exercise 2.13, but use cosine series instead.

2.15 Repeat Exercises 2.13 and 2.14, but now use 4096 points. You should observe less discrepancy between the two graphs.

3

The Fast Fourier Transform (FFT)

A direct calculation of an N-point DFT requires $(N-1)^2$ multiplications. For large N, say $N > 1000$, this would require too much computer time. The term *Fast Fourier Transform* (FFT) is used to describe a computer algorithm that reduces the amount of computer time enormously. The FFT that we describe in this chapter reduces the calculation time by a factor of 200 when $N = 1024$. The FFT is one of the greatest contributions to numerical analysis made in this century.

Besides the FFT we shall also describe a Fast Sine Transform (FST) and a Fast Cosine Transform (FCT).

See Appendix B for computer programs related to the material in this chapter.

1 Decimation in time, radix 2, FFT

In this section we shall describe one of the most widely used FFT algorithms, the decimation in time, radix 2, FFT. *Throughout this chapter we will assume that N is a power of 2, say $N = 2^R$, where R is a positive integer.* Also, we will continue with the notation introduced in the last chapter for the N-point DFT:

$$H_k = \sum_{j=0}^{N-1} h_j W^{jk} \tag{1.1}$$

where W stands for either $e^{i2\pi/N}$ or $e^{-i2\pi/N}$. The base 2, for $N = 2^R$, is often called radix 2. In this book we shall work with radix 2 exclusively. Other types of radices, including mixed radices, are discussed in the references.

We begin our FFT algorithm by halving the N-point DFT in (1.1) into two sums, each of which is a $(1/2)N$-point DFT

$$H_k = \sum_{j=0}^{\frac{1}{2}N-1} h_{2j}(W^2)^{jk} + \sum_{j=0}^{\frac{1}{2}N-1} h_{2j+1}(W^2)^{jk}W^k. \qquad (1.2)$$

Based on (1.2) we write H_k as

$$H_k = H_k^0 + W^k H_k^1$$

$$H_k^0 = \sum_{j=0}^{\frac{1}{2}N-1} h_{2j}(W^2)^{jk}$$

$$H_k^1 = \sum_{j=0}^{\frac{1}{2}N-1} h_{2j+1}(W^2)^{jk} \qquad (k = 0, 1, \ldots, N-1). \qquad (1.3)$$

Notice that the $(1/2)N$-point DFTs $\{H_k^0\}$ and $\{H_k^1\}$ use weight W^2 (not W). The periods of $\{H_k^0\}$ and $\{H_k^1\}$ are $(1/2)N$; they are the $(1/2)N$-point DFTs of $\{h_0, h_2, \ldots, h_{N-2}\}$ and $\{h_1, h_3, \ldots, h_{N-1}\}$, respectively. It is important to realize that the binary (base 2) expansions of the indices of $h_0, h_2, \ldots, h_{N-2}$ all end with 0 (since the indices are even). This is the reason for the superscript 0 in $\{H_k^0\}$. Similarly, the binary expansions of the indices of $h_1, h_3, \ldots, h_{N-1}$ all end with 1 and that's the reason for the superscript 1 in $\{H_k^1\}$. Using the prefix DFT to stand for "DFT of," we can represent this as

$$\{H_k^0\} = DFT\{h_{\ldots 0}\}, \quad \{H_k^1\} = DFT\{h_{\ldots 1}\}. \qquad (1.4)$$

Since $N = 2^R$ we can divide N evenly by 2, and we have

$$W^{\frac{1}{2}N} = -1. \qquad (1.5)$$

Using (1.5) we will write the first equation in (1.3) as

$$H_k = H_k^0 + W^k H_k^1, \qquad H_{k+(1/2)N} = H_k^0 - W^k H_k^1 \qquad (1.6)$$

for $k = 0, 1, \ldots, (1/2)N - 1$. The calculations in (1.6) can be diagrammed as

$$
\begin{array}{lll}
H_k^0 & \longrightarrow & H_k^0 + W^k H_k^1 \\
& \times & \\
H_k^1 & \longrightarrow & H_k^0 - W^k H_k^1 \qquad \left(k = 0, 1, \ldots, \frac{1}{2}N - 1\right).
\end{array} \qquad (1.7)
$$

Diagram (1.7) is called a *butterfly*. There are $(1/2)N$ butterflies for this stage of the FFT and each butterfly requires one multiplication, W^k times H_k^1. Figure 3.1 illustrates the butterflies needed for this stage when $N = 16$.

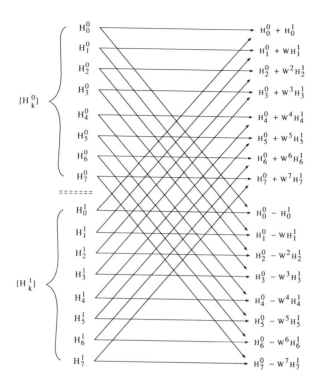

FIGURE 3.1
First reduction in FFT algorithm for $N = 16$.

The splitting of $\{H_k\}$ into two half-size DFTs, $\{H_k^0\}$ and $\{H_k^1\}$, can be repeated on $\{H_k^0\}$ and $\{H_k^1\}$ themselves. We get

$$H_k^0 = H_k^{00} + (W^2)^k H_k^{01}, \qquad H_{k+\frac{1}{4}N}^0 = H_k^{00} - (W^2)^k H_k^{01}$$

$$H_k^1 = H_k^{10} + (W^2)^k H_k^{11}, \qquad H_{k+\frac{1}{4}N}^1 = H_k^{10} - (W^2)^k H_k^{11} \qquad (1.8)$$

for $k = 0, 1, \ldots, (1/4)N - 1$. In (1.8), $\{H_k^{00}\}$ is the $(1/4)N$-point DFT of $\{h_0, h_4, h_8, \ldots, h_{N-4}\}$, $\{H_k^{01}\}$ is the $(1/4)N$-point DFT of $\{h_2, h_6, \ldots, h_{N-2}\}$, $\{H_k^{10}\}$ is the $(1/4)N$-point DFT of $\{h_1, h_5, \ldots, h_{N-3}\}$, $\{H_k^{11}\}$ is the $(1/4)N$-point DFT of $\{h_3, h_7, \ldots, h_{N-1}\}$. Or, in terms of the binary expansions of h_j, we have

$$\{H_k^{00}\} = DFT\{h_{...00}\}, \qquad \{H_k^{01}\} = DFT\{h_{...10}\}$$

$$\{H_k^{10}\} = DFT\{h_{...01}\}, \qquad \{H_k^{11}\} = DFT\{h_{...11}\}.$$

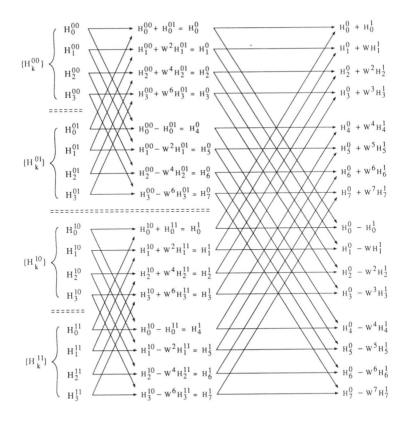

FIGURE 3.2
Two stages of FFT, $N = 16$.

Notice that there is a *reversal* of the last two digits (*bits*) in the binary expansions of the indices j in $\{h_j\}$ for which we calculate the $(1/4)N$-point DFTs $\{H_k^{00}\}$, $\{H_k^{01}\}$, $\{H_k^{10}\}$, $\{H_k^{11}\}$. In Figure 3.2 we have diagrammed the two stages of the FFT that we have described.

If we continue with this process of halving the order of the DFTs, then after $R = \log_2 N$ stages we reach a point where we are performing N one-point DFTs. A one-point DFT of a number h_j is just the identity $h_j \longrightarrow h_j$. Moreover, since the process of indexing the smaller size DFTs by superscripts written *by reversing the order of the bits* in the indices of $\{h_j\}$ will continue, by the time the one-point stage is reached *all* bits in the binary expansion of j will be arranged in reverse order. See Figure 3.3. Therefore, to begin the FFT calculation one must first *rearrange* $\{h_j\}$ so it is listed in *bit reverse order*.

In Figure 3.4 we have shown the calculations needed for performing an 8-point FFT. The result, of course, matches with the definition of an 8-point DFT.

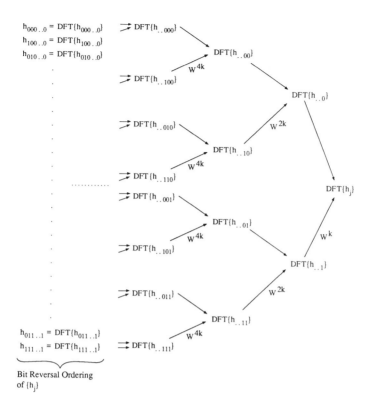

$h_{000\,.\,.0} = DFT\{h_{000\,.\,.0}\}$

$h_{100\,.\,.0} = DFT\{h_{100\,.\,.0}\}$

$h_{010\,.\,.0} = DFT\{h_{010\,.\,.0}\}$

$h_{011\,.\,.1} = DFT\{h_{011\,.\,.1}\}$

$h_{111\,.\,.1} = DFT\{h_{111\,.\,.1}\}$

Bit Reversal Ordering
of $\{h_j\}$

FIGURE 3.3
Reductions of DFTs and bit reversal. The indices are in binary form.

In the exercises the reader is asked to perform a similar calculation of a 16-point FFT. Ultimately, this work is essential to understanding the FFT algorithm.

The FFT results in an enormous savings in computing time. For each of the $\log_2 N$ stages there are $(1/2)N$ multiplications, hence there are $(1/2)N \log_2 N$ multiplications needed for the FFT. This takes much less time than the $(N-1)^2$ multiplications needed for a direct DFT calculation. When $N = 1024$, the FFT requires 5120 multiplications, while the DFT requires $1,046,529$ multiplications. Therefore, for $N = 1024$, the FFT results in a savings, in multiplication time, by a factor of almost 200.

In the next few sections we shall analyze in detail the components of the FFT:

(a) bit reversal reordering of the initial data

(b) rotations involved in butterfly computations

(c) computation of the sines needed to perform (b).

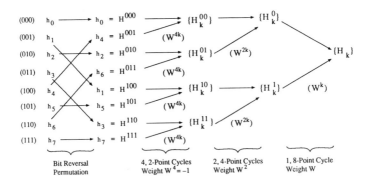

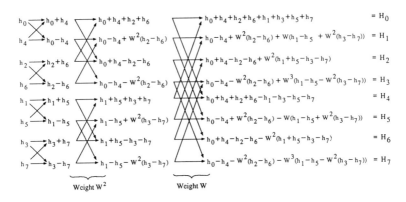

FIGURE 3.4
Decimation in Time FFT for $N = 8$. [*Note:* $W^4 = -1$ and $W^8 = 1$.]

REMARK 1.9   In this section we have described a decimation in time FFT. The other major type of FFT is known as a *decimation in frequency* FFT. This type of FFT is described in [Wa, Chapter 7.13]. ∎

2   Bit reversal

In this section we shall examine in detail the first component of the FFT, bit reversal reordering of the initial data. We shall describe two methods of performing this bit reversal permutation. These methods are *Buneman's algorithm* and another, faster, algorithm.

Buneman's algorithm is the simplest method for performing the bit reversal permutation. It is based on a simple pattern. If we have permuted the N

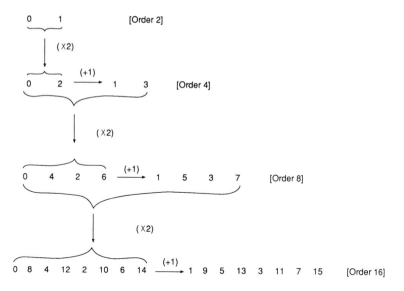

FIGURE 3.5
Buneman's algorithm for generating lists of bit reversed numbers.

numbers

$$\{0, 1, \ldots, N - 1\}$$

by bit reversing their binary expansions, then the permutation of the $2N$ numbers

$$\{0, 1, \ldots, 2N - 1\}$$

is obtained by doubling the numbers in the permutation of $\{0, 1, \ldots, N - 1\}$ to get the first N numbers and then adding 1 to these doubled numbers to get the last N numbers. See Figure 3.5.

That Buneman's algorithm works is easily proved by looking at binary expansions. Suppose m has the binary expansion

$$m = a_1 a_2 \ldots a_R \quad \text{(base 2)} \tag{2.1}$$

where each a_j is either 0 or 1. The number m is mapped to $P_N(m)$, its bit reversed image, hence

$$P_N(m) = a_R \ldots a_2 a_1 \quad \text{(base 2)}. \tag{2.2}$$

If we double $P_N(m)$, then

$$2P_N(m) = a_R \ldots a_2 a_1 0 \quad \text{(base 2)} \tag{2.3}$$

and we see that $2P_N(m)$ is the bit reversed image of

$$m = 0\, a_1\, a_2\, \ldots\, a_R \quad \text{(base 2)} \tag{2.4}$$

where m is considered as an element of $\{0, 1, \ldots, 2N - 1\}$. Note that (2.4) describes the first N numbers in the set $\{0, 1, \ldots, 2N - 1\}$. Also,

$$2P_N(m) + 1 = a_R \ldots a_2\, a_1\, 1 \quad \text{(base 2)} \tag{2.5}$$

is the bit reversal of

$$m = 1\, a_1\, a_2\, \ldots\, a_R \quad \text{(base 2)} \tag{2.6}$$

which accounts for the last N numbers in the list $\{0, 1, \ldots, 2N - 1\}$.

Buneman's algorithm proceeds inductively, at each stage doubling the number of indices for which bit reversals can be done and *swapping the data having these indices*. This is easy to program. We leave this task to the reader as an exercise.

While Buneman's method is admirably simple, it has the defect that it performs *unneccesary* swaps. It can be improved upon by more efficient algorithms that perform only those swaps that are absolutely necessary. We will now describe one such algorithm, the one that *FAS* uses.

To see how this algorithm works, let's suppose that $N = 2^R$ *where $R = 2Q$ is even*. (The case of R odd will be discussed afterwards.) In this case, a number m from the list $\{0, 1, \ldots, N - 1\}$ has a binary expansion

$$m = a_Q \ldots a_1\, b_Q \ldots b_1 \quad \text{(base 2)} \tag{2.7}$$

where a_j, b_j are either 0 or 1. The number $P_N(m)$ then has as its binary expansion

$$P_N(m) = b_1 \ldots b_Q\, a_1 \ldots a_Q \quad \text{(base 2)}. \tag{2.8}$$

Equations (2.7) and (2.8) can be written as

$$m = KM + L$$
$$P_N(m) = P_M(L)M + P_M(K)$$
$$(K = 0, 1, \ldots, M - 1 \text{ and } L = 0, 1, \ldots, M - 1) \tag{2.9}$$

where $M = N^{1/2} = 2^Q$, the number K has the binary expansion $a_Q \ldots a_1$, and L has the binary expansion $b_Q \ldots b_1$.

Equation (2.9) shows how the bit reversal permutation P_N can be split into two bit reversal permutations *of square root size*, P_M, where $M = N^{1/2}$.

Using the facts that $P_M \circ P_M$ is the identity permutation and that P_M permutes $\{0, 1, 2, \ldots, M-1\}$ we can express the set of numbers described in (2.9) in the following new way:

$$n = P_M(K)M + L$$
$$P_N(n) = P_M(L)M + K$$
$$(K = 0, 1, \ldots, M-1 \text{ and } L = 0, 1, \ldots, M-1). \quad (2.10)$$

Equation (2.10) describes all the bit reversals, $n \longrightarrow P_N(n)$. However, each swap of data

$$h_n \iff h_{P_N(n)} \quad (2.11)$$

covers two of the bit reversals described in (2.10). Namely, $n \longrightarrow P_N(n)$ and $P_N(n) \longrightarrow n$. Also, for those n for which $P_N(n) = n$, we do not need to perform the swaps in (2.11). If we were to make a table of K versus $P_M(L)$ for $K = 0, 1, \ldots, M-1$ and $L = 0, 1, \ldots, M-1$, then the unnecessary swaps would be characterized by the *diagonal elements* $K = J$ and $L = J$. For, in this case, (2.10) becomes

$$n = P_M(J)M + J$$
$$P_N(n) = P_M(J)M + J \quad (2.12)$$

so $n = P_N(n)$. The *lower triangle* of this table corresponds to

$$n = P_M(K)M + L$$
$$P_N(n) = P_M(L)M + K$$
$$(K = 1, 2, \ldots, M-1 \text{ and } L = 0, \ldots, K-1) \quad (2.13)$$

and describes $(1/2)M(M-1) = (1/2)N^{1/2}(N^{1/2}-1)$ distinct swaps. This follows from the fact that there are no repeats in the list

$$\{n\} = \{P_M(K) + L\}_{K=1,L=0}^{M-1,K-1}.$$

For, if two swaps $h_n \iff h_{P_N(n)}$ and $h_{n'} \iff h_{P_N(n')}$ were identical, then we would have

$$n = P_M(J)M + I = n' = P_M(J')M + I' \quad (2.14)$$

where $1 \le J, J' < M$, $0 \le I \le J-1 < M$, and $0 \le I' \le J'-1 < M$. Since

$$0 \le P_M(J) < M \quad \text{and} \quad 0 \le P_M(J') < M$$

equation (2.14) implies that $I = I'$ and $P_M(J) = P_M(J')$, hence $J = J'$ as well.

Therefore, we have shown that equation (2.13) describes, *in a one-to-one fashion*, $(1/2)N^{1/2}(N^{1/2}-1)$ swaps. However, that is precisely the number of swaps needed to perform the bit reversal permutation P_N. This last fact follows from the fact that there are $N^{1/2}$ numbers of the form

$$a_1 a_2 \ldots a_Q \, a_Q \ldots a_2 \, a_1 \quad \text{(base 2)} \tag{2.15}$$

which do *not* require swaps, since they are their own bit reversals. Subtracting these $N^{1/2}$ elements from the N total elements and multiplying by $1/2$ (since there is just one swap needed for each pair swapped) we get $(1/2)(N - N^{1/2}) = (1/2)N^{1/2}(N^{1/2} - 1)$. Thus, *equation (2.13) completely describes all bit reversals when the power R in $N = 2^R$ is even.*

When the power R is odd, a general bit reversal will look like

$$n = a_Q \ldots a_1 \, c \, b_Q \ldots b_1$$
$$P_N(n) = b_1 \ldots b_Q \, c \, a_1 \ldots a_Q \tag{2.16}$$

where c equals 0 or 1. For $M = ((1/2)N)^{1/2}$, we can express (2.16) as

$$\text{(a)} \quad \begin{aligned} n &= P_M(K)(2M) + L \\ P_N(n) &= P_M(L)(2M) + K \end{aligned} \qquad \text{(b)} \quad \begin{aligned} & n + M \\ & P_N(n+M) = P_N(n) + M \end{aligned} \tag{2.17}$$

for $K = 1, 2, \ldots, M - 1$ and $L = 0, 1, \ldots, K - 1$.

To accomplish a bit reversal permutation one must program the equations in (2.13) for the power R being even and program the equations in (2.17) for the power R being odd, the swaps in each case being carried out according to (2.11). A computer procedure that accomplishes this, called BITREV, can be found in Appendix B. Part of the program uses Buneman's method to compute the *lower order* permutation P_M.

While Buneman's method is simple and relatively fast, the algorithm programmed in BITREV runs faster by a considerable percentage.

---

3   Rotations in FFTs

The second major component of the FFT is the succession of butterflies performed during each stage. Each butterfly calculation requires a multiplication of the form

$$W^m H \tag{3.1}$$

where H is a complex number and so is W^m.

We now show that (3.1) can be interpreted as a rotation. The complex number H can be written as

$$H = A + iB \tag{3.2}$$

where A and B are real numbers. The weight factor W^m can be written, if $W = e^{i2\pi/N}$, as

$$W^m = \cos\left(\frac{2\pi m}{N}\right) + i\sin\left(\frac{2\pi m}{N}\right), \qquad \left(m = 0, 1, \ldots, \frac{1}{2}N - 1\right). \quad (3.3)$$

For simplicity of notation we define $C(m)$ and $S(m)$ by

$$C(m) = \cos\left(\frac{2\pi m}{N}\right), \qquad S(m) = \sin\left(\frac{2\pi m}{N}\right) \quad (3.4)$$

and, for later use, we define $T(m)$ by

$$T(m) = \tan\left(\frac{\pi m}{N}\right). \quad (3.5)$$

Using (3.4), formula (3.3) becomes

$$W^m = C(m) + iS(m). \quad (3.6)$$

And, using (3.2) and (3.6), we express $W^m H$ in (3.1) as

$$W^m H = \left[C(m)A - S(m)B\right] + i\left[C(m)B + S(m)A\right]. \quad (3.7)$$

Or,

$$Re(W^m H) = C(m)A - S(m)B, \qquad Im(W^m H) = S(m)A + C(m)B. \quad (3.8)$$

Because of (3.4), we can intepret (3.8) as describing a *rotation* of the vector (A, B) through the angle $2\pi m/N$.

Formula (3.8) requires four real multiplications. An algorithm due to Buneman reduces the number of multiplications needed to three, while increasing the number of additions from two to three. Since multiplications take up a considerable part of FFT computation time, usually taking up much more time than the additions, this reduction of multiplications yields a considerable time savings.

Buneman's algorithm is based on the trigonometric identity

$$\tan\frac{1}{2}\theta = \frac{1 - \cos\theta}{\sin\theta} \quad (3.9)$$

which follows from using the double angle identities for cosine and sine, where $\theta = 2((1/2)\theta)$, on the right side of (3.9). From (3.9), with a little algebra, we get

$$\cos\theta = 1 - \sin\theta\tan\frac{1}{2}\theta. \quad (3.10)$$

And, multiplying both sides of (3.9) by $1 + \cos\theta$, yields (after applying a trigonometric identity)

$$\sin\theta = (1 + \cos\theta)\tan\frac{1}{2}\theta. \quad (3.11)$$

Replacing θ by $2\pi m/N$ in (3.10) and (3.11) yields

$$C(m) = 1 - S(m)T(m) \tag{3.12}$$

$$S(m) = [1 + C(m)]T(m). \tag{3.13}$$

Using (3.12) in the first equation in (3.8) we get

$$Re(W^m H) = A - S(m)[B + T(m)A]. \tag{3.14}$$

If we define the auxiliary quantity V by

$$V = B + T(m)A \tag{3.15}$$

then (3.14) becomes

$$Re(W^m H) = A - S(m)V. \tag{3.16}$$

Using (3.13) and (3.12) in the second equation in (3.8) we get

$$
\begin{aligned}
Im(W^m H) &= [1 + C(m)]T(m)A + [1 - S(m)T(m)]B \\
&= [B + T(m)A] + [C(m)A - S(m)B]T(m) \\
&= V + Re(W^m H)T(m). \tag{3.17}
\end{aligned}
$$

Summarizing these results, we have from formulas (3.15)–(3.17) a three-step process for computing $Re(W^m H)$ and $Im(W^m H)$ [when $W = e^{i2\pi/N}$]

$$
\begin{aligned}
V &= B + T(m)A \\
Re(W^m H) &= A - S(m)V \\
Im(W^m H) &= V + Re(W^m H)T(m). \tag{3.18}
\end{aligned}
$$

The three-step process in (3.18) requires only three real multiplications instead of the four real multiplications required for (3.8).

We saw in Section 1 that the FFT requires a total of $(1/2)N \log_2 N$ complex multiplications. There are $2N \log_2 N$ *real* multiplications needed if formula (3.8) is used. Using formula (3.18) requires only $(3/2)N \log_2 N$ real multiplications. This represents a savings of about 25% when a computer is used that takes *significantly longer to perform multiplications than additions* [formula (3.18) requires three additions while (3.8) requires two]. Such a savings represents a significant reduction in computation time.

4 Computation of sines

In the previous section we saw that, using (3.18), the butterflies in the FFT require the computation of $S(m)$ and $T(m)$ for $m = 0, \ldots, (1/2)N - 1$, where

$S(m)$ and $T(m)$ are defined by

$$S(m) = \sin\left(\frac{2\pi m}{N}\right),$$

$$T(m) = \tan\left(\frac{\pi m}{N}\right), \qquad \left(m = 0, \ldots, \frac{1}{2}N - 1\right). \tag{4.1}$$

Buneman's method requires that we compute the quantities $S(m)$ and $T(m)$ in advance of performing the butterflies in the FFT. This saves time because one does not have to have the computer calculate a particular weight factor over and over again from one stage of the FFT to the next. There is also a time savings when several FFTs of the same size (N points) are computed in a row; the same values of $S(m)$ and $T(m)$ can be used again and again. In fact, this is what happens in *FAS* when several partial sums for Fourier series of a given function are computed in a row.

In this section we will show that only the values of $S(m)$ for $m = 0, \ldots,$ $(1/4)N$ need to be precalculated, and we will discuss Buneman's method for efficient computation of these values of $S(m)$. They are *not* computed by repeatedly invoking the sine function in the computer; there is a much more efficient way of computing these very special sine values.

First, we show why only the values of $S(m)$ for $m = 0, \ldots, (1/4)N$ need to be computed. This reduction saves memory storage. If $N = 1024$, then we only need a 257-point auxiliary array to store these values of $S(m)$. This is a tolerable increase in extra memory.

The elementary trigonometric identity

$$\sin(\pi - \theta) = \sin\theta \tag{4.2}$$

implies that

$$S\left(\frac{1}{2}N - m\right) = S(m), \qquad \left(m = 1, \ldots, \frac{1}{4}N\right). \tag{4.3}$$

Formula (4.3) implies that only $S(m)$ for $m = 0, \ldots, (1/4)N$ needs to be calculated; the rest of the sines $S(m)$, for $m = (1/4)N + 1, \ldots, (1/2)N - 1$, can be obtained using formula (4.3).

The factors $T(m)$ for $m = 0, \ldots, (1/2)N - 1$ can all be computed from the sine values $S(m)$. For $m = 0$, we note that $T(0) = S(0) = 0$. For $m > 0$, we proceed as follows. From (3.9), we have

$$T(m) = \frac{1 - C(m)}{S(m)}. \tag{4.4}$$

Since

$$\cos\theta = \sin\left(\frac{1}{2}\pi - \theta\right) \tag{4.5}$$

we also have

$$C(m) = S\left(\frac{1}{4}N - m\right), \qquad \left(m = 1, 2, \dots, \frac{1}{4}N\right). \tag{4.6}$$

Combining (4.6) and (4.4) yields

$$T(m) = \frac{1 - S(\frac{1}{4}N - m)}{S(m)}, \qquad \left(m = 1, 2, \dots, \frac{1}{4}N\right). \tag{4.7}$$

Formula (4.7) shows how to produce $T(m)$ for $m = 1, \dots, (1/4)N$ from the values of $S(m)$ for the same range of m. The rest of the values of $T(m)$ for $m = (1/4)N + 1$, $\dots$, $(1/2)N - 1$ can be produced using the following formula:

$$T\left(\frac{1}{2}N - m\right) = \frac{1 + S(\frac{1}{4}N - m)}{S(m)}, \qquad \left(m = 1, \dots, \frac{1}{4}N - 1\right). \tag{4.8}$$

We leave the verification of (4.8) to the reader as an exercise.

We will now discuss Buneman's method for efficiently calculating the values of $S(m)$ for $m = 0, 1, \dots, (1/4)N$. As we can see from (4.1), we do not need to compute arbitrary values of the sine function. Rather, we must compute $\sin(2\pi m/N)$ for $m = 0, 1, \dots, (1/4)N$. Buneman devised a clever method for inductively constructing these values.

The method is based on the following trigonometric identity:

$$\sin\theta = \frac{\sin(\theta + \phi) + \sin(\theta - \phi)}{2\cos\phi}. \tag{4.9}$$

Formula (4.9) follows from the sine addition and subtraction identities applied to the numerator on the right side of (4.9). Using (4.9) it is possible to compute $S(m)$ for $m = 0, 1, \dots, (1/4)N$ by an inductive process. The first step is to put $S(0) = 0$ and $S((1/4)N) = 1$. Then, we put $\phi_1 = (1/4)\pi$ and define $z_1 = 2^{1/2}$. Using (4.9) we have, since $z_1 = 2\cos\phi_1$,

$$S\left(\frac{1}{8}N\right) = \frac{S(\frac{1}{4}N) + S(0)}{z_1}.$$

The next step is to multiply ϕ_1 by $1/2$ and put $\phi_2 = (1/2)\phi_1$. Then, by the cosine half-angle formula,

$$2\cos\phi_2 = 2\cos\left(\frac{1}{2}\phi_1\right) = (2 + 2\cos\phi_1)^{\frac{1}{2}}$$

$$= (2 + z_1)^{\frac{1}{2}}. \tag{4.10}$$

Defining z_2 to be $2\cos\phi_2$, we rewrite (4.10) as

$$z_2 = (2 + z_1)^{\frac{1}{2}}. \tag{4.11}$$

Combining (4.10) and (4.11), we get from (4.9) that

$$S\left(\frac{1}{16}N\right) = \frac{S(\frac{1}{8}N) + S(0)}{z_2}, \qquad S\left(\frac{3}{16}N\right) = \frac{S(\frac{1}{4}N) + S(\frac{1}{8}N)}{z_2}.$$

The third step in the method is to define z_3 by

$$z_3 = (2 + z_2)^{\frac{1}{2}}. \tag{4.12}$$

Using z_3 in place of $2\cos\phi$ in (4.9) and $\phi_3 = (1/2)\phi_2$ in place of ϕ in (4.9), we obtain

$$S\left(\frac{1}{32}N\right) = \frac{S(\frac{1}{16}N) + S(0)}{z_3}, \qquad S\left(\frac{3}{32}N\right) = \frac{S(\frac{1}{8}N) + S(\frac{1}{16}N)}{z_3}$$

$$S\left(\frac{5}{32}N\right) = \frac{S(\frac{3}{16}N) + S(\frac{1}{8}N)}{z_3}, \qquad S\left(\frac{7}{32}N\right) = \frac{S(\frac{1}{4}N) + S(\frac{3}{16}N)}{z_3}.$$

This completes the first three stages of Buneman's method. We have shown how to compute $S(jN/32)$ for $j = 0, 1, \ldots, 8$. If $N = 32$, then we are done. If $N = 64$ we would need to go through one more stage. Buneman's method is to continue the process we have begun. For $N = 2^R$, Buneman's method takes $R - 2$ stages to compute all the sines $S(m)$ for $m = 0, 1, \ldots, (1/4)N$. For each new stage we compute

$$\phi_{k+1} = \frac{1}{2}\phi_k \ , \quad z_{k+1} = (2 + z_k)^{\frac{1}{2}} \tag{4.13}$$

and use (4.9) in the form

$$\sin\theta = \frac{\sin(\theta + \phi_{k+1}) + \sin(\theta - \phi_{k+1})}{z_{k+1}}$$

to compute the values of $S(m)$ where the angles $2\pi m/N$ lie midway between the angles previously computed.

A computer procedure, called SINES, which executes the method just described, can be found in Appendix B.

Bracewell has shown that Buneman's methods of handling trigonometric operations in the FFT, discussed in this section and the previous section, result in significant time savings over more conventional methods (see [Ba]).

We have now finished our discussion of the basic elements of the FFT algorithm. See Appendix B for computer programs that implement this algorithm.

---

5 Computing two real FFTs simultaneously

In this section and the next, we shall discuss FFTs of *real* data. Most FFTs are performed on real-valued data sequences. For real data there are some useful devices that reduce the number of calculations involved and, consequently, the computation time.

In this section, we show how to compute two real FFTs simultaneously by performing one complex FFT. This time-saving device will be used in the next section to reduce the amount of calculation needed to perform an FFT of real data by about one-half. It will also be used in Chapter 4 to facilitate the computation of the convolution of two real data sequences.

Suppose $\{f_j\}_{j=0}^{N-1}$ and $\{g_j\}_{j=0}^{N-1}$ are two sequences of N *real* numbers. Define the sequence of N complex numbers $\{h_j\}$ by

$$h_j = f_j + ig_j, \qquad (j = 0, 1, \ldots, N - 1). \tag{5.1}$$

Taking the N-point FFT of $\{h_j\}$, we get $\{H_k\}$, where

$$H_k = \sum_{j=0}^{N-1}(f_j + ig_j)W^{jk}, \qquad (k = 0, 1, \ldots, N - 1). \tag{5.2}$$

and $W = e^{i2\pi/N}$ (or $W = e^{-i2\pi/N}$). Using algebra we rewrite (5.2) as

$$H_k = F_k + iG_k, \qquad (k = 0, 1, \ldots, N - 1). \tag{5.3}$$

where

$$F_k = \sum_{j=0}^{N-1} f_j W^{jk}, \qquad G_k = \sum_{j=0}^{N-1} g_j W^{jk} \tag{5.4}$$

are N-point FFTs of $\{f_j\}$ and $\{g_j\}$, respectively.

There is a crucial symmetry property for the DFT of a real-valued sequence that we need to utilize. It is an analogue of the symmetry relation for Fourier series coefficients of a real-valued function. We take the complex conjugate F_{N-k}^* of F_{N-k} and we get

$$F_{N-k}^* = \left[\sum_{j=0}^{N-1} f_j W^{j(N-k)}\right]^* = \sum_{j=0}^{N-1} f_j^* (W^{jN})^*(W^{-jk})^*. \tag{5.5}$$

Because f_j is real we have $f_j^* = f_j$ and, because $W^N = 1$, we have $(W^{jN})^* = 1$. Finally, because $W^{-jk} = (W^{jk})^*$, we have $(W^{-jk})^* = W^{jk}$. From these facts, we observe that (5.5) simplifies to

$$F_{N-k}^* = F_k \qquad (\text{or } F_{N-k} = F_k^*). \tag{5.6}$$

Similarly, we have

$$G_{N-k}^* = G_k \qquad (\text{or } G_{N-k} = G_k^*) \tag{5.7}$$

because $\{g_j\}$ are real numbers, too. Using (5.6) and (5.7), we obtain from (5.3)

$$H^*_{N-k} = F_k - iG_k.\tag{5.8}$$

By combining (5.3) and (5.8) we get

$$F_k = \frac{1}{2}\left[H_k + H^*_{N-k}\right], \qquad G_k = \frac{i}{2}\left[H^*_{N-k} - H_k\right].\tag{5.9}$$

From formula (5.9) one can obtain $\{F_k\}_{k=0}^{N-1}$ and $\{G_k\}_{k=0}^{N-1}$.

We have shown how to obtain the FFTs of two real-valued N-point sequences $\{f_j\}$ and $\{g_j\}$. The method consists of the following steps:

Step 1. Form the sequence $h_j = f_j + ig_j$ for $j = 0, 1, \ldots, N - 1$.
Step 2. Perform the N-point FFT of $\{h_j\}$ to get $\{H_k\}$.
Step 3. Use formula (5.9) to get F_k and G_k for $k = 0, 1, \ldots, N - 1$.

6   Computing a real FFT

In this section we shall apply the method of the previous section, along with formula (5.6), to compute the FFT of a real data sequence. This method will reduce the number of computations involved, from simply computing an FFT in the form described in Section 1, by about one-half.

We begin by considering again the first reduction involved in computing the FFT of the sequence of N *real* points $\{f_j\}$

$$
\begin{aligned}
F_k &= \sum_{j=0}^{\frac{1}{2}N-1} f_{2j}(W^2)^{jk} + \sum_{j=0}^{\frac{1}{2}N-1} f_{2j+1}(W^2)^{jk}W^k \\
&= F_k^0 + F_k^1 W^k
\end{aligned}\tag{6.1}
$$

where $\{F_k^0\}$ and $\{F_k^1\}$ are the $(1/2)N$-point FFTs of the real sequences $\{f_{2j}\}$ and $\{f_{2j+1}\}$, respectively. To calculate *simultaneously* the $(1/2)N$-point FFTs of $\{f_{2j}\}$ and $\{f_{2j+1}\}$, we use the method summarized at the end of Section 5. First, we form the sequence $\{h_j\}$ where $h_j = f_{2j} + if_{2j+1}$. Second, we calculate $\{F_k^0\}$ and $\{F_k^1\}$ by

$$
\begin{aligned}
F_k^0 &= \frac{1}{2}\left[H_k + H^*_{\frac{1}{2}N-k}\right] \\
F_k^1 &= \frac{i}{2}\left[H^*_{\frac{1}{2}N-k} - H_k\right] \qquad \left(k = 0, 1, \ldots, \frac{1}{2}N - 1\right).
\end{aligned}\tag{6.2}
$$

Now, for $k = 0, 1, \ldots, (1/2)N - 1$, we obtain F_k from (6.1):

$$F_k = F_k^0 + F_k^1 W^k, \qquad \left(k = 0, 1, \ldots, \frac{1}{2}N - 1 \right). \tag{6.3}$$

To get F_k for $k = (1/2)N + 1, \ldots, N - 1$ we make use of (5.6). Namely,

$$F_{N-k} = F_k^*, \qquad \left(k = 1, \ldots, \frac{1}{2}N - 1 \right). \tag{6.4}$$

To obtain $F_{(1/2)N}$ we note that $\{F_k^0\}$ and $\{F_k^1\}$ have period $(1/2)N$. Therefore, $F_{(1/2)N}^0 = F_0^0$ and $F_{(1/2)N}^1 = F_0^1$. Furthermore, $W^{(1/2)N} = -1$, hence we obtain from (6.1)

$$F_{\frac{1}{2}N} = F_0^0 - F_0^1. \tag{6.5}$$

Formulas (6.3)–(6.5) give a complete description of $\{F_k\}_{k=0}^{N-1}$.

The method we have just described for computing an FFT of a real sequence is considerably faster than just applying the FFT described in Section 1. For example, the method of this section requires $1/2((1/2)N) \log_2((1/2)N) + (1/2)N$ *complex* multiplications. This results in $(3/4)N \log_2((1/2)N) + (3/2)N$ *real* multiplications if we use Buneman's method for performing butterflies (which was described in Section 3). This represents a reduction by approximately one-half of the number of real multiplications, $(3/2)N \log_2 N$, needed for the complex FFT described in Section 1. This reduction by one-half is what we should expect based on the symmetry inherent in formula (5.6).

A computer procedure, called REALFFT, that performs the algorithm above can be found in Appendix B.

We can also apply the symmetry in formula (5.6) to reduce the number of calculations needed for an inverse FFT of the FFT of a real sequence. We will describe our approach to this problem after the next section, since it involves fast sine and cosine transforms.

7 Fast sine and cosine transforms

To compute discrete sine transforms (DSTs) and discrete cosine transforms (DCTs) efficiently we make use of the work of the previous section and we use some important symmetry properties of finite sequences. An efficient method of computing a DST is called a *fast sine transform* (FST). After discussing an FST we will then discuss an efficient method for computing a DCT, which is called a *fast cosine transform* (FCT).

We will need to make use of some fundamental symmetry properties of finite sequences. These symmetry properties are closely related to the properties of evenness and oddness of functions. A sequence $\{C_j\}_{j=1}^{N-1}$ is called *even about* $(1/2)N$ if

$$C_{N-j} = C_j, \qquad (j = 1, \ldots, N-1). \tag{7.1}$$

The sequence is called *odd about* $(1/2)N$ if

$$C_{N-j} = -C_j, \qquad (j = 1, \ldots, N-1). \tag{7.2}$$

If a sequence is even about $(1/2)N$, then

$$\sum_{j=1}^{N-1} C_j = 2 \sum_{j=1}^{\frac{1}{2}N-1} C_j + C_{\frac{1}{2}N}. \tag{7.3}$$

If a sequence is odd about $(1/2)N$, then

$$\sum_{j=1}^{N-1} C_j = 0. \tag{7.4}$$

The proofs of (7.3) and (7.4) are left to the reader as exercises.

Here are a few examples. The sequence $\{\sin(j\pi/N)\}_{j=1}^{N-1}$ is even about $(1/2)N$. For *fixed* k, the sequence $\{\sin(2\pi jk/N)\}_{j=1}^{N-1}$ is odd about $(1/2)N$, while the sequence $\{\cos(2\pi jk/N)\}_{j=1}^{N-1}$ is even about $(1/2)N$.

The algebra of even and odd sequences about $(1/2)N$ is the same as the algebra of even and odd functions from calculus. For example, if $\{A_j\}_{j=1}^{N-1}$ is odd about $(1/2)N$ and $\{S_j\}_{j=1}^{N-1}$ is even about $(1/2)N$, then $\{A_j S_j\}_{j=1}^{N-1}$ is odd about $(1/2)N$. The general pattern is (even)(even) = even, (odd)(odd) = even, and (odd)(even) = odd.

Now, to compute an FST of a real sequence $\{h_j\}_{j=1}^{N-1}$, we define a new sequence $\{f_j\}_{j=0}^{N-1}$ by

$$f_0 = 0$$
$$f_j = s_j + a_j, \qquad (j = 1, \ldots, N-1) \tag{7.5}$$

where

$$s_j = [h_j + h_{N-j}] \sin \frac{j\pi}{N},$$

$$a_j = \frac{1}{2} [h_j - h_{N-j}].$$

The sequence $\{s_j\}_{j=1}^{N-1}$ defined in (7.5) is even about $(1/2)N$, while the sequence $\{a_j\}_{j=1}^{N-1}$ is odd about $(1/2)N$. We now compute an FFT (using weight $W = e^{i2\pi/N}$) of $\{f_j\}$. This FFT $\{F_k\}$ has real parts $\{R_k\}$ defined by

$$R_k = \sum_{j=0}^{N-1} f_j \cos \frac{2\pi jk}{N} = \sum_{j=1}^{N-1} f_j \cos \frac{2\pi jk}{N} . \tag{7.6}$$

The imaginary parts $\{I_k\}$ of $\{F_k\}$ are defined by

$$I_k = \sum_{j=0}^{N-1} f_j \sin \frac{2\pi jk}{N} = \sum_{j=1}^{N-1} f_j \sin \frac{2\pi jk}{N} . \tag{7.7}$$

From the symmetry relation (7.4) we get [using (7.5) and (7.6)]

$$\begin{aligned}
R_k &= \sum_{j=1}^{N-1} s_j \cos \frac{2\pi jk}{N} + \sum_{j=1}^{N-1} a_j \cos \frac{2\pi jk}{N} \\
&= \sum_{j=1}^{N-1} s_j \cos \frac{2\pi jk}{N} \\
&= \sum_{j=1}^{N-1} h_j \sin \frac{j\pi}{N} \cos \frac{2\pi jk}{N} + \sum_{j=1}^{N-1} h_{N-j} \sin \frac{j\pi}{N} \cos \frac{2\pi jk}{N} .
\end{aligned} \tag{7.8}$$

By substituting $N - j$ in place of j we find that, because $\{\sin(j\pi/N)\}$ is even about $(1/2)N$,

$$\sum_{j=1}^{N-1} h_{N-j} \sin \frac{j\pi}{N} \cos \frac{2\pi jk}{N} = \sum_{j=1}^{N-1} h_j \sin \frac{j\pi}{N} \cos \frac{2\pi jk}{N} . \tag{7.9}$$

Substituting from (7.9) back into (7.8) we get

$$R_k = \sum_{j=1}^{N-1} 2h_j \sin \frac{j\pi}{N} \cos \frac{2\pi jk}{N} . \tag{7.10}$$

Applying the trigonometric identity $2 \sin \phi \cos \theta = \sin(\theta + \phi) - \sin(\theta - \phi)$ we obtain from (7.10)

$$\begin{aligned}
R_k &= \sum_{j=1}^{N-1} h_j \left[\sin \left(\frac{2k+1}{N} j\pi \right) - \sin \left(\frac{2k-1}{N} j\pi \right) \right] \\
&= H_{2k+1}^S - H_{2k-1}^S .
\end{aligned} \tag{7.11}$$

Similarly, for I_k we have

$$
\begin{aligned}
I_k &= \sum_{j=1}^{N-1} s_j \sin \frac{2\pi jk}{N} + \sum_{j=1}^{N-1} a_j \sin \frac{2\pi jk}{N} \\
&= \sum_{j=1}^{N-1} a_j \sin \frac{2\pi jk}{N} \\
&= \sum_{j=1}^{N-1} \frac{1}{2} h_j \sin \frac{2\pi jk}{N} - \sum_{j=1}^{N-1} \frac{1}{2} h_{N-j} \sin \frac{2\pi jk}{N} \\
&= \sum_{j=1}^{N-1} h_j \sin \frac{2\pi jk}{N} \ .
\end{aligned}
$$

Thus,

$$
I_k = H_{2k}^S. \tag{7.12}
$$

Formula (7.12) says that the *even-indexed* numbers H_{2k}^S are given by I_k, the imaginary part of F_k, for $k = 0, 1, \ldots, (1/2)N - 1$. The *odd-indexed* numbers H_{2k+1}^S are gotten from the recursion formula

$$
H_{2k+1}^S = H_{2k-1}^S + R_k, \qquad \left(k = 1, \ldots, \frac{1}{2}N - 1 \right), \tag{7.13}
$$

obtained from (7.11), where R_k is the real part of F_k. To get the recursion in (7.13) started, the value of H_1^S is needed. But,

$$
H_1^S = \sum_{j=1}^{N-1} h_j \sin \frac{j\pi}{N} \tag{7.14}
$$

can be calculated first (for instance, when creating $\{f_j\}$), and then used to begin the recursion in (7.13).

The calculation of a fast cosine transform (FCT) follows a pattern similar to that of the FST. In this case, we define the sequence $\{f_j\}$ by

$$
\begin{aligned}
f_0 &= h_0 \\
f_j &= s_j + a_j, \qquad (j = 1, \ldots, N - 1)
\end{aligned} \tag{7.15}
$$

where

$$
\begin{aligned}
s_j &= \frac{1}{2} \left[h_j + h_{N-j} \right], \\
a_j &= \left[h_{N-j} - h_j \right] \sin \frac{j\pi}{N} \ .
\end{aligned}
$$

We leave it as an exercise for the reader to show that

$$H_{2k}^C = R_k, \qquad \left(k = 0, 1, \ldots, \frac{1}{2}N - 1\right) \tag{7.16}$$

and

$$H_{2k+1}^C = H_{2k-1}^C + I_k, \qquad \left(k = 1, \ldots, \frac{1}{2}N - 1\right)$$

$$H_1^C = \sum_{j=1}^{N-1} h_j \cos \frac{j\pi}{N} \tag{7.17}$$

where $\{R_k\}$ and $\{I_k\}$ are the real and imaginary parts of the FFT of $\{f_j\}$ using weight $W = e^{i2\pi/N}$.

Computer procedures, called COSTRAN and SINETRAN, that perform the algorithms described in this section can be found in Appendix B.

8 Inversion of discrete sine and cosine transforms

In this section we will describe how to invert DSTs and DCTs, thereby recovering the original sequences before they were transformed. First, we show that a DST has the nice property that *it is its own inverse* (up to multiplication by a constant). More precisely, we will show that the inversion of a DST $\{H_k^C\}$ is given by

$$\frac{2}{N} \sum_{k=1}^{N-1} H_k^S \sin \frac{\pi j k}{N} = h_j, \qquad (j = 1, \ldots, N - 1). \tag{8.1}$$

Formula (8.1) follows from

$$\sum_{k=1}^{N-1} \sin \frac{\pi m k}{N} \sin \frac{\pi j k}{N} = \begin{cases} \frac{1}{2}N & \text{if } m = j \\ 0 & \text{if } m \neq j. \end{cases} \tag{8.2}$$

We will prove (8.2) in a moment. First, we show that (8.1) follows from it.

The left side of (8.1) can be rewritten as

$$\frac{2}{N} \sum_{k=1}^{N-1} \left[\sum_{m=1}^{N-1} h_m \sin \frac{\pi m k}{N}\right] \sin \frac{\pi j k}{N} = \sum_{m=1}^{N-1} h_m \left[\frac{2}{N} \sum_{k=1}^{N-1} \sin \frac{\pi m k}{N} \sin \frac{\pi j k}{N}\right]$$

which, by applying (8.2), yields (8.1).

Now, to prove (8.2), we note that for *fixed* j and m, the sequence

$$\left\{\sin \frac{\pi m k}{N} \sin \frac{\pi j k}{N}\right\}_{k=1}^{2N-1}$$

is even about N. Hence, from formula (7.3), using N in place of $(1/2)N$, and noting that the middle element $\sin(\pi mN/N)\sin(\pi jN/N)$ equals 0, we have

$$
\begin{aligned}
\sum_{k=1}^{N-1} \sin\frac{\pi mk}{N}\sin\frac{\pi jk}{N} &= \frac{1}{2}\sum_{k=0}^{2N-1}\sin\frac{\pi mk}{N}\sin\frac{\pi jk}{N}\\
&= \frac{1}{2}\sum_{k=0}^{2N-1}\sin\frac{2\pi mk}{2N}\sin\frac{2\pi jk}{2N}\\
&= \frac{1}{4}\sum_{k=0}^{2N-1}\cos\frac{2\pi(m-j)k}{2N}\\
&\quad - \frac{1}{4}\sum_{k=0}^{2N-1}\cos\frac{2\pi(m+j)k}{2N}\,.
\end{aligned}
\tag{8.3}
$$

Moreover,

$$
\begin{aligned}
\sum_{k=0}^{2N-1}\cos\frac{2\pi(m-j)k}{2N} &= Re\left[\sum_{k=0}^{2N-1}e^{i2\pi(m-j)k/2N}\right]\\
&= \begin{cases} 0 & \text{for } m \neq j\\ 2N & \text{for } m = j\end{cases}
\end{aligned}
\tag{8.4}
$$

where the last equality follows from formula (1.10), Chapter 2, with $2N$ in place of N. [The argument being similar to the one used to prove Theorem 2.2(c) in Chapter 2.]

Similarly, we have

$$
\sum_{k=0}^{2N-1}\cos\frac{2\pi(m+j)k}{2N} = 0.
\tag{8.5}
$$

Substituting from (8.4) and (8.5) back into (8.3), we obtain formula (8.2).

There is also an inversion procedure for discrete cosine transforms. It is, however, more complicated than the inversion procedure for discrete sine transforms. First, we need to prove the following formula:

$$
\sum_{k=0}^{N-1}\cos\frac{\pi jk}{N}\cos\frac{\pi mk}{N} = \begin{cases} N & \text{for } m = j = 0\\[2mm] \frac{1}{2}N & \text{for } m = j \neq 0\\[2mm] \frac{1}{2}-\frac{1}{2}(-1)^{j+m} & \text{for } m \neq j.\end{cases}
\tag{8.6}
$$

Since, for fixed j and m, the sequence $\{\cos(\pi jk/N)\cos(\pi mk/N)\}_{k=1}^{2N-1}$ is even about N, we have

$$\sum_{k=0}^{N-1} \cos\frac{\pi jk}{N}\cos\frac{\pi mk}{N} = \frac{1}{2}\sum_{k=0}^{2N-1}\cos\frac{\pi jk}{N}\cos\frac{\pi mk}{N} + \frac{1}{2} - \frac{1}{2}(-1)^{j+m}. \quad (8.7)$$

That (8.7) implies (8.6) is shown by a similar argument to the one used above to show (8.2). The details are left to the reader as an exercise.

From (8.6) we find that

$$\frac{2}{N}\sum_{k=0}^{N-1}H_k^C\cos\frac{\pi jk}{N} = \frac{2}{N}\sum_{m=0}^{N-1}h_m\left[\sum_{k=0}^{N-1}\cos\frac{\pi jk}{N}\cos\frac{\pi mk}{N}\right]$$

$$= \begin{cases} 2h_0 + \frac{1}{N}\sum_{m=1}^{N-1}h_m - \frac{1}{N}\sum_{m=1}^{N-1}h_m(-1)^m & \text{for } j = 0 \\ h_j + \frac{1}{N}\sum_{m=0}^{N-1}h_m - \frac{1}{N}\sum_{m=0}^{N-1}h_m(-1)^{m+j} & \text{for } j \neq 0. \end{cases} \quad (8.8)$$

Consequently, since

$$\frac{1}{N}\sum_{m=0}^{N-1}h_m - \frac{1}{N}\sum_{m=0}^{N-1}h_m(-1)^{m+j} = \frac{1}{N}\sum_{m=0}^{N-1}h_m\left[1 - (-1)^{m+j}\right]$$

we obtain from (8.8)

$$\frac{2}{N}\sum_{k=0}^{N-1}H_k^C\cos\frac{\pi jk}{N} = \begin{cases} h_j + \frac{2}{N}\sum_{m\ \text{odd}}^{N-1}h_m & \text{for } j \text{ even} \neq 0 \\ h_j + \frac{2}{N}\sum_{m\ \text{even}}^{N-1}h_m & \text{for } j \text{ odd} \\ 2h_0 + \frac{2}{N}\sum_{m\ \text{odd}}^{N-1}h_m & \text{for } j = 0. \end{cases} \quad (8.9)$$

Using (8.9) it is possible to recover $\{h_j\}$ from $\{H_k^C\}$ *provided* the sums

$$s_1 = \frac{2}{N}\sum_{m\ \text{odd}}^{N-1}h_m \qquad \text{and} \qquad s_2 = \frac{2}{N}\sum_{m\ \text{even}}^{N-1}h_m$$

are known. To find these unknown sums, we introduce the sequence $\tilde{h}_j$ defined by

$$\tilde{h}_j = \frac{2}{N}\sum_{k=0}^{N-1}H_k^C\cos\frac{\pi jk}{N}, \qquad (j = 0, 1, \ldots, N-1).$$

Then, from (8.9) we obtain

$$\sum_{\substack{j\ \text{even}}}^{N-2}\tilde{h}_j = h_0 + \sum_{j=0}^{N-1}h_j \quad (8.10)$$

and

$$\sum_{\substack{j \text{ odd}}}^{N-1} \tilde{h}_j = \sum_{j=0}^{N-1} h_j. \tag{8.11}$$

From (8.10) and (8.11) it follows that

$$h_0 = \sum_{\substack{j \text{ even}}}^{N-2} \tilde{h}_j - \sum_{\substack{j \text{ odd}}}^{N-1} \tilde{h}_j. \tag{8.12}$$

Hence, since (8.9) implies

$$\tilde{h}_0 = 2h_0 + \frac{2}{N} \sum_{\substack{m \text{ odd}}}^{N-1} h_m$$

we get

$$s_1 = \tilde{h}_0 - 2 \left[\sum_{\substack{j \text{ even}}}^{N-2} \tilde{h}_j - \sum_{\substack{j \text{ odd}}}^{N-1} \tilde{h}_j \right]. \tag{8.13}$$

Furthermore, it also follows from (8.11), and the original definitions of s_1 and s_2, that

$$s_2 = \frac{2}{N} \sum_{\substack{j \text{ odd}}}^{N-1} \tilde{h}_j - s_1. \tag{8.14}$$

To summarize, we can invert $\{H_k^C\}$ to recover $\{h_j\}$ by the following steps:

Step 1. Form the sequence $\{\tilde{h}_j\}$ defined by

$$\tilde{h}_j = \frac{2}{N} \sum_{k=0}^{N-1} H_k^C \cos \frac{\pi j k}{N}, \qquad (j = 0, 1, \ldots, N-1).$$

Step 2. Calculate the quantities s_1 and s_2 by

$$s_1 = \tilde{h}_0 - 2 \left[\sum_{\substack{j \text{ even}}}^{N-2} \tilde{h}_j - \sum_{\substack{j \text{ odd}}}^{N-1} \tilde{h}_j \right]$$

$$s_2 = \frac{2}{N} \sum_{\substack{j \text{ odd}}}^{N-1} \tilde{h}_j - s_1.$$

Step 3. The sequence $\{h_j\}_{j=0}^{N-1}$ is then obtained by

$$
h_j = \begin{cases}
\frac{1}{2}[\tilde{h}_0 - s_1] & \text{for } j = 0 \\[2mm]
\tilde{h}_j - s_1 & \text{for } j \text{ even and } \neq 0 \\[2mm]
\tilde{h}_j - s_2 & \text{for } j \text{ odd.}
\end{cases}
$$

9   Inversion of the FFT of a real sequence

In this section we show how the symmetry property (5.6) of the FFT of a real sequence can be used to reduce by about one-half the computations involved in inverting the FFT. Although there are other methods of doing this than the method described below, this method is of interest because it uses no extra sines other than the set $\{S(m)\}_{m=0}^{(1/4)N}$ that was used for performing the FFT itself.

We will assume in this section that $W = e^{i2\pi/N}$. Let R_k and I_k stand for the real and imaginary parts of the FFT $\{F_k\}$. That is,

$$
F_k = R_k + iI_k, \qquad (k = 0, 1, \ldots, N - 1) \tag{9.1}
$$

where F_k is defined by

$$
F_k = \sum_{j=0}^{N-1} f_j W^{jk}.
$$

Then, by the formula for DFT inversion,

$$
f_j = \frac{1}{N} g_j, \qquad (j = 0, 1, \ldots, N - 1) \tag{9.2}
$$

where

$$
g_j = \sum_{k=0}^{N-1} F_k (W^{jk})^*. \tag{9.3}
$$

Using $W = e^{i2\pi/N}$ and (9.1) we have

$$
g_j = \sum_{j=0}^{N-1} (R_k + iI_k) e^{-i2\pi jk/N}
$$

$$
= \sum_{k=0}^{N-1} (R_k + iI_k) \left(\cos \frac{2\pi jk}{N} - i \sin \frac{2\pi jk}{N} \right). \tag{9.4}
$$

Since $\{f_j\}$ consists of real numbers, so does $\{g_j\}$ because of (9.2). Consequently, only the real part of the sum in (9.4) is nonzero. Therefore, we must have

$$g_j = \sum_{k=0}^{N-1} R_k \cos \frac{2\pi jk}{N} + \sum_{k=0}^{N-1} I_k \sin \frac{2\pi jk}{N}. \tag{9.5}$$

Because of (5.6) we have

$$R_{N-k} = R_k, \qquad I_{N-k} = -I_k, \qquad (k = 1, \ldots, N-1). \tag{9.6}$$

It follows from (9.6) that, for each fixed j, the sequences

$$\left\{ R_k \cos \frac{2\pi jk}{N} \right\}_{k=1}^{N-1} \qquad \text{and} \qquad \left\{ I_k \sin \frac{2\pi jk}{N} \right\}_{k=1}^{N-1}$$

are even about $(1/2)N$. From formula (7.3) we obtain

$$\sum_{k=1}^{N-1} R_k \cos \frac{2\pi jk}{N} = R_{\frac{1}{2}N}(-1)^j + 2 \sum_{k=1}^{\frac{1}{2}N-1} R_k \cos \frac{2\pi jk}{N} \tag{9.7}$$

and

$$\sum_{k=1}^{N-1} I_k \sin \frac{2\pi jk}{N} = 2 \sum_{k=1}^{\frac{1}{2}N-1} I_k \sin \frac{2\pi jk}{N}. \tag{9.8}$$

Using (9.7) and (9.8) in (9.5) yields

$$g_j = R_0 + R_{\frac{1}{2}N}(-1)^j + 2 \sum_{k=1}^{\frac{1}{2}N-1} R_k \cos \frac{\pi jk}{\frac{1}{2}N} + 2 \sum_{k=1}^{\frac{1}{2}N-1} I_k \sin \frac{\pi jk}{\frac{1}{2}N}. \tag{9.9}$$

Formula (9.9) shows that $\{g_j\}_{j=0}^{N-1}$, and consequently $\{f_j\}_{j=0}^{N-1}$, can be generated from a $(1/2)N$-point fast cosine transform of

$$\{0, 2R_1, 2R_2, \ldots, 2R_{\frac{1}{2}N-1}\}$$

and a fast sine transform of

$$\{2I_1, 2I_2, \ldots, 2I_{\frac{1}{2}N-1}\}.$$

These fast cosine and fast sine transforms are even and odd about $(1/2)N$, respectively. Taking this symmetry into account, formula (9.9) applies to $j = 0$, $1, \ldots, N-1$.

As we noted above, an interesting feature of (9.9) is that to compute the fast cosine and fast sine transforms, *of order* $(1/2)N$, by the methods of Section 7, one needs only the *same* sines $\{S(m)\}_{m=0}^{(1/4)N}$ generated to calculate the FFT $\{F_k\}_{k=0}^{N-1}$. Therefore, inverting the FFT using (9.9) requires only these same sines. This is a useful memory savings.

A computer procedure, called InvRFFT, that carries out the algorithm described in this section can be found in Appendix B.

References

For further discussion of DFTs and FFTs, see [Bri], [El-R], [Nu], [Op-S], [Pr-e], [Ra], and [Ra-G]. An important set of foundational papers is collected in [Ra-R]. Computer programs can be found in [Pr-e]. Those programs do not, however, include Buneman's improvements (see [Br]).

Exercises

Section 1

3.1 Make a diagram, like the one in Figure 3.4, for the 16-point FFT.

3.2 Construct tables of bit reversal numbers for $N = 4$, 8, 16, and 32.

3.3 Suppose that $N = 4^R$. Derive an FFT based on successive divisions of N by 4. [*Remark*: This is called a radix-4 FFT.]

Section 2

3.4 Construct tables of base 4 digit reversals for $N = 4$ and $N = 16$. [This is needed because of Exercise 3.3.]

3.5 Describe the analog of Buneman's algorithm for digit reversal, base 4. [If $n = 1\,3\,2\,0\,1$ (base 4), then the digit reversal of n is $1\,0\,2\,3\,1$.]

3.6 Repeat Exercise 3.5 for base r, where r is an integer greater than 1.

3.7 Generalize the second algorithm described in Section 2, using base 4.

3.8 Generalize the second algorithm described in Section 2, but use base r.

3.9 Show that $S((1/2)N - m) = S(m)$ for $m = 0, 1, \ldots, (1/4)N$.

Section 3

3.10 Show that, for $m = 0, 1, \ldots, (1/4)N$

$$C(m) = S\left(\frac{1}{4}N - m\right), \qquad C\left(\frac{1}{2}N - m\right) = -S\left(\frac{1}{4}N - m\right).$$

3.11 What sines (tangents) are needed for performing rotations in a radix-4 FFT ($N = 4^R$)?

Section 4

3.12 Use the first three stages of Buneman's method, along with a hand calculator, to find $\sin(2\pi j/32)$ for $j = 0, 1, \ldots, 8$. Check your results by computing the same values directly on a calculator.

3.13 Develop the fourth stage of Buneman's method to find $\sin(2\pi j/64)$ for $j = 0, 1, \ldots, 16$.

3.14 Suppose that $\{F_k\}$ is the N-point DFT of $\{f_j\}$, defined by

$$F_k = \sum_{j=0}^{N-1} f_j W^{jk}.$$

Prove that, for each $j = 0, 1, \ldots, N - 1$,

$$f_j = \frac{1}{N} \left[\sum_{k=0}^{N-1} F_k^* W^{jk} \right]^*. \tag{3.15}$$

Also, explain why formula (3.15) shows that only the sines $\{S(j)\}_{j=0}^{(1/4)N}$ are needed for inverting an FFT, when that FFT used weight $W = e^{i2\pi/N}$. (See Sample Program 2 in Appendix B.)

Section 7

3.16 Prove (7.3) and (7.4).

3.17 Show that (7.16) and (7.17) are true.

Section 8

3.18 Show that (8.6) follows from (8.7).

3.19 Prove that, for $m, k = 0, 1, \ldots, N - 1$,

$$\sum_{n=0}^{N-1} \cos \frac{(n + \frac{1}{2})(m + \frac{1}{2})\pi}{N} \cos \frac{(n + \frac{1}{2})(k + \frac{1}{2})\pi}{N} = \begin{cases} \frac{1}{2}N & \text{if } m = k \\ 0 & \text{if } m \neq k. \end{cases} \tag{a}$$

Also, prove that a formula just like (a) holds if sines are used instead of cosines.

3.20 Using the results of Exercise 3.19, formulate a definition of an N-point discrete cosine transform *that is (up to a constant multiple) its own inverse.* Do the same for sines instead of cosines. How would you program a fast version of this discrete cosine (sine) transform?

4

Some Applications

In this chapter we will describe computer modeling of solutions to a few fundamental equations from mathematical physics. We shall discuss one-dimensional versions of the heat equation, the wave equation, and the Schrödinger wave equation for an enclosed freely moving particle. These problems will help introduce some of the filtered Fourier series that are used in signal processing.

---

1 Heat equation

In this section we shall describe computer modeling of a simple problem in heat conduction. Let the function $u(x,t)$, where $0 \leq x \leq L$ and $t \geq 0$, be interpreted as a temperature, 0 at the ends, and the function $g(x)$ be interpreted as the *initial temperature* $u(x,0) = g(x)$. In Chapter 3.1 of [Wa] it is shown that

$$\frac{\partial u}{\partial t} = a^2 \frac{\partial^2 u}{\partial x^2} \qquad \text{(heat equation)}$$

$$u(0,t) = 0 , \quad u(L,t) = 0 \qquad \text{(boundary conditions)} \qquad (1.1)$$

$$u(x,0) = g(x) \qquad \text{(initial condition)}.$$

The constant a^2 is called the *diffusion constant* for the material of the rod. Its units are commonly cm^2/sec. Copper, for example, has a diffusion constant of 1.14, while granite has a diffusion constant of 0.011.

Problem (1.1) is well known in mathematical physics. It was first solved by Fourier. Its derivation and solution by Fourier series is described in many other books besides [Wa]. See, for example, [Fo], [Ch-B], and [We].

A second interpretation of problem (1.1) is that the function u describes the relative concentration of a solvent within a surrounding solute. The function $g(x)$ describes the initial relative concentration. This interpretation is described in many physical chemistry textbooks.

Usually, the method of separation of variables is used to solve (1.1). Since this method is so well known, we will describe another method. We begin by expanding $u(x, t)$ in a Fourier series in the spatial variable x. Since all of the functions of the system $\{\sin(n\pi x/L)\}_{n=1}^{\infty}$ satisfy the *zero boundary conditions* at $x = 0$ and $x = L$, we expand $u(x, t)$ in a *Fourier sine series*

$$u(x, t) = \sum_{n=1}^{\infty} B_n(t) \sin \frac{n\pi x}{L} ,$$

$$\left(B_n(t) = \frac{2}{L} \int_0^L u(x, t) \sin \frac{n\pi x}{L} \, dx \right). \tag{1.2}$$

To find the functions $B_n(t)$ we substitute into the heat equation in (1.1) and, differentiating term by term, get

$$\sum_{n=1}^{\infty} B_n'(t) \sin \frac{n\pi x}{L} = \sum_{n=1}^{\infty} -\left(\frac{n\pi a}{L} \right)^2 B_n(t) \sin \frac{n\pi x}{L}. \tag{1.3}$$

Expressing (1.3) in terms of one sine series yields

$$\sum_{n=1}^{\infty} \left[B_n'(t) + \left(\frac{n\pi a}{L} \right)^2 B_n(t) \right] \sin \frac{n\pi x}{L} = 0. \tag{1.4}$$

Since the sine coefficients of the 0-function are all 0, we infer from (1.4) that

$$B_n'(t) + \left(\frac{n\pi a}{L} \right)^2 B_n(t) = 0 , \qquad (n = 1, 2, 3, \ldots). \tag{1.5}$$

Substituting $t = 0$ into $B_n(t)$ in (1.2) we also have

$$B_n(0) = \frac{2}{L} \int_0^L u(x, 0) \sin \frac{n\pi x}{L} \, dx , \qquad (n = 1, 2, 3, \ldots). \tag{1.6}$$

The right side of (1.6) is a constant for each n; it represents the sine coefficients of the initial temperature $u(x, 0)$. If we define b_n by

$$b_n = \frac{2}{L} \int_0^L u(x, 0) \sin \frac{n\pi x}{L} \, dx , \qquad (n = 1, 2, 3, \ldots) \tag{1.7}$$

then (1.6) becomes

$$B_n(0) = b_n , \qquad (n = 1, 2, 3, \ldots). \tag{1.8}$$

The solution to the differential equation (1.5) with initial condition (1.8) is

$$B_n(t) = b_n e^{-(n\pi a/L)^2 t}. \tag{1.9}$$

Combining (1.9) with the first equation in (1.2) yields the Fourier sine series solution to problem (1.1)

$$u(x,t) = \sum_{n=1}^{\infty} \left[e^{-(n\pi a/L)^2 t} \right] b_n \sin \frac{n\pi x}{L} \tag{1.10}$$

where

$$u(x,0) = \sum_{n=1}^{\infty} b_n \sin \frac{n\pi x}{L} \ .$$

In other words, *the temperature $u(x,t)$ for $t > 0$ is obtained by putting damping factors*

$$e^{-(n\pi a/L)^2 t} \qquad \text{(Gaussian damping factor)}$$

on the coefficients of the Fourier sine series for the initial temperature $u(x,0)$.

Example 1.11

Suppose that $a^2 = 1.14$ cm^2/sec, the diffusion constant of copper, and $L = 10$ cm. Graph $u(x,t)$ for $t = 0.1$, 0.5, 1.0, and 2.0 sec; given an initial temperature of

$$u(x,0) = \begin{cases} 0 & \text{for } 0 \le x \le 4 \text{ and } 6 \le x \le 10 \\ 20x - 80 & \text{for } 4 \le x \le 5 \\ 120 - 20x & \text{for } 5 \le x \le 6. \end{cases} \quad \blacksquare$$

SOLUTION   In Figure 4.1 we show the graph of the 25 harmonic partial sum for the Fourier sine series of $u(x,0)$ using 512 points. We will use this partial sum as an approximation to $u(x,0)$. Although the graph of this partial sum appears noticeably different than the graph of the initial function $u(x,0)$, we shall see below that it will still provide good approximations for the positive values of t that we are using. For the constants given, we have damping factors

$$e^{-(n\pi a/L)^2 t}.$$

*Using **FAS**, these factors can be put as coefficients on the terms of the 25 harmonic partial sum by choosing to do a filter and then choosing **Gauss** as the type of filter. For $t = 0.1$ the damping constant T asked for by FAS is $T = a^2 0.1 = 0.114$. And, for $t = 0.5$, 1.0, and 2.0 sec, we use $T = 0.57$, 1.14, and 2.28, respectively. The graphs of $u(x,t)$ for these values of t are shown in Figure 4.2.*

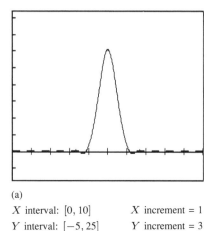

(a)

X interval: $[0, 10]$	X increment = 1
Y interval: $[-5, 25]$	Y increment = 3

FIGURE 4.1
Graph of the 25 harmonic Fourier sine series partial sum for the function $u(x, 0)$ in Example 1.11.

We will now show that using a 25 harmonic partial sum gives good approximations to $u(x, t)$ for the times $t = 0.1$, 0.5, 1.0, and 2.0 that we used above. First, we observe that the sine coefficients b_n satisfy

$$|b_n| = \left| \frac{1}{5} \int_0^{10} u(x, 0) \sin \frac{n\pi x}{10} \, dx \right| \leq \frac{1}{5} \int_0^{10} |u(x, 0)| \, dx.$$

Thinking of the integral of $|u(x, 0)|$ as an area, we get

$$|b_n| \leq 4, \qquad (n = 1, 2, 3, \ldots). \tag{1.12}$$

From (1.12) we obtain $(T = a^2 t)$

$$\left| u(x, t) - \sum_{n=1}^{25} b_n e^{-n^2(\pi/10)^2 T} \sin \frac{n\pi x}{10} \right| = \left| \sum_{n=26}^{\infty} b_n e^{-n^2(\pi/10)^2 T} \sin \frac{n\pi x}{10} \right|$$

$$\leq \sum_{n=26}^{\infty} |b_n| e^{-n^2(\pi/10)^2 T}$$

$$\leq \sum_{n=26}^{\infty} 4 e^{-n^2(\pi/10)^2 T}.$$

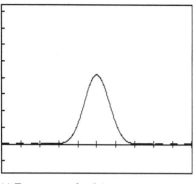

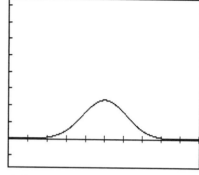

(a) Temperature after 0.1 sec

X interval: $[0, 10]$ X increment = 1

Y interval: $[-5, 25]$ Y increment = 3

(b) Temperature after 0.5 sec

X interval: $[0, 10]$ X increment = 1

Y interval: $[-5, 25]$ Y increment = 3

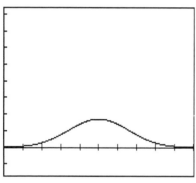

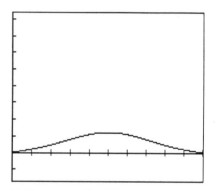

(c) Temperature after 1.0 sec

X interval: $[0, 10]$ X increment = 1

Y interval: $[-5, 25]$ Y increment = 3

(d) Temperature after 2.0 sec

X interval: $[0, 10]$ X increment = 1

Y interval: $[-5, 25]$ Y increment = 3

FIGURE 4.2
Graphs of the temperature $u(x, t)$ evolving from the initial temperature shown in Figure 4.1.

Now, if we write out the terms of the last series, we have

$$\sum_{n=26}^{\infty} 4e^{-n^2(\pi/10)^2 T} = 4e^{-26^2(\pi/10)^2 T} + 4e^{-27^2(\pi/10)^2 T} + 4e^{-28^2(\pi/10)^2 T} + \cdots$$

$$\leq 4e^{-26^2(\pi/10)^2 T} \left[1 + e^{-53(\pi/10)^2 T} + e^{-(2)53(\pi/10)^2 T} + \cdots\right]$$

$$= 4e^{-26^2(\pi/10)^2 T} \sum_{k=0}^{\infty} e^{-k53(\pi/10)^2 T} = \frac{4e^{-26^2(\pi/10)^2 T}}{1 - e^{-53(\pi/10)^2 T}}.$$

Combining this last result with our previous estimates yields

$$\left| u(x,t) - \sum_{n=1}^{25} b_n e^{-n^2(\pi/10)^2 T} \sin \frac{n\pi x}{10} \right| \leq \frac{4e^{-26^2(\pi/10)^2 T}}{1 - e^{-53(\pi/10)^2 T}} \,. \tag{1.13}$$

Substituting the values of $T = 0.114$, 0.57, 1.14, and 2.28 into the right side of (1.13), we see that the maximum differences between $u(x,t)$ and the 25 harmonic partial sums used for each T are about 0.00443, 1.284×10^{-16}, 3.725×10^{-33}, and 3.451×10^{-66}, respectively. This shows what *excellent* approximations the 25 harmonic partial sums are to the actual temperature $u(x,t)$ for the last three values of t. For $t = 0.1$ more harmonics must be used to obtain such precision.

∎

We will end this section with a brief description of how *FAS* graphs the filtered sine series partial sums like the ones just described. First, the Fourier sine coefficients b_n are approximated by a uniform, left endpoint, sum

$$b_n \approx \frac{2}{N} \sum_{j=1}^{N-1} u(jL/N, 0) \sin \frac{\pi j n}{N} \,. \tag{1.14}$$

Thus, at least for $1 \leq n \leq (1/4)N$, we will have b_n approximated by $2/N$ times the discrete sine transform of $\{u(jL/N, 0)\}_{j=1}^{N-1}$. Denoting this DST by $\{U_n\}$, we substitute into an M-harmonic partial sum of the series for $u(x,t)$ in (1.10) and we obtain (putting $a^2 t = T$)

$$u(x,t) \approx \frac{2}{N} \sum_{n=1}^{M} \left[U_n e^{-(n\pi/L)^2 T} \right] \sin \frac{n\pi x}{L} \,, \qquad (T = a^2 t). \tag{1.15}$$

Assuming that $M \leq (1/4)N$ (as in Chapter 2, Section 5) we substitute mL/N, for $m = 0, 1, \ldots, N-1$, into (1.15) obtaining

$$u(mL/N, t) \approx \frac{2}{N} \sum_{n=1}^{N-1} V_n \sin \frac{\pi n m}{N} \tag{1.16}$$

where

$$V_n = \begin{cases} U_n e^{-(n\pi/L)^2 T} & \text{for } n = 1, 2, \ldots, M \\ 0 & \text{for } n = M+1, \ldots, N-1. \end{cases}$$

Based on (1.16) we have $u(x,t)$ approximated by the following steps:

Step 1. A DST $\{U_n\}_{n=1}^{N-1}$ is computed from the set of sample values $\{u(jL/N, 0)\}_{j=1}^{N-1}$ of the initial function $u(x,0)$.

Step 2. The DST $\{U_n\}$ is lopped off by setting to 0 the values for the indices $n = M + 1, \ldots, N - 1$ where $M \leq (1/4)N$. The remaining values of U_n are multiplied by the damping factors $e^{-(n\pi/L)^2 T}$ where T is a nonnegative constant (called the damping constant) entered by the user. For solving (1.1), T should be given the value $a^2 t$ for each specific time t. The new sequence obtained by this procedure is called $\{V_n\}$.

Step 3. An inverse DST is applied to V_n to obtain approximations to $\{u(mL/N, t)\}_{m=1}^{N-1}$. Also, since $u(x, t)$ equals the sine series in (1.10) we have $u(0, t) = 0$ and $u(L, t) = 0$.

Step 4. The values of $\{u(mL/N, t)\}_{m=0}^{N}$ are connected by straight line segments to give approximations to $u(x, t)$ for $0 \leq x \leq L$.

2   The wave equation

In this section we describe the computer modeling of solutions to the problem of describing the motion of a vibrating string.

Consider a string of length L with both its ends fixed (for example, a guitar string or a violin string). It can be shown (see [Wa, Chapter 3.1]) that in the absence of any externally applied force, the position $y(x, t)$ of the string for $0 \leq x \leq L$ and $t \geq 0$ satisfies

$$
\begin{aligned}
&c^2 \frac{\partial^2 y}{\partial x^2} = \frac{\partial^2 y}{\partial t^2} && \text{(wave equation)} \\
&y(0, t) = 0 \ , \quad y(L, t) = 0 && \text{(boundary conditions)} \\
&y(x, 0) = f(x) \ , \quad \frac{\partial y}{\partial t}(x, 0) = g(x) && \text{(initial conditions).}
\end{aligned}
\tag{2.1}
$$

The function $f(x)$ specifies the *initial position* of the string, while the function $g(x)$ specifies the *initial velocity* of the string.

The system of units involved is the CGS system. The constant c^2 is equal to T/ρ where T is the constant tension along the string and ρ is the constant linear density of the string. In the CGS system c has the units of *velocity*, cm/sec.

Expanding $y(x, t)$ in a Fourier sine series

$$
y(x, t) = \sum_{n=1}^{\infty} B_n(t) \sin \frac{n\pi x}{L} \ ,
$$

$$
\left(B_n(t) = \frac{2}{L} \int_0^L y(x, t) \sin \frac{n\pi x}{L} \, dx \right)
\tag{2.2}
$$

and substituting this series for $y(x,t)$ into the wave equation in (2.1) and into the initial conditions in (2.1), we obtain

$$\sum_{n=1}^{\infty} \left[B_n''(t) + (nc\pi/L)^2 B_n(t) \right] \sin \frac{n\pi x}{L} = 0$$

$$B_n(0) = \frac{2}{L} \int_0^L f(x) \sin \frac{n\pi x}{L} \qquad (n = 1, 2, 3, \ldots)$$

$$B_n'(0) = \frac{2}{L} \int_0^L g(x) \sin \frac{n\pi x}{L} \, dx \qquad (n = 1, 2, 3, \ldots). \tag{2.3}$$

Letting a_n and b_n stand for the nth sine coefficients of f and g, respectively, we obtain from (2.3)

$$B_n''(t) + (nc\pi/L)^2 B_n(t) = 0 \,, \qquad B_n(0) = a_n \,, \qquad B_n'(0) = b_n \tag{2.4}$$

for $n = 1, 2, 3, \ldots$. For each n, the solution to (2.4) is

$$B_n(t) = a_n \cos (nc\pi/L)t + b_n \frac{\sin (nc\pi/L)t}{nc\pi/L} \,. \tag{2.5}$$

Using the notation

$$\operatorname{sinc} v = \frac{\sin \pi v}{\pi v} \,, \qquad (\operatorname{sinc} 0 = 1) \tag{2.6}$$

we can express (2.5) as

$$B_n(t) = a_n \cos (nc\pi/L)t + b_n t \operatorname{sinc} (nc/L)t. \tag{2.7}$$

Combining (2.7) with (2.2) we have found the following series solution to problem (2.1):

$$y(x,t) = \sum_{n=1}^{\infty} \left[a_n \cos (nc\pi/L)t + b_n t \operatorname{sinc} (nc/L)t \right] \sin \frac{n\pi x}{L}$$

$$a_n = \frac{2}{L} \int_0^L y(x,0) \sin \frac{n\pi x}{L} \, dx,$$

$$b_n = \frac{2}{L} \int_0^L \frac{\partial y}{\partial t}(x,0) \sin \frac{n\pi x}{L} \, dx \tag{2.8}$$

Let's consider an example.

Example 2.9

Suppose that $L = 50$ cm, $c^2 = 10,000$ cm²/sec², the initial position is $f(x) = 0.1x(50 - x)$, and the initial velocity g is zero. Graph $y(x,t)$ for the times $t = 0, 0.1, 0.2$, and 0.3 sec. ▯

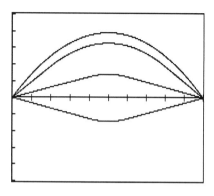

X interval: $[0, 50]$	X increment $= 5$
Y interval: $[-80, 80]$	Y increment $= 16$

FIGURE 4.3
Graphs of the string position for Example 2.9.

SOLUTION Since $g(x) = 0$ for $0 \leq x \leq L$, it follows that $b_n = 0$ for $n = 1$, 2, 3, ... Using a 200 harmonic partial sum for $y(x, t)$ yields

$$y(x, t) \approx \sum_{n=1}^{200} a_n \cos(n\pi ct/L) \sin \frac{n\pi x}{L} \qquad (2.10)$$

where $L = 50$ and $c = 100$. The sine series partial sum in (2.10) is obtained by applying *filter coefficients*

$$\cos(n\pi ct/L) , \qquad (n = 1, 2, 3, \ldots) \qquad (2.11)$$

to the terms of the 200 harmonic partial sum of the sine series for $y(x, 0) = f(x)$. This can be done by first calculating the sine series partial sum, using 1024 points. Then, after pressing y in response to the question about using a filter, one presses the *Tab* key to display the second list of filter choices. The filter needed in this case is the *cos* filter. You will be asked for the value of c, called the *wave constant*, which is 100 for this example. Then you will be asked for the value of t, called the *time constant*. By successively performing these steps and entering the three values of t for the time constant, you can obtain graphs like the ones shown in Figure 4.3. ∎

The approximation in (2.10) is a good one. Calculating the sine coefficients $\{a_n\}$ for $f(x) = 0.1x(50 - x)$, we get

$$a_n = \frac{1000}{n^3 \pi^3} \left[1 - (-1)^n \right] , \qquad (n = 1, 2, 3, \ldots). \qquad (2.12)$$

Therefore,

$$\left| y(x,t) - \sum_{n=1}^{200} a_n \cos(n\pi ct/L) \sin \frac{n\pi x}{L} \right| = \left| \sum_{n=201}^{\infty} a_n \cos(n\pi ct/L) \sin \frac{n\pi x}{L} \right|$$

$$\leq \sum_{n=201}^{\infty} |a_n| = \sum_{k=101}^{\infty} \frac{2000}{(2k - 1)^3 \pi^3} .$$

The last sum above can be estimated using the integral test from calculus

$$\sum_{k=101}^{\infty} \frac{2000}{(2k - 1)^3 \pi^3} \leq \int_{100}^{\infty} \frac{2000}{(2x - 1)^3 \pi^3} \, dx = \frac{500}{(199)^2 \pi^3}$$

$$< 0.001.$$

Therefore, the approximation in (2.10) involves an error of less than 0.001 for $0 \leq x \leq 50$ and all $t \geq 0$.

Example 2.13

Suppose that L and c^2 have the same values from the previous example, but that the initial velocity is described by $20x(50 - x)$ while the initial position is zero. Graph $y(x,t)$ for times $t = 0.001$, 0.002, and 0.003 sec. □

SOLUTION In this case, the coefficients $\{a_n\}$ are all 0. Again using 200 harmonics in a partial sum for $y(x,t)$ we have

$$y(x,t) \approx \sum_{n=1}^{200} \left[t\operatorname{sinc}(nct/L) \right] b_n \sin \frac{n\pi x}{L} . \qquad (2.14)$$

In this example, one computes the filter factors

$$t\operatorname{sinc}(nct/L) , \qquad (n = 1, 2, 3, \ldots). \qquad (2.15)$$

One chooses the filter *sinc* on the second filter menu (press the *Tab* key when the first filter menu appears). You are asked to enter the wave constant, which is $c = 100$ as in the previous example, and the time constants (0.001, 0.002, and 0.003). This is how one obtains the graphs shown in Figure 4.4. ▮

The approximation in (2.14) is as accurate as in the previous example. Since the values of b_n are 200 times the values of a_n in the previous example (since

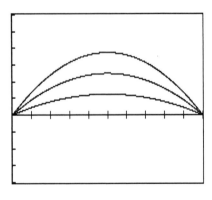

X interval: [0, 50]   X increment = 5
Y interval: [−40, 60]   Y increment = 10

FIGURE 4.4
Graphs of the string position for Example 2.13.

$g = 200f$ where f is as in the previous example), we obtain

$$\left| y(x,t) - \sum_{n=1}^{200} t\,\text{sinc}\,\frac{nct}{L}\, b_n \sin\frac{n\pi x}{L} \right|$$

$$= \left| \sum_{n=201}^{\infty} \frac{200\sin(nc\pi t/L)}{nc\pi/L} \frac{1000[1-(-1)^n]}{n^3\pi^3} \sin\frac{n\pi x}{L} \right|$$

$$\leq \sum_{k=101}^{\infty} \frac{1}{2\pi} \frac{2000}{(2k-1)^3\pi^3} < \frac{0.001}{2\pi}\ .$$

Therefore, the approximation in (2.14) involves an error of no more than $0.001/2\pi$ for $0 \leq x \leq 50$ and all $t \geq 0$.

3   Schrödinger's equation for a free particle

In this section we shall discuss computer modeling of the filtered Fourier series solution to the following problem

$$i\frac{\partial\psi}{\partial t} = \frac{-\hbar}{2m}\frac{\partial^2\psi}{\partial x^2} \qquad \text{(Schrödinger's equation)}$$

$$\psi(x,t) = 0\ , \quad \text{for } |x| > L \qquad \text{(boundary conditions)}$$

$$\psi(x,0) = f(x)\ , \quad \text{for } |x| < L \qquad \text{(initial condition).} \qquad (3.1)$$

The function ψ is interpreted in quantum mechanics as a generator of a probability distribution, $|\psi|^2$ $(= \psi\psi^*)$, governing the position x of a particle of mass m at time $t \geq 0$. Later, we will show that

$$\|\psi\|_2 = \left[\int_{-L}^{L} |\psi(x,t)|^2 \, dx \right]^{\frac{1}{2}} = 1 \tag{3.2}$$

provided that the given initial function f satisfies

$$\|f\|_2 = \left[\int_{-L}^{L} |f(x)|^2 \, dx \right]^{\frac{1}{2}} = 1. \tag{3.3}$$

Formulas (3.2) and (3.3) are the *normalization conditions* that are imposed on the function ψ whose evolution from f is governed by Schrödinger's equation. The constant $\hbar$ is Planck's constant divided by 2π; $\hbar$ has the value 1.054×10^{-27} erg-sec. By comparison, an electron has mass 0.911×10^{-27} g. Because of the absence of a potential term in Schrödinger's equation in (3.1), the particle is said to move *freely*. The boundary conditions in (3.1) say that the particle is constrained to move along the x-axis between $-L$ and L. Problem (3.1) is a one-dimensional version of the *particle in a box* problem.

To find the solution ψ to (3.1) we proceed as in the previous two sections. Only this time, we expand $\psi(x,t)$ and $f(x)$ in complex Fourier series in x (period $2L$)

$$\psi(x,t) = \sum_{n=-\infty}^{\infty} C_n(t) e^{i\pi nx/L} \,, \quad \left[C_n(t) = \frac{1}{2L} \int_{-L}^{L} \psi(x,t) e^{-i\pi nx/L} \right]$$

$$f(x) = \sum_{n=-\infty}^{\infty} c_n e^{i\pi nx/L} \,, \quad \left[c_n = \frac{1}{2L} \int_{-L}^{L} f(x) e^{-i\pi nx/L} \, dx \right]. \tag{3.4}$$

Substituting the series for $\psi(x,t)$ into Schrödinger's equation in (3.1), differentiating term by term, and combining the two series into one, yields

$$\sum_{n=-\infty}^{\infty} \left[iC_n'(t) - \frac{\hbar}{2m} \left(\frac{n\pi}{L} \right)^2 C_n(t) \right] e^{i\pi nx/L} = 0. \tag{3.5}$$

Setting $t = 0$ in the first equation in (3.4) yields [by comparison to the second equation in (3.4)]

$$C_n(0) = c_n \,, \qquad (n = 0, \pm 1, \pm 2, \dots). \tag{3.6}$$

Setting each coefficient in (3.5) equal to 0, we get

$$C_n'(t) = -\frac{i\hbar}{2m} \left(\frac{n\pi}{L} \right)^2 C_n(t) \,, \qquad (n = 0, \pm 1, \pm 2, \dots). \tag{3.7}$$

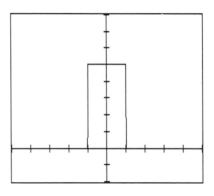

X interval: $[-10, 10]$ X increment = 2
Y interval: $[-.2, .8]$ Y increment = .1

FIGURE 4.5
Graph of the initial function in Example 3.10. 4096 points were used.

Solving (3.7), subject to the initial condition (3.6), for each n yields

$$C_n(t) = c_n e^{-i\frac{\hbar}{2m}(\frac{n\pi}{L})^2 t}, \qquad (n = 0, \pm 1, \pm 2, \ldots). \qquad (3.8)$$

Combining (3.8) with (3.4) we have our solution to (3.1):

$$\psi(x,t) = \sum_{n=-\infty}^{\infty} \left[e^{\frac{-i\hbar}{2m}(\frac{n\pi}{L})^2 t} \right] c_n e^{i\pi nx/L} \qquad (3.9a)$$

where

$$\psi(x,0) = \sum_{n=-\infty}^{\infty} c_n e^{i\pi nx/L}. \qquad (3.9b)$$

Our solution to (3.1) expresses $\psi(x,t)$ as a *filtered Fourier series* obtained from the Fourier series for the initial function $\psi(x,0)$. The filter factors are the complex, time dependent exponentials given in equation (3.8).

Example 3.10
Suppose that the initial function $\psi(x,0)$ is given by

$$\psi(x,0) = \begin{cases} 0.5 & \text{for } |x| < 2 \\ 0 & \text{for } 2 < |x| < 32 \end{cases}$$

(see Figure 4.5) and the particle has a mass of 0.911×10^{-27} g (mass of electron). Graph $\psi(x,t)$ for $t = 0.1, 0.2$, and 0.3 sec.   ☐

SOLUTION The initial function $\psi(x,0)$ is *real valued*. Consequently, the Fourier series for $\psi(x,t)$ can be split into real and imaginary parts

$$\psi(x,t) = \psi^R(x,t) - i\psi^I(x,t) \tag{3.11}$$

where

$$\psi^R(x,t) = \sum_{n=-\infty}^{\infty} \left[\cos \left[\frac{\hbar}{2m} \left(\frac{n\pi}{L} \right)^2 t \right] \right] c_n e^{i\pi nx/L}$$

$$\psi^I(x,t) = \sum_{n=-\infty}^{\infty} \left[\sin \left[\frac{\hbar}{2m} \left(\frac{n\pi}{L} \right)^2 t \right] \right] c_n e^{i\pi nx/L}. \tag{3.12}$$

Because $\psi(x,0)$ is real valued, it follows that $c^*_{-n} = c_n$ and, therefore

$$\left[c_{-n} \cos \left[\frac{\hbar}{2m} \left(\frac{-n\pi}{L} \right)^2 t \right] \right]^* = c_n \cos \left[\frac{\hbar}{2m} \left(\frac{n\pi}{L} \right)^2 t \right]$$

$$\left[c_{-n} \sin \left[\frac{\hbar}{2m} \left(\frac{-n\pi}{L} \right)^2 t \right] \right]^* = c_n \sin \left[\frac{\hbar}{2m} \left(\frac{n\pi}{L} \right)^2 t \right] \tag{3.13}$$

From (3.13) it follows that ψ^R and ψ^I are real valued. *FAS* can be used to graph partial sum approximations to ψ^R and ψ^I, and use these approximations to approximate

$$|\psi| = \left[(\psi^R)^2 + (\psi^I)^2 \right]^{\frac{1}{2}}. \tag{3.14}$$

To graph a 500 harmonic partial sum approximation to $\psi^R(x,t)$, call it $S_{500}^R(x,t)$, one computes a 500 harmonic partial sum for $\psi(x,0)$, using *FAS* (4096 points were used for these examples). Then, one chooses the *FresCos* filter from the second filter menu. The filter procedure asks you to input the value of $\hbar/2m$ which it calls the *wave constant*. For an electron, the wave constant has a value of 0.578. The *FresCos* procedure then asks you for the *time constant*, for which you enter one of the values of t. *FAS* then computes filter factors

$$\cos \left[\frac{\hbar}{2m} \left(\frac{n\pi}{L} \right)^2 t \right] , \qquad (n = 0, \pm 1, \pm 2, \ldots)$$

and graphs S_{500}^R, the 500 harmonic partial sum approximation to ψ^R.

To graph the 500 harmonic partial sum approximation to $\psi^I(x,t)$, call it $S_{500}^I(x,t)$, you choose *FresSin* from the filter menu and then enter the same wave constant and the same time constants as for ψ^R. If you graph S_{500}^R and S_{500}^I simultaneously for $t = 0.1$, 0.2, and 0.3 seconds, then you obtain graphs like the ones shown in Figure 4.6.

To better visualize things, in Figure 4.7 we show graphs of

$$|S_{500}(x,t)| = \left[(S_{500}^R(x,t))^2 + (S_{500}^I(x,t))^2 \right]^{\frac{1}{2}}$$

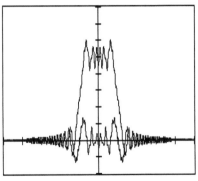

(a) $t = 0.1$ sec
X interval: $[-10, 10]$ X increment $= 2$
Y interval: $[-.2, .8]$ Y increment $= .1$

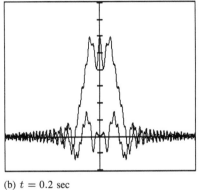

(b) $t = 0.2$ sec
X interval: $[-10, 10]$ X increment $= 2$
Y interval: $[-.2, .8]$ Y increment $= .1$

(c) $t = 0.3$ sec
X interval: $[-10, 10]$ X increment $= 2$
Y interval: $[-.2, .8]$ Y increment $= .1$

FIGURE 4.6
Graphs of approximations to the real and imaginary parts of $\psi(x,t)$ in
Example 3.10 for $t = 0.1$, 0.2, and 0.3 sec.

which is the 500 harmonic approximation to $|\psi(x,t)|$. These graphs were ob-
tained by choosing *Graphs* on the display menu and then choosing *Sum of*
Squares and pressing y when asked whether to normalize or not (if n is pressed,
then

$$(S_{500}^R(x,t))^2 + (S_{500}^I(x,t))^2$$

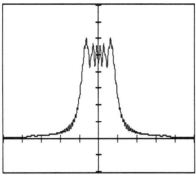

(a) $t = 0.1$ sec

X interval: $[-10, 10]$ X increment $= 2$

Y interval: $[-.2, .8]$ Y increment $= .1$

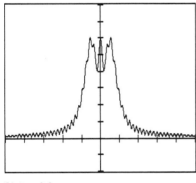 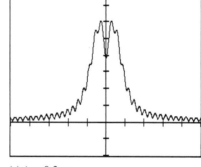

(b) $t = 0.2$ sec

X interval: $[-10, 10]$ X increment $= 2$

Y interval: $[-.2, .8]$ Y increment $= .1$

(c) $t = 0.3$ sec

X interval: $[-10, 10]$ X increment $= 2$

Y interval: $[-.2, .8]$ Y increment $= .1$

FIGURE 4.7
Graphs of approximations to $|\psi(x,t)|$ in Example 3.10 for $t = 0.1, 0.2,$ and 0.3 sec.

is graphed). This sequence of steps is performed after both S_{500}^{R} are S_{500}^{I} are displayed simultaneously.

The similarity between these graphs of approximations to $|\psi(x,t)|$ and one-dimensional Fresnel diffraction patterns from optics will be obvious to anyone familiar with such diffraction patterns (see Chapter 6, Section 2).

The graphs of $|S_{500}|$ $(\approx |\psi(x,t)|)$ can be compared to the initial function by doing another partial sum, but entering f when asked for how many harmonics. Then *FAS* does an inverse FFT on the FFT of the initial function data, thereby

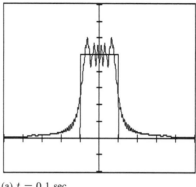

(a) $t = 0.1$ sec
X interval: $[-10, 10]$ X increment = 2
Y interval: $[-.2, .8]$ Y increment = .1

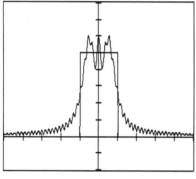

(b) $t = 0.2$ sec
X interval: $[-10, 10]$ X increment = 2
Y interval: $[-.2, .8]$ Y increment = .1

(c) $t = 0.3$ sec
X interval: $[-10, 10]$ X increment = 2
Y interval: $[-.2, .8]$ Y increment = .1

FIGURE 4.8
Comparison between the initial function in Example 3.1 and $|\psi(x,t)|$ for $t = 0.1$, 0.2, and 0.3 sec.

recovering this function data. The comparisons between $\psi(x,0)$ and $|S_{500}(x,t)|$ are shown in Figure 4.8. ∎

It is interesting to see what happens if the initial probability distribution is narrowed, as in our next example.

Example 3.15

Suppose that the initial function $\psi(x,0)$ is

$$\psi(x,0) = \begin{cases} 2^{\frac{1}{2}} & \text{for } |x| < 0.25 \\ 0 & \text{for } 0.25 < |x| < 32. \end{cases}$$

Graph approximations to $|\psi(x,t)|$ for the same times as in the previous example and for the same mass, too. ◻

SOLUTION The graphs of $|S_{500}|$ ($\approx |\psi(x,t)|$) are shown in Figure 4.9. The similarity between the graphs for $t = 0.2$ and 0.3 seconds and *Fraunhofer diffraction patterns* from optics will be obvious to anyone familiar with such diffraction patterns (see Chapter 6, Section 3). ∎

We close this section by showing that, given (3.3), equation (3.2) must hold. And, we also show that the 500 harmonic partial sum approximations used in the last two examples are fairly good if the 2-Norm

$$\|g\|_2 = \left[\int_{-L}^{L} |g(x)|^2 \, dx \right]^{\frac{1}{2}} \tag{3.16}$$

is used to measure the magnitudes of the differences between functions.

We need the following theorem, whose proof is beyond the scope of this text (see [Ru, Chapter 4]).

THEOREM 3.17

The series

$$\sum_{n=-\infty}^{\infty} a_n e^{i\pi n x / L}$$

*is a Fourier series for a function g, where $\|g\|_2 < \infty$, if and only if $\sum_{n=-\infty}^{\infty} |a_n|^2$ converges. Moreover, **Parseval's equality** holds*

$$\sum_{n=-\infty}^{\infty} |a_n|^2 = \frac{1}{2L} \int_{-L}^{L} |g(x)|^2 \, dx = \frac{1}{2L} \|g\|_2^2$$

*and we have the **completeness relation***

$$\lim_{M \to \infty} \|g - S_M\|_2 = \lim_{M \to \infty} \left[2L \sum_{|n|>M}^{\infty} |a_n|^2 \right]^{\frac{1}{2}} = 0$$

where S_M is the M-harmonic partial sum of the Fourier series for g.

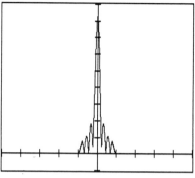

(a) $t = 0.1$ sec
X interval: $[-32, 32]$ X increment = 6.4
Y interval: $[-.1, .9]$ Y increment = .1

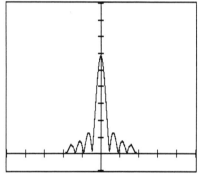

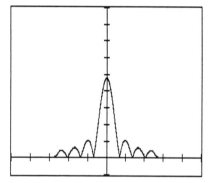

(b) $t = 0.2$ sec
X interval: $[-32, 32]$ X increment = 6.4
Y interval: $[-.1, .9]$ Y increment = .1

(c) $t = 0.3$ sec
X interval: $[-32, 32]$ X increment = 6.4
Y interval: $[-.1, .9]$ Y increment = .1

FIGURE 4.9
Graphs of approximations to $|\psi(x, t)|$ in Example 3.15 for $t = 0.1$, 0.2, and 0.3 sec.

Using Theorem (3.17) we have, based on (3.3) and the second equation in (3.4),

$$1 = \|\psi(x, 0)\|_2^2 = 2L \sum_{n=-\infty}^{\infty} |c_n|^2. \qquad (3.18)$$

However, we also have

$$\left| e^{i \frac{\hbar}{2m} \left(\frac{n\pi}{L} \right)^2 t} c_n \right| = |c_n|$$

so (3.18) implies

$$1 = 2L \sum_{n=-\infty}^{\infty} \left| e^{i\frac{\hbar}{2m}(\frac{n\pi}{L})^2 t} c_n \right|^2. \tag{3.19}$$

Using Theorem (3.17) again, we conclude from (3.19) that the series in the first equation in (3.4) is the Fourier series for $\psi(x,t)$ and (3.2) holds.

Theorem (3.17) can also be used to show that the 500 harmonic partial sums used in the examples above are good approximations. Letting $S_M(x,t)$ stand for the M-harmonic partial sum of the first series in (3.9), it follows that $\psi(x,t) - S_M(x,t)$ has the Fourier series expansion

$$\psi(x,t) - S_M(x,t) = \sum_{|n|>M}^{\infty} \left[e^{i\frac{\hbar}{2m}(\frac{n\pi}{L})^2 t} c_n \right] e^{i\pi nx/L} \tag{3.20}$$

and, for $t = 0$ in particular,

$$\psi(x,0) - S_M(x,0) = \sum_{|n|>M}^{\infty} c_n e^{i\pi nx/L}. \tag{3.21}$$

By Theorem 3.17 applied to (3.20) and (3.21) we obtain, as we did above,

$$\|\psi(x,0) - S_M(x,0)\|_2 = \left[2L \sum_{|n|>M}^{\infty} |c_n|^2 \right]^{\frac{1}{2}}$$

$$= \|\psi(x,t) - S_M(x,t)\|_2. \tag{3.22}$$

Equation (3.22) says that the 2-Norm measure of the magnitude of the difference between $\psi(x,t)$ and $S_M(x,t)$ is a constant for all $t \geq 0$. Using the choices *Graphs* (on the display menu) and *Norm Difference* (entering 2 for the *Power norm*), one can estimate the 2-Norm magnitude of the difference between the initial function $\psi(x,0)$ and its M-harmonic Fourier series $S_M(x,t)$ *when both graphs are displayed.* When $M = 500$, the magnitude of the 2-Norm difference obtained for Example 3.10 is approximately 0.04 while for Example 3.15 it is approximately 0.11. In both cases, $S_{500}(x,t)$ is a reasonable approximation, in terms of 2-Norm, to $\psi(x,t)$. This is especially true if the 2-Norm difference is thought of as measuring the distance between two vectors of length 1 (corresponding to an *angular*, or more precisely, *chordal*, distance).

4 Filters used in signal processing

In each of the preceding sections an initial Fourier series (or Fourier sine series) was *filtered* by multiplying its coefficients by *filter factors*. These factors were time dependent. In this section we will give an introduction to some of the common *time independent* filters, most of which are used in signal processing.

Cesàro Filter

Cesàro filtering is also known as the method of *arithmetic means*. To motivate this procedure, consider Figure 4.10. In parts (b)–(d) of Figure 4.10, we can see how the successive partial sums of the Fourier series seem to *interlace* around the graph of the step function. By interlace we mean not only that the partial sums oscillate about the function, but that *at most points* they also change from being above the step function to below the step function (or from below to above), at least approximately, as one passes from one partial sum to the next. It makes sense then to form an average, an arithmetic mean, of partial sums in order to better approximate the function.

DEFINITION 4.1 *Given a function with Fourier series partial sums* $\{S_n\}_{n=0}^{\infty}$, *the* Mth **arithmetic mean**, *or* **Cesàro filtered Fourier series using** M **harmonics**, *is denoted by* σ_M *where*

$$\sigma_M = \frac{1}{M}\left[S_0 + S_1 + \ldots + S_{M-1}\right].$$

By replacing S_k by $\sum_{j=-k}^{k} c_j e^{i\pi jx/L}$, we get

$$\sigma_M(x) = \frac{1}{M}\sum_{k=0}^{M-1}\left[\sum_{j=-k}^{k} c_j e^{i\pi jx/L}\right]. \tag{4.2}$$

By fixing a value of n, for some $n = 0, \pm 1, \pm 2, \ldots, \pm M$, and counting how often c_n appears in the sums in (4.2), we obtain

$$\sigma_M(x) = \sum_{n=-M}^{M}\left(1 - \left|\frac{n}{M}\right|\right) c_n e^{i\pi nx/L}. \tag{4.3}$$

If we compare σ_M in (4.3) to S_M, where

$$S_M(x) = \sum_{n=-M}^{M} c_n e^{i\pi nx/L} \tag{4.4}$$

we see that σ_M is obtained from S_M by multiplying the coefficients in S_M by the *filter factors* $\{1 - |n/M|\}_{n=-M}^{n=M}$. These factors are often called *convergence factors* since they will usually help improve the pointwise convergence of the Cesàro filtered Fourier series to the original function. One form of this improved convergence is *suppression of Gibbs' phenomenon* (see Figure 4.11). Another aspect of Cesàro filtering is that the filter coefficients are near 0 for $|n|$ near M, consequently the *higher frequency harmonics in* S_M *are damped down in* σ_M.

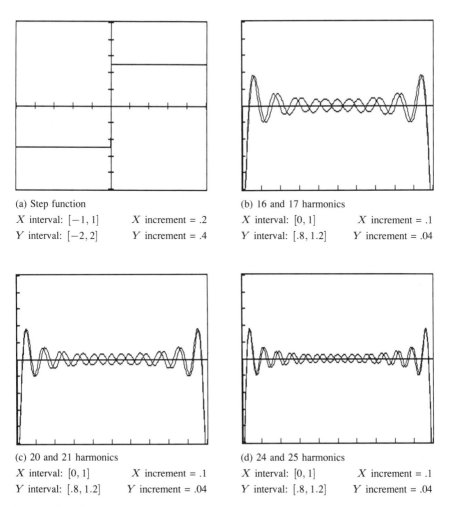

(a) Step function

X interval: $[-1, 1]$ X increment = .2

Y interval: $[-2, 2]$ Y increment = .4

(b) 16 and 17 harmonics

X interval: $[0, 1]$ X increment = .1

Y interval: $[.8, 1.2]$ Y increment = .04

(c) 20 and 21 harmonics

X interval: $[0, 1]$ X increment = .1

Y interval: $[.8, 1.2]$ Y increment = .04

(d) 24 and 25 harmonics

X interval: $[0, 1]$ X increment = .1

Y interval: $[.8, 1.2]$ Y increment = .04

FIGURE 4.10
Behavior of Fourier series partial sums for a step function.

This is a common feature of many of the filtering procedures used in signal processing.

de la Vallee Poussin filter

Another filter that is closely related to the Cesàro filter is the de la Vallee Poussin filter (dlVP filter, for short). To motivate the use of the dlVP filter, we note that for small values of n the Fourier series partial sum S_n is often not a good approximation to the original function (especially if that function is

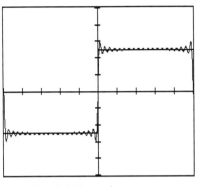 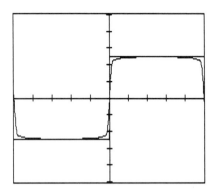

(a) Unfiltered, 40 harmonics

X interval: $[-1, 1]$ X increment = .2

Y interval: $[-2, 2]$ Y increment = .4

(b) Cesàro filtering, 40 harmonics

X interval: $[-1, 1]$ X increment = .2

Y interval: $[-2, 2]$ Y increment = .4

FIGURE 4.11
Suppression of Gibb's phenomenon by Cesàro filtering.

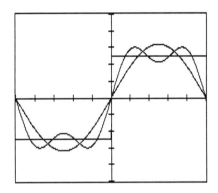

X interval: $[-1, 1]$ X increment = .2

Y interval: $[-2, 2]$ Y increment = .4

FIGURE 4.12
2 and 3 harmonics Fourier series partial sums for a step function.

a step function). For example, graphs of the 2 and 3 harmonic Fourier series partial sums for the step function shown in Figure 4.12 are not close to that step function at all. Putting this another way, notice that in Figure 4.10(a)–(d), we chose partial sums that were intertwined about the step function *and this intertwining becomes denser with larger number of harmonics*. Therefore, it might be more advantageous to average only the upper half of the partial sums.

If we have an even number of harmonics, say $2M$, then we define V_{2M} by

$$V_{2M} = \frac{1}{M} \sum_{k=M}^{2M-1} S_k. \tag{4.5}$$

This function V_{2M} is called the *dlVP filtered partial sum containing $2M$ harmonics*.

We now show that

$$V_{2M}(x) = \sum_{n=-2M}^{2M} v_n c_n e^{i\pi nx/L} \tag{4.6}$$

where

$$v_n = \begin{cases} 1 & \text{if } |n| \le M \\ 2(1 - |\frac{n}{2M}|) & \text{if } M \le |n| \le 2M. \end{cases} \tag{4.7}$$

The first step is to rewrite (4.5) using the definition of S_k:

$$V_{2M}(x) = \frac{1}{M} \sum_{k=M}^{2M-1} \left[\sum_{j=-k}^{k} c_j e^{i\pi jx/L} \right]. \tag{4.8}$$

Second, we find the coefficient of $e^{i\pi nx/L}$ for each $n = 0, \pm1, \ldots, \pm2M$. There are two cases to consider.

Case 1. $(|n| \le M)$. For this case c_n appears in each term in brackets in (4.8) for $k = M$ to $2M - 1$. Therefore, the exponential $e^{i\pi nx/L}$ has coefficient

$$\frac{1}{M} \sum_{k=M}^{2M-1} c_n = c_n \left[\frac{1}{M} \sum_{k=M}^{2M-1} 1 \right] = c_n.$$

Hence $v_n = 1$ as in (4.7).

Case 2. $(M < |n| \le 2M)$. For this case, c_n appears in only those terms in the brackets in (4.8) where $k = |n|, \ldots, 2M - 1$. Therefore, the coefficient of $e^{i\pi nx/L}$ is

$$\frac{1}{M} \sum_{k=|n|}^{2M-1} c_n = c_n \left[\frac{1}{M} \sum_{k=|n|}^{2M-1} 1 \right] = c_n \frac{2M - |n|}{M}$$

$$= \left(2 - \left| \frac{n}{M} \right| \right) c_n = 2 \left(1 - \left| \frac{n}{2M} \right| \right) c_n.$$

Hence, $v_n = 2(1 - |n/2M|)$ as in (4.7).

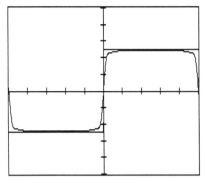
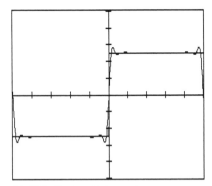

(a) Cesàro filtering, 30 harmonics

X interval: $[-1, 1]$ X increment = .2

Y interval: $[-2, 2]$ Y increment = .4

(b) dlVP filtering, 30 harmonics

X interval: $[-1, 1]$ X increment = .2

Y interval: $[-2, 2]$ Y increment = .4

FIGURE 4.13
Comparison of Cesàro and dlVP filtering.

In *FAS*, the dlVP filter is found on the second filter menu. *FAS* will calculate the dlVP filtered Fourier series

$$V_M(x) = \sum_{n=-M}^{M} v_n c_n e^{i\pi n x/L} \tag{4.9}$$

where

$$v_n = \begin{cases} 1 & \text{if } 2|n| \leq M \\ 2\left(1 - \frac{|n|}{M}\right) & \text{if } M \leq 2|n| \leq 2M \end{cases} \tag{4.10}$$

which, *for $2M$ in place of M*, matches the description of V_{2M} in (4.6) and (4.7). Formula (4.9) has the advantage of allowing the calculation of a dlVP filter for an odd number of harmonics as well as an even number of harmonics.

In Figure 4.13 we have graphed a dlVP filtered Fourier series partial sum for a step function. For this example we can see that dlVP filtering gives a closer approximation to the original step function than Cesàro filtering does.

Hamming and hanning filters

We close this section by mentioning two other filters frequently used in signal processing. When the function $H_M(x)$ is computed by the formula

$$H_M(x) = \sum_{n=-M}^{M} A_n c_n e^{i\pi n x/L} \tag{4.11}$$

where

$$A_n = 0.5 + 0.5\cos\frac{n\pi}{M} \tag{4.12}$$

then the Fourier series

$$\sum_{n=-\infty}^{\infty} c_n e^{i\pi nx/L}$$

is said to be *hanning filtered*. The function $H_M(x)$ is called the *hanning filtered Fourier series using M harmonics*.

If, instead of formula (4.12), one uses

$$A_n = 0.54 + 0.46\cos\frac{n\pi}{M} \tag{4.13}$$

then the Fourier series is said to be *Hamming filtered*. And the function $H_M(x)$ is called the *Hamming filtered Fourier series using M harmonics*.

REMARK 4.14   (a) The similarity between the standard names for these filters, hanning and Hamming, is a cause for confusion. The hanning filter (note the lowercase h) is named in honor of the Austrian mathematician von Hann. The Hamming filter (note the capital H) is named after Richard W. Hamming, who is from the USA. (b) The underlying motivation for hanning and Hamming filtering is not as easily explained as our two previous examples. Their description requires an examination of *point spread functions*, which we discuss in Section 6. ∎

5   Designing filters

All of the filters described in Section 4 fit into one common framework. Understanding this framework will allow you to design your own filters, using the function creation procedure of *FAS*.

Each filtering procedure used a modified Fourier series partial sum

$$\sum_{n=-M}^{M} F\left(\frac{n}{M}\right) c_n e^{i\pi nx/L} \tag{5.1}$$

where the *unfiltered* Fourier series partial sum was

$$\sum_{n=-M}^{M} c_n e^{i\pi nx/L}. \tag{5.2}$$

In (5.1) we shall assume that the function F is a continuous, even function over the interval $[-1, 1]$. For three of the filters described in Section 4, we have the following formulas for $F(n/M)$ and $F(x)$:

Filter	$F(n/M)$	$F(x)$
Cesàro	$1 - \left|\frac{n}{m}\right|$	$1 - |x|$
hanning	$0.5 + 0.5\cos(\pi n/M)$	$0.5 + 0.5\cos(\pi x)$
Hamming	$0.54 + 0.46\cos(\pi n/M)$	$0.54 + 0.46\cos(\pi x)$

(5.3)

Since $F(x)$ is even, it really only needs to be defined over the interval $[0,1]$. In which case, the functions in (5.3) can be expressed as

Filter	$F(x)$, x in $[0,1]$
Cesàro	$1 - x$
hanning	$0.5 + 0.5\cos(\pi x)$
Hamming	$0.54 + 0.46\cos(\pi x)$

(5.4)

The *user* choice on the filter menu of *FAS* allows you to design you own filter function $F(x)$ over the interval $[0,1]$. *FAS* then automatically takes care of computing $\{F(n/M)\}$, using the even extension of F for negative n, for $n = 0$, $\pm1, \pm2, \ldots, \pm M$ and using these filter coefficients in (5.1). Here is an example.

Example 5.5
If you choose *user* on the filter menu and then *Piece* on the function menu, you can create the following function:

$$F(x) = \begin{cases} 1 & \text{if } 0 \le x \le 0.2 \\ \cos\left(\pi\frac{x-0.2}{1.6}\right) & \text{if } 0.2 \le x \le 1. \end{cases}$$

(5.6)

If you create this filter function for $M = 30$ harmonics (using, say 512 points), then the graph shown in Figure 4.14 results when the initial function is $f(x) = x$ over $[-\pi, \pi]$. If the initial function is

$$f(x) = \begin{cases} -1 & \text{for } -\pi < x < 0 \\ 1 & \text{for } 0 < x < \pi \end{cases}$$

(5.7)

then the graph shown in Figure 4.15 results for $M = 30$ harmonics.   ☐

Earlier in this chapter we described filtering of Fourier sine series. A filtered Fourier sine series partial sum has the form

$$\sum_{n=1}^{M} F\left(\frac{n}{M}\right) b_n \sin\frac{n\pi x}{L}$$

(5.8)

where $F(x)$ is a continuous function over the interval $[0,1]$, and

$$\sum_{n=1}^{M} b_n \sin\frac{n\pi x}{L}$$

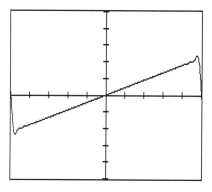

X interval: $[-\pi, \pi]$ X increment $= \pi/5$
Y interval: $[-7, 7]$ Y increment $= 1.4$

FIGURE 4.14
User-filtered Fourier series.

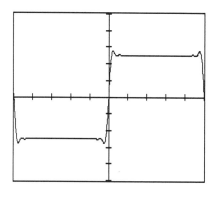

X interval: $[-\pi, \pi]$ X increment $= \pi/5$
Y interval: $[-2, 2]$ Y increment $= .4$

FIGURE 4.15
User-filtered Fourier series.

is the M-harmonic partial sum of the Fourier sine series. By choosing *user* on the filter menu when doing sine series, you can create your own filter function $F(x)$, just as for Fourier series. Here is an example of an application that requires the creation of such a filter.

Example 5.9   INHOMOGENEOUS WAVE EQUATION
It can be shown (see Chapter 3.1 of [Wa]) that when a force $F(x,t)$ *per unit length* is applied to an elastic string with fixed ends and constant tension, then the string height $y(x,t)$ satisfies (ρ is the linear density of the string, L is the length of the string)

$$c^2 \frac{\partial^2 y}{\partial x^2} + \frac{F(x,t)}{\rho} = \frac{\partial^2 y}{\partial t^2} \qquad \text{(inhomogenous wave equation)}$$

$$y(0,t) = 0 , \quad y(L,t) = 0 \qquad \text{(boundary conditions)}$$

$$y(x,0) = f(x), \quad \frac{\partial y}{\partial t}(x,0) = g(x) \quad \text{(initial conditions)}. \qquad (5.10)$$

Problem (5.10) is solved by a method similar to the one used to solve (2.1). Expanding $y(x,t)$ and $F(x,t)/\rho$ in Fourier sine series for $0 \le x \le L$ we have

$$y(x,t) = \sum_{n=1}^{\infty} B_n(t) \sin \frac{n\pi x}{L},$$

$$\left(B_n(t) = \frac{2}{L} \int_0^L y(x,t) \sin \frac{n\pi x}{L} \, dx \right) \qquad (5.11)$$

and

$$\frac{F(x,t)}{\rho} = \sum_{n=1}^{\infty} F_n(t) \sin \frac{n\pi x}{L},$$

$$\left(F_n(t) = \frac{2}{L} \int_0^L \frac{F(x,t)}{\rho} \sin \frac{n\pi x}{L} \, dx \right). \qquad (5.12)$$

Since $F(x,t)$ will be given to us, we assume that the functions $F_n(t)$ are also known. Substituting the series from (5.11) and (5.12) into the inhomogeneous wave equation in (5.10), differentiating term by term, and rearranging into a single series, yields

$$\sum_{n=1}^{\infty} \left[F_n(t) - \left(\frac{nc\pi}{L} \right)^2 B_n(t) - B_n''(t) \right] \sin \frac{n\pi x}{L} = 0. \qquad (5.13)$$

Since the 0-function has sine coefficients all equal to 0, we set the expression in brackets in (5.13) equal to 0, obtaining

$$B_n''(t) + \left(\frac{nc\pi}{L} \right)^2 B_n(t) = F_n(t), \qquad (n = 1, 2, 3, \ldots). \qquad (5.14)$$

For each n, we also have initial conditions,

$$B_n(0) = a_n, \quad \left(a_n = \frac{2}{L} \int_0^L y(x,0) \sin \frac{n\pi x}{L} \, dx \right)$$

$$B_n'(0) = b_n, \quad \left(b_n = \frac{2}{L} \int_0^L \frac{\partial y}{\partial t}(x,0) \sin \frac{n\pi x}{L} \, dx \right). \tag{5.15}$$

For notational purposes, we define ω_n *to be* $nc\pi/L$. Our problem is then to solve, for each n, each of the problems

$$B_n''(t) + \omega_n^2 B_n(t) = F_n(t), \quad \left(\omega_n = \frac{nc\pi}{L} \right)$$

$$B_n(0) = a_n, \quad B_n'(0) = b_n \tag{5.16}$$

for the unknown function $B_n(t)$.

For instance, suppose that we have a string of length $L = 10$ and that a force of $0.2\rho \sin(\omega t)$ is applied to the string between $x = 2$ and $x = 3$. Suppose also that $c^2 = 10,000$ and that the string is initially at rest in the horizontal position. *We will show how a filtered Fourier sine series can be used to describe the motion of the string.*

For this example, we have

$$y(x,0) = 0, \quad \frac{\partial y}{\partial t}(x,0) = 0 \tag{5.17}$$

since the string is initially horizontal and at rest. Consequently, using (5.17) in (5.15), we have

$$a_n = 0, \quad b_n = 0, \quad (n = 1, 2, 3, \ldots). \tag{5.18}$$

Now, the function $F(x,t)/\rho$ has the form

$$\frac{F(x,t)}{\rho} = G(x) \sin \omega t$$

where

$$G(x) = \begin{cases} 0.2 & \text{for } 2 < x < 3 \\ 0 & \text{for } 0 < x < 2 \text{ and } 3 < x < 10. \end{cases} \tag{5.19}$$

Consequently,

$$F_n(t) = (\sin \omega t) \frac{2}{L} \int_0^L G(x) \sin \frac{n\pi x}{L} \, dx$$

$$= K_n \sin \omega t, \quad \left(K_n = \frac{2}{L} \int_0^L G(x) \sin \frac{n\pi x}{L} \, dx \right). \tag{5.20}$$

We will see that for our computer work, the exact expression for K_n is *not* needed. The important thing is that $\{K_n\}$ is the set of Fourier sine coefficients for the function G in (5.19). That is,

$$G(x) = \sum_{n=1}^{\infty} K_n \sin \frac{n\pi x}{L},$$

$$\left(K_n = \frac{2}{L} \int_0^L G(x) \sin \frac{n\pi x}{L} \, dx \right). \tag{5.21}$$

Now, after substituting the form for $F_n(t)$ from (5.20) and the values for a_n and b_n from (5.18) back into (5.16), we have to solve

$$B_n''(t) + \omega_n^2 B_n(t) = K_n \sin \omega t$$

$$B_n(0) = 0, \qquad B_n'(0) = 0. \tag{5.22}$$

If we assume that $\omega \neq \omega_n$ for any n, then the solution is found to be

$$B_n(t) = \frac{K_n}{\omega_n^2 - \omega^2} \sin \omega t - \frac{K_n \omega / \omega_n}{\omega_n^2 - \omega^2} \sin \omega_n t.$$

After simplifying this last expression for $B_n(t)$ and substituting into the series for $y(x, t)$ in (5.11), we obtain

$$y(x, t) = \sum_{n=1}^{\infty} \left[\frac{\omega_n \sin \omega t - \omega \sin \omega_n t}{\omega_n (\omega_n^2 - \omega^2)} \right] K_n \sin \frac{n\pi x}{L} \tag{5.23}$$

where $L = 10$ and $\omega_n = nc\pi / L = 10 n\pi$.

Formula (5.23) expresses $y(x, t)$ as a *time dependent* filtered Fourier sine series, where the initial (unfiltered) series is the Fourier sine series for the function G in (5.21). Approximating the series in (5.23) using M harmonics yields

$$y(x, t) \approx \sum_{n=1}^{M} \left[\frac{\omega_n \sin \omega t - \omega \sin \omega_n t}{\omega_n (\omega_n^2 - \omega^2)} \right] K_n \sin \frac{n\pi x}{L}. \tag{5.24}$$

To create a filter function we need to express the filter coefficients

$$\frac{\omega_n \sin \omega t - \omega \sin \omega_n t}{\omega_n (\omega_n^2 - \omega^2)} \tag{5.25}$$

in the form $F(n/M)$ for a function $F(x)$ over the interval $[0, 1]$. The key thing to observe is that only ω_n depends on n. In fact, $\omega_n = 10 n\pi = 10 \pi M(n/M)$. Consequently, using x in place of n/M for each occurence of ω_n in (5.25) yields the filter function

$$F(x) = \frac{(10\pi M x) \sin \omega t - \omega \sin[(10\pi M x)t]}{(10\pi M x)[(10\pi M x)^2 - \omega^2]}.$$

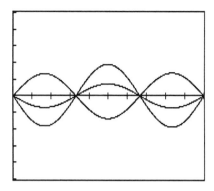

X interval: $[0, 10]$ X increment $=1$
Y interval: $[-.001, .001]$ Y increment $=.0002$

FIGURE 4.16
String motion under driving force, frequency $\omega = 29.9\pi$.

Or, more simply, using the sinc function defined in (2.6)

$$F(x) = \frac{\sin \omega t - \omega t \operatorname{sinc}(10Mxt)}{(10\pi Mx)^2 - \omega^2}.\tag{5.26}$$

Suppose we use 1024 points, $M = 200$ harmonics, $\omega = 29.9\pi$, and times $t = 1$, $t = 3$, and $t = 12$ sec. Then, computing a 200 harmonic sine series for $G(x)$ using formula (5.19), over the interval $[0, 10]$, and creating a filter function using formula (5.26) *with the values of M, ω, and each t that we have chosen*, we obtain the graphs shown in Figure 4.16. If, instead, we use $\omega = 19.9\pi$, then we obtain the graphs shown in Figure 4.17 (using the same number of harmonics and the same times). □

6 Convolution and point spread functions

In this section we continue our examination of filters. We will examine the concepts of convolution and point spread functions, which are essential for a solid understanding of filtered Fourier series. In the preceding sections we applied filters to Fourier series in the following way. From an initial Fourier series for a function f

$$\sum_{n=-\infty}^{\infty} c_n e^{i\pi nx/L}, \qquad \left(c_n = \frac{2}{L}\int_{-L}^{L} f(x)e^{-i\pi nx/L}\,dx\right)\tag{6.1}$$

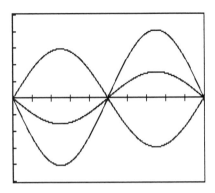

X interval: $[0, 10]$ X increment $= 1$
Y interval: $[-.001, .001]$ Y increment $= .0002$

FIGURE 4.17
String motion under driving force, frequency $\omega = 19.9\pi$.

we obtain a filtered Fourier series

$$\sum_{n=-\infty}^{\infty} c_n F_n e^{i\pi nx/L}. \tag{6.2}$$

To understand this filtering process, we need the following theorem.

THEOREM 6.3   FOURIER SERIES CONVOLUTION THEOREM
If f and $\mathcal{P}$ have period $2L$, and Fourier series expansions

$$f(x) = \sum_{n=-\infty}^{\infty} c_n e^{i\pi nx/L}, \qquad \left(c_n = \frac{1}{2L} \int_{-L}^{L} f(x) e^{-i\pi nx/L}\, dx \right) \tag{6.3a}$$

$$\mathcal{P}(x) = \sum_{n=-\infty}^{\infty} F_n e^{i\pi nx/L}, \qquad \left(F_n = \frac{1}{2L} \int_{-L}^{L} \mathcal{P}(x) e^{-i\pi nx/L}\, dx \right) \tag{6.3b}$$

then there is a function, denoted by $f * \mathcal{P}$, with Fourier series expansion

$$(f * \mathcal{P})(x) = \sum_{n=-\infty}^{\infty} c_n F_n e^{i\pi nx/L}. \tag{6.3c}$$

This function $f * \mathcal{P}$, called the **convolution over** $[-L, L]$ of f and $\mathcal{P}$, is defined by

$$(f * \mathcal{P})(x) = \frac{1}{2L} \int_{-L}^{L} f(s)\mathcal{P}(x - s)\, ds. \tag{6.3d}$$

PROOF Given the existence of this function $f * \mathcal{P}$, we compute its nth Fourier coefficient (call it p_n)

$$p_n = \frac{1}{2L} \int_{-L}^{L} (f * \mathcal{P})(x)e^{-i\pi nx/L}\, dx$$

$$= \frac{1}{2L} \int_{-L}^{L} \left[\frac{1}{2L} \int_{-L}^{L} f(s)\mathcal{P}(x - s)\, ds \right] e^{-i\pi nx/L}\, dx. \tag{6.4}$$

Since $e^{-i\pi nx/L}$ is independent of s we bring it inside the integral in brackets and reverse the order of integration, obtaining

$$p_n = \frac{1}{2L} \int_{-L}^{L} f(s) \left[\frac{1}{2L} \int_{-L}^{L} \mathcal{P}(x - s)e^{-i\pi nx/L}\, dx \right] ds. \tag{6.5}$$

Replacing x by $(x - s) + s$ in $e^{-i\pi nx/L}$ and doing some algebra, we get (after factoring out $e^{-in\pi s/L}$ from the inner integral)

$$p_n = \frac{1}{2L} \int_{-L}^{L} f(s)e^{-in\pi s/L} \left[\frac{1}{2L} \int_{-L}^{L} \mathcal{P}(x - s)e^{-in\pi(x-s)/L}\, dx \right] ds. \tag{6.6}$$

Changing variables, the inner integral in (6.6) becomes

$$\frac{1}{2L} \int_{-L}^{L} \mathcal{P}(x - s)e^{-in\pi(x-s)/L}\, dx = \frac{1}{2L} \int_{-L-s}^{L-s} \mathcal{P}(v)e^{-in\pi v/L}\, dv$$

$$= \frac{1}{2L} \int_{-L}^{L} \mathcal{P}(v)e^{-in\pi v/L}\, dv. \tag{6.7}$$

The second equality holding because $\mathcal{P}(v)e^{-in\pi v/L}$ has period $2L$.

The last integral in (6.7) is the nth Fourier coefficient for $\mathcal{P}$, which we called F_n in equation (6.3b). Therefore, returning to (6.6), we have

$$p_n = \frac{1}{2L} \int_{-L}^{L} f(s)e^{-in\pi s/L} F_n\, ds = c_n F_n.$$

Thus, the nth Fourier coefficient for $f * \mathcal{P}$ is $c_n F_n$ so (6.3c) holds. ∎

Before we examine the application of this convolution theorem to filtering of Fourier series, we give an example of a convolution.

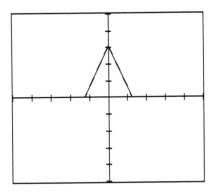

X interval: $[-4, 4]$	X increment = .8
Y interval: $[-.2, .2]$	Y increment = .04

FIGURE 4.18
Convolution described in Example 6.8.

Example 6.8
Compute $f * g$ over the interval $[-4, 4]$ where

$$f(x) = g(x) = \begin{cases} 1 & \text{for } |x| < 0.5 \\ 0 & \text{for } 0.5 < |x| < 4. \end{cases} \quad \Box$$

SOLUTION   Applying (6.3d) to compute $f * g$, with g in place of $\mathcal{P}$, we have

$$(f * g)(x) = \frac{1}{8} \int_{-4}^{4} f(s) g(x - s) \, ds = \frac{1}{8} \int_{-0.5}^{0.5} g(x - s) \, ds.$$

We leave it as an exercise for the reader to finish the calculation, showing that

$$(f * g)(x) = \begin{cases} (1 - |x|)/8 & \text{for } |x| < 1 \\ 0 & \text{for } 1 < |x| < 4. \end{cases} \tag{6.9}$$

With *FAS* we can easily check (6.9). First, choose *Conv* on the initial menu, then choose the interval $[-4, 4]$ and 1024 points. *FAS* then asks you to choose a first function and a second function for convolution. In both cases, you could create the rec function for both f and g. Or, you could construct f as the first function, then save it to a data file and recall this saved file as the second function. In any case, after doing its computations, *FAS* asks you if you want to divide by the length of the interval, and you should press y for this example (in Chapter 5, we will discuss another type of convolution for which we do not divide by the length of the interval). The result is shown in Figure 4.18.   ∎

REMARK It is an important fact that convolution is *commutative*. That is, if f and g have period $2L$, then

$$f * g = g * f.$$

The proof of this is not hard. We leave it to the reader as an exercise. ∎

The convolution theorem gives us a tool for studying filtering of Fourier series. If we look at a partial sum of the filtered Fourier series in (6.2), containing M harmonics, then we have

$$\sum_{n=-M}^{M} c_n F_n e^{i\pi nx/L}. \tag{6.10}$$

By Theorem 6.3), we then have

$$\sum_{n=-M}^{M} c_n F_n e^{i\pi nx/L} = (f * \mathcal{P}_M)(x) \tag{6.11}$$

where

$$\mathcal{P}_M(x) = \sum_{n=-M}^{M} F_n e^{i\pi nx/L}. \tag{6.12}$$

Indeed, (6.11) follows immediately from (6.3c) since the Fourier coefficients of $\mathcal{P}_M$ are all 0 for $|n| > M$.

We now give a name to this important function $\mathcal{P}_M$.

DEFINITION 6.13 *Given a sequence $\{F_n\}$ of filter coefficients, the function $\mathcal{P}_M$, defined in (6.12), is called the **point spread function** (PSF) or **kernel** for the filter process.*

We will now describe how *FAS* plots filtered Fourier series and how it plots *PSFs*. This method is similar to the creation of sampled Fourier series described in Chapter 2, Section 4.

First, since $f * \mathcal{P}_M$ has period $2L$, it can be sampled over the interval $[0, 2L]$. Replacing x in formula (6.11) by $x_j = 2Lj/N$ for $j = 0, 1, \ldots, N-1$, we get

$$(f * \mathcal{P}_M)(x_j) = \sum_{n=-M}^{M} c_n F_n e^{i2\pi nj/N}, \qquad (j = 0, 1, \ldots, N-1). \tag{6.14}$$

With some index shifting we can view (6.14) as an N-point DFT (provided $M < (1/2)N$). Splitting the sum in (6.14) into two sums, one over positive indices and the other over negative indices, we get

$$(f * \mathcal{P}_M)(x_j) = \sum_{n=0}^{M} c_n F_n e^{i2\pi nj/N} + \sum_{n=1}^{M} c_{-n} F_{-n} e^{-i2\pi nj/N}. \tag{6.15}$$

Hence, because $e^{-i2\pi nj/N} = e^{i2\pi(N-n)j/N}$,

$$(f * \mathcal{P}_M)(x_j) = \sum_{n=0}^{M} c_n F_n e^{i2\pi nj/N} + \sum_{n=1}^{M} c_{-n} F_{-n} e^{-i2\pi(N-n)j/N}. \qquad (6.16)$$

Now, recall that the Fourier series coefficients of f are approximated by a DFT (see Remark 1.5(a) in Chapter 2). Denoting this DFT by $\{G_n\}_{n=0}^{N-1}$, we have (for $M \leq (1/8)N$)

$$c_n \approx G_n/N \qquad \text{and} \qquad c_{-n} \approx G_{N-n}/N. \qquad (6.17)$$

Using (6.17), we rewrite (6.16) as the following approximation:

$$(f * \mathcal{P}_M)(x_j) \approx \frac{1}{N} \sum_{n=0}^{M} G_n F_n e^{i2\pi nj/N} + \frac{1}{N} \sum_{n=1}^{M} G_{N-n} F_{-n} e^{-i2\pi(N-n)j/N}.$$

$$(6.18)$$

If we define the sequence $\{H_n\}$ by

$$H_n = \begin{cases} G_n F_n/N & \text{for } n = 0,\, 1,\, \ldots,\, M \\ 0 & \text{for } n = M+1,\, \ldots,\, N-M-1 \\ G_n F_{n-N}/N & \text{for } n = N-M,\, \ldots,\, N-1 \end{cases} \qquad (6.19)$$

then we can express (6.18) as

$$(f * \mathcal{P}_M)(x_j) \approx \sum_{n=0}^{N-1} H_n e^{i2\pi nj/N}. \qquad (6.20)$$

Formula (6.20) shows how we can approximate $f * \mathcal{P}_M$ at the points

$$x_j = 2Lj/N, \qquad (j = 0,\, 1,\, \ldots,\, N-1)$$

by doing an FFT, with weight $e^{i2\pi/N}$, on the sequence $\{H_n\}$ defined in (6.19). The value $(f * \mathcal{P}_M)(x_N) = (f * \mathcal{P}_M)(2L)$ can be obtained by periodicity:

$$(f * \mathcal{P}_M)(2L) = (f * \mathcal{P}_M)(0).$$

By connecting all of these values by line segments, $(f * \mathcal{P}_M)(x)$ is approximated for every x in the interval $[0, 2L]$. (When the interval used is $[-L, L]$, then *FAS* makes use of periodicity. It just converts to $[0, 2L]$, does the calculations described above, and converts back to $[-L, L]$.)

Using the method just described, it is also possible to graph the *PSF* $\mathcal{P}_M$. All we need for this is a sequence $\{\Delta_N(j)\}$ whose FFT $\{G_n\}$ is the constant sequence $\{N\}$, since then (6.19) will reduce to the Fourier coefficients of $\mathcal{P}_M$. If we define Δ_N by

$$\Delta_N(j) = \begin{cases} N & \text{if } j = 0 \\ 0 & \text{if } j = 1,\, 2,\, \ldots,\, N-1 \end{cases} \qquad (6.21)$$

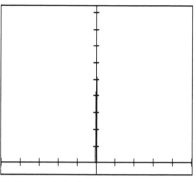

(a) Discrete delta function (1024 points)
X interval: $[-1, 1]$ X increment = .2
Y interval: $[-100, 1400]$ Y increment = 150

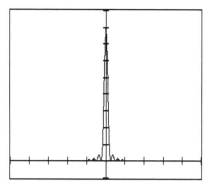

(b) Cesàro *PSF*, 40 harmonics
X interval: $[-\pi, \pi]$ X increment = $\pi/5$
Y interval: $[-5, 45]$ Y increment = 5

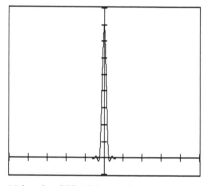

(c) hanning *PSF*, 40 harmonics
X interval: $[-\pi, \pi]$ X increment = $\pi/5$
Y interval: $[-5, 45]$ Y increment = 5

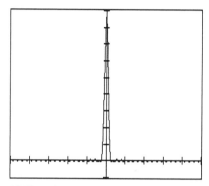

(d) Hamming *PSF*, 40 harmonics
X interval: $[-\pi, \pi]$ X increment = $\pi/5$
Y interval: $[-5, 45]$ Y increment = 5

FIGURE 4.19
Graphs of *PSFs*.

then the N-point DFT of Δ_N is the constant sequence $\{N\}_{j=0}^{N-1}$. The function Δ_N is called a *discrete delta function*. The procedure *Point* on the function menu of *FAS* is designed to insure that Δ_N is obtained *if you proceed as follows*. Enter 1 for the number of integers, enter 0 for the integer 1 out of 1, and for the function value enter the number of points you chose at the start. For example, if $N = 1024$, the graphs of Δ_{1024}, and $\mathcal{P}_{40}$ for Cesàro, hanning, and Hamming *PSFs* are shown in Figure 4.19. These *PSFs* were obtained by successively applying Cesàro, hanning, and Hamming filters to the 40 harmonic Fourier series for the discrete delta function Δ_{1024}.

7   Discrete convolutions using FFTs

In the previous section we defined the concept of convolution of two functions. In this section we describe the method by which the computer can be used to approximate the convolution

$$(f * g)(x) = \frac{1}{2L} \int_{-L}^{L} f(s)g(x - s) \, ds \tag{7.1}$$

where f and g both have period $2L$.

The method used to approximate (7.1) is very similar to the method described at the end of Section 6. It basically consists of multiplying Fourier coefficients for f and g and then doing an inverse FFT.

Dividing the interval $[-L, L]$ into N subintervals of *equal length* $2L/N$ with left endpoints

$$\{s_m\}_{m=0}^{N-1} = \left\{-L + m\frac{2L}{N}\right\}_{m=0}^{N-1}$$

and using a left-endpoint sum for the integral in (7.1) yields

$$(f * g)(x) \approx \frac{1}{2L} \sum_{m=0}^{N-1} f(s_m)g(x - s_m) \, \Delta s_m. \tag{7.2}$$

Since $\Delta s_m = 2L/N$ for each m, we have

$$(f * g)(x) \approx \frac{1}{N} \sum_{m=0}^{N-1} f(s_m)g(x - s_m). \tag{7.3}$$

Now, *since g has period $2L$*, if we extend the sequence $\{s_m\}$ beyond the interval $[-L, L]$ using the formula

$$s_m = -L + m\frac{2L}{N}, \qquad (m = 0, \pm 1, \pm 2, \ldots)$$

then we will have for all integers j and m

$$g(x_j - s_m) = g(s_{j-m}) \tag{7.4}$$

where x_j is defined by

$$x_j = j\frac{2L}{N}, \qquad (j = 0, \pm 1, \pm 2, \ldots).$$

Using (7.4), we can rewrite (7.3) with x_j in place of x

$$(f * g)(x_j) \approx \frac{1}{N} \sum_{m=0}^{N-1} f(s_m)g(s_{j-m}). \tag{7.5}$$

The right side of (7.5) is a discrete convolution (multiplied by $1/N$).

DEFINITION 7.6 *Let $\{u_j\}$ and $\{v_j\}$ be two sequences of numbers, each having period N. The **cyclic convolution** of $\{u_j\}$ and $\{v_j\}$ is the sequence $\{(u * v)_j\}$ defined by*

$$(u * v)_j = \sum_{m=0}^{N-1} u_m v_{j-m}.$$

For example, formula (7.5) shows that the sequence $\{(f * g)(x_j)\}$ is approximated by $1/N$ times the cyclic convolution of the sequences $\{f(s_j)\}$ and $\{g(s_j)\}$.

A direct calculation of $(u*v)_j$ for $j = 0, 1, \ldots, N-1$ would seem to require $4N^2$ real multiplications. The following theorem allows this number to be reduced to $(9/2)N \log_2 N + 6N$ real multiplications by taking advantage of the FFT. The FFT provides a very efficient way of performing cyclic convolution.

THEOREM 7.7   DISCRETE CONVOLUTION
If $\{u_j\}$ and $\{v_j\}$ are sequences of period N with DFTs $\{U_k\}$ and $\{V_k\}$, respectively, then the DFT of $\{(u * v)_j\}$ is $\{U_k V_k\}$.

PROOF   If we put $W = e^{-i2\pi/N}$, then the DFT of $\{(u * v)_j\}$ is

$$\sum_{j=0}^{N-1} (u * v)_j W^{jk}.$$

Replacing $(u * v)_j$ by the sum that defines it, we have upon rearranging sums

$$\sum_{j=0}^{N-1} (u * v)_j W^{jk} = \sum_{j=0}^{N-1} \left[\sum_{m=0}^{N-1} u_m v_{j-m} \right] W^{jk}$$

$$= \sum_{m=0}^{N-1} u_m \left[\sum_{j=0}^{N-1} v_{j-m} W^{jk} \right].$$

Replacing W^{jk} by $W^{(j-m)k}W^{mk}$ in the last sum yields

$$\sum_{j=0}^{N-1}(u*v)_j W^{jk} = \sum_{m=0}^{N-1} u_m W^{mk}\left[\sum_{j=0}^{N-1} v_{j-m}W^{(j-m)k}\right]$$

$$= \sum_{m=0}^{N-1} u_m W^{mk}\left[\sum_{p=-m}^{N-m-1} v_p W^{pk}\right]. \qquad (7.8)$$

Now, the last sum in brackets can be rewritten as follows:

$$\sum_{p=-m}^{N-m-1} v_p W^{pk} = \sum_{p=0}^{N-m-1} v_p W^{pk} + \sum_{p=-m}^{-1} v_p W^{pk}$$

$$= \sum_{p=0}^{N-m-1} v_p W^{pk} + \sum_{p=N-m}^{N-1} v_{p-N}W^{(p-N)k}.$$

Because $\{v_p\}$ has period N, we have $v_{p-N} = v_p$ for all p. Also, $W^{-N} = 1$. Therefore,

$$\sum_{p=-m}^{N-m-1} v_p W^{pk} = \sum_{p=0}^{N-1} v_p W^{pk} = V_k.$$

Hence (7.8) becomes

$$\sum_{j=0}^{N-1}(u*v)_j\, W^{jk} = \sum_{m=0}^{N-1} u_m W^{mk}V_k = U_k V_k$$

which proves that the DFT of $\{(u*v)_j\}$ is $\{U_k V_k\}$. ∎

    Theorem 7.7 shows how to efficiently compute the cyclic convolution $\{(u*v)_j\}_{j=0}^{N-1}$. First, compute the FFTs of $\{u_j\}_{j=0}^{N-1}$ and $\{v_j\}_{j=0}^{N-1}$, which are $\{U_k\}_{k=0}^{N-1}$ and $\{V_k\}_{k=0}^{N-1}$. Then, multiply each element of the FFTs, obtaining $\{U_k V_k\}_{k=0}^{N-1}$. Finally, take the inverse FFT of $\{U_k V_k\}_{k=0}^{N-1}$ and divide by N. Since each FFT and the inverse FFT take $(3/2)N\log_2 N$ real multiplications and multiplying each element of the two FFTs takes $4N$ real multiplications, we find that the whole process takes $(9/2)N\log_2 N + 6N$ real multiplications. This indicates how efficient the FFT method is, as compared to a direct computation, which would take $4N^2$ real multiplications. For instance, when $N = 1024$, the FFT method takes 80 times fewer multiplications than a direct computation.

8 Kernels for some common filters

In this section we will examine kernels (*PSFs*) for the Cesàro, dlVP, hanning, and Hamming filters.

To begin our discussion we must first examine the kernel for an unfiltered Fourier series

$$\sum_{n=-\infty}^{\infty} c_n e^{inx}, \qquad \left(c_n = \frac{1}{2\pi} \int_{-\pi}^{\pi} f(x) e^{-inx}\, dx \right)$$

for a function f having period 2π. For simplicity, *we shall assume that all functions discussed in this section have period* 2π; this results in no loss of generality. The partial sum S_M for the Fourier series above is

$$S_M(x) = \sum_{n=-M}^{M} c_n e^{inx}. \tag{8.1}$$

This partial sum can be thought of as a filtering of a Fourier series, in which case its kernel D_M is defined by

$$D_M(x) = \sum_{n=-M}^{M} 1 \cdot e^{inx}. \tag{8.2}$$

Using this kernel, we have

$$S_M(x) = (f * D_M)(x) = \frac{1}{2\pi} \int_{-\pi}^{\pi} f(s) D_M(x - s)\, ds. \tag{8.3}$$

A closed form expression for D_M can be found. First, we split the sum for D_M in (8.2) into two sums

$$D_M(x) = \sum_{n=0}^{M} e^{inx} + \sum_{n=1}^{M} e^{-inx}. \tag{8.4}$$

Using formula (1.10) from Chapter 2, with M in place of $N - 1$, e^{ix} in place of r for the first sum in (8.4), and e^{-ix} in place of r for the second sum in (8.4), we obtain

$$
\begin{aligned}
D_M(x) &= \frac{1 - e^{i(M+1)x}}{1 - e^{ix}} + \frac{e^{-ix} - e^{-i(M+1)x}}{1 - e^{-ix}} \\
&= \frac{-e^{i(M+1)x} + e^{iMx} - e^{-i(M+1)x} + e^{-iMx}}{(1 - e^{ix})(1 - e^{-ix})} \\
&= \frac{2\cos Mx - 2\cos(M+1)x}{2 - 2\cos x}.
\end{aligned}
\tag{8.5}
$$

Thus,

$$D_M(x) = \frac{\cos Mx - \cos (M+1)x}{1 - \cos x}. \tag{8.6}$$

Formula (8.6) can be further simplified using trigonometric identities. First, since

$$\cos Mx = \cos \left(M + \frac{1}{2} - \frac{1}{2} \right) x$$

it follows that

$$\cos Mx = \cos \left(M + \frac{1}{2} \right) x \cos \frac{1}{2}x + \sin \left(M + \frac{1}{2} \right) x \sin \frac{1}{2}x$$

and, similarly,

$$\cos (M+1)x = \cos \left(M + \frac{1}{2} \right) x \cos \frac{1}{2}x - \sin \left(M + \frac{1}{2} \right) x \sin \frac{1}{2}x.$$

Hence, by subtraction

$$\cos Mx - \cos (M+1)x = 2 \sin \left(M + \frac{1}{2} \right) x \sin \frac{1}{2}x. \tag{8.7}$$

For $M = 0$, formula (8.7) becomes

$$1 - \cos x = 2 \sin^2 \frac{1}{2}x. \tag{8.8}$$

Combining formulas (8.6)–(8.8) yields the following closed form expression for D_M:

$$D_M(x) = \frac{\sin (M + \frac{1}{2})x}{\sin \frac{1}{2}x}. \tag{8.9}$$

Formula (8.9) is the classic expression for *Dirichlet's kernel* D_M. Another expression that is more amenable to graphing with *FAS* is

$$D_M(x) = (2M+1)\mathrm{sinc}((M+0.5)x/\pi)/\mathrm{sinc}(0.5x/\pi) \tag{8.9a}$$

where sinc is the function defined in (2.6). Graphs of Dirichlet's kernel using formula (8.9a) are shown in Figure 4.20.

To derive the kernel for Cesàro filtering, we use Definition 4.1 and formula (8.3) to obtain

$$\sigma_M(x) = \frac{1}{M} \sum_{k=0}^{M-1} (f * D_k)(x) = f * \left(\frac{1}{M} \sum_{k=0}^{M-1} D_k(x) \right). \tag{8.10}$$

Denoting the kernel for Cesàro filtering by C_M, we have from (8.10)

$$C_M(x) = \frac{1}{M} \sum_{k=0}^{M-1} D_k(x) \tag{8.11}$$

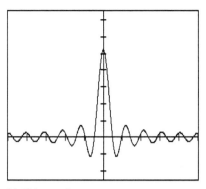

(a) 10 harmonics
X interval: $[-\pi, \pi]$ X increment = $\pi/5$
Y interval: $[-10, 30]$ Y increment = 4

(b) 21 harmonics
X interval: $[-\pi, \pi]$ X increment = $\pi/5$
Y interval: $[-20, 60]$ Y increment = 8

FIGURE 4.20
Dirichlet's kernel.

and

$$\sigma_M(x) = (f * C_M)(x). \tag{8.12}$$

To obtain a closed form for C_M we use (8.9) in (8.11), obtaining

$$C_M(x) = \frac{1}{M} \sum_{k=0}^{M-1} \frac{\sin\left(k + \frac{1}{2}\right)x}{\sin \frac{1}{2}x}. \tag{8.13}$$

We now show that

$$C_M(x) = \frac{1}{M} \frac{\sin^2\left(\frac{1}{2}Mx\right)}{\sin^2\left(\frac{1}{2}x\right)}. \tag{8.14}$$

First, multiply (8.13) by $2\sin^2((1/2)x)$, and get

$$2\sin^2\left(\frac{1}{2}x\right) C_M(x) = \frac{1}{M} \sum_{k=0}^{M-1} 2\sin\left(k + \frac{1}{2}\right)x \sin \frac{1}{2}x. \tag{8.15}$$

Using a trigonometric identity

$$2\sin\theta \sin\phi = \cos(\theta - \phi) - \cos(\theta + \phi) \tag{8.16}$$

we obtain from (8.15)

$$2\sin^2\left(\frac{1}{2}x\right)C_M(x) = \frac{1}{M}\sum_{k=0}^{M-1}\left[\cos kx - \cos(k+1)x\right]$$

$$= \frac{1}{M}\left[1 - \cos Mx\right]. \tag{8.17}$$

The last equality holding because of the telescoping of the finite series. Putting $\theta = \phi = (1/2)Mx$ in (8.16), we obtain

$$1 - \cos Mx = 2\sin^2\left(\frac{1}{2}Mx\right)$$

which used in (8.17) yields

$$2\sin^2\left(\frac{1}{2}x\right)C_M(x) = \frac{2}{M}\sin^2\left(\frac{1}{2}Mx\right). \tag{8.18}$$

Dividing (8.18) by $2\sin^2((1/2)x)$ yields (8.14), our desired formula.

REMARK 8.19 The kernel C_M is usually called *Fejér's kernel*. Sometimes we will refer to it as *Cesàro's kernel* in order to remind the reader that it is the kernel for Cesàro filtering. ∎

Another form of Fejér's kernel, more amenable to graphing by *FAS*, is

$$C_M(x) = M[\mathrm{sinc}(Mx/(2\pi))/\mathrm{sinc}(x/(2\pi))] \wedge 2. \tag{8.20}$$

We now obtain a closed form for the dlVP kernel using $2M$ harmonics. By formula (4.7) the kernel V_{2M} for the dlVP filter must be (for $L = 2\pi$)

$$V_{2M}(x) = \sum_{n=-2M}^{2M} v_n e^{inx} \tag{8.21}$$

where

$$v_n = \begin{cases} 1 & \text{if } |n| \leq M \\ 2(1 - |\frac{n}{2M}|) & \text{if } M \leq |n| \leq 2M. \end{cases} \tag{8.22}$$

The second part of formula (8.22) reminds us of the coefficients for the Cesàro kernel C_{2M}, multiplied by 2. Once we realize this, then with some algebra we obtain

$$V_{2M}(x) = 2C_{2M}(x) - C_M(x). \tag{8.23}$$

Using either formula (8.14) or (8.20), formula (8.23) gives a closed form expression for the dlVP kernel V_{2M}. For graphing with *FAS*, however, formula (8.20) is the better choice (if one desires to use a formula, the method described at the end of Section 6 is much easier in this case).

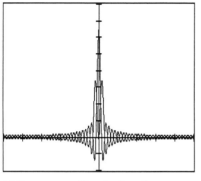
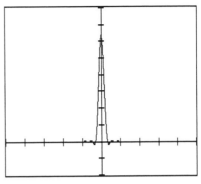

(a) Components of hanning kernel
X interval: $[-\pi, \pi]$ X increment $= \pi/5$
Y interval: $[-10, 40]$ Y increment $= 5$

(b) hanning kernel
X interval: $[-\pi, \pi]$ X increment $= \pi/5$
Y interval: $[-10, 40]$ Y increment $= 5$

FIGURE 4.21
Construction of hanning kernel.

We close this section by describing closed forms for the hanning and Hamming kernels. The *hanning kernel* h_M is defined by [see (4.12)]

$$h_M(x) = \sum_{n=-M}^{M} \left[0.5 + 0.5 \cos \frac{n\pi}{M} \right] e^{inx}. \tag{8.24}$$

We leave it as an exercise for the reader to verify that

$$h_M(x) = \frac{1}{2} D_M(x) + \frac{1}{4} D_M \left(x + \frac{\pi}{M} \right) + \frac{1}{4} D_M \left(x - \frac{\pi}{M} \right) \tag{8.25}$$

where D_M is Dirichlet's kernel. In Figure 4.21(a) we show graphs of

$$\frac{1}{2} D_M(x) \quad \text{and} \quad \frac{1}{4} D_M \left(x + \frac{\pi}{M} \right) + \frac{1}{4} D_M \left(x - \frac{\pi}{M} \right)$$

for $M = 32$. The effect of adding the second function to the first is a large amount of cancellation of the oscillations of D_M. This is evident in Figure 4.21(b).

The *Hamming kernel* H_M is defined by [see (4.13)]

$$H_M(x) = \sum_{n=-M}^{M} \left[0.54 + 0.46 \cos \frac{n\pi}{M} \right] e^{inx}. \tag{8.26}$$

By comparing formulas (8.2), (8.24), and (8.26) it follows that

$$H_M(x) = 0.08 D_M(x) + 0.92 h_M(x). \tag{8.27}$$

In Figure 4.22(a) we have graphed $0.08 D_{32}(x)$ and $0.92 h_{32}(x)$. It can be seen that the similarity between the figures lies in the nearly perfect match (with

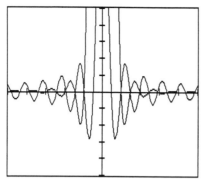

(a) Components of Hamming kernel

X interval: $[-\pi, \pi]$ X increment = $\pi/5$

Y interval: $[-2, 2]$ Y increment = .4

(b) Hamming kernel

X interval: $[-\pi, \pi]$ X increment = $\pi/5$

Y interval: $[-2, 2]$ Y increment = .4

FIGURE 4.22
Construction of Hamming kernel.

opposite signs) between the two side lobes just to the right and left of the two center lobes. Consequently, as can be seen in Figure 4.22(b), which is a graph of H_{32}, the effect of summing $0.08D_{32}$ and $0.92h_{32}$ is to *cancel out these sidelobes*. The main advantage of this is to *suppress Gibbs' phenomenon* in Fourier series expansions.

9   Convergence of filtered Fourier series

In this section we will prove a theorem that has wide applicability to the convergence of filtered Fourier series. It covers many (but not all) of the filters commonly used in mathematics, signal processing, and other areas.

We begin with the following definition of a summation kernel.

DEFINITION 9.1 *A **PSF**, $\mathcal{P}_M$, defined on $[-L, L]$ is called a **summation kernel** if it satisfies the following conditions:*

(a)   For each M,

$$\frac{1}{2L} \int_{-L}^{L} \mathcal{P}_M(x)\, dx = 1.$$

(b)   There is a positive constant C for which

$$\frac{1}{2L} \int_{-L}^{L} |\mathcal{P}_M(x)|\, dx \leq C \quad \text{for all } M.$$

(c) *Suppose $\epsilon > 0$ is given, no matter how small. Then, given $\delta > 0$, satisfying $0 < \delta < L$, we will have for $\delta \leq |x| \leq L$*

$$|\mathcal{P}_M(x)| < \epsilon$$

provided M *is chosen sufficiently large.*

REMARK 9.2   Condition (c) in Definition 9.1 can also be stated as

$$\lim_{M \to \infty} \left[\sup_{\delta \leq |x| \leq L} |\mathcal{P}_M(x)| \right] = 0$$

for each $\delta > 0$.   ∎

The Cesàro, dlVP, and hanning kernels are all summation kernels over $[-\pi, \pi]$. We will show this for the Cesàro kernel, leaving the verification for the dlVP kernel to the reader as an exercise. Showing that the hanning kernel is a summation kernel is easier for us to do by an indirect argument utilizing the Fourier transform, so we won't do it until Section 10 of Chapter 5.

Example 9.3
Show that the Cesàro kernel is a summation kernel.   ⬚

SOLUTION   We must check that the properties (a)–(c) in Definition 9.1 are satisfied.
(a) From formula (8.11) we obtain

$$\frac{1}{2\pi} \int_{-\pi}^{\pi} C_M(x)\, dx = \frac{1}{M} \sum_{k=0}^{M-1} \frac{1}{2\pi} \int_{-\pi}^{\pi} D_k(x)\, dx. \tag{9.4}$$

Using formula (8.2) we have, because of the orthogonality of $\{e^{inx}\}$,

$$\frac{1}{2\pi} \int_{-\pi}^{\pi} D_k(x)\, dx = \sum_{n=-k}^{k} \frac{1}{2\pi} \int_{-\pi}^{\pi} e^{inx}\, dx$$

$$= \frac{1}{2\pi} \int_{-\pi}^{\pi} 1\, dx = 1.$$

Hence, formula (9.4) becomes

$$\frac{1}{2\pi} \int_{-\pi}^{\pi} C_M(x)\, dx = \frac{1}{M} \sum_{k=0}^{M-1} 1 = 1 \tag{9.5}$$

so (a) is true.

(b) From formula (8.14) we see that $C_M(x) \geq 0$ for all M, hence $|C_M(x)| = C_M(x)$ for all M. Therefore, because of (a),

$$\frac{1}{2\pi} \int_{-\pi}^{\pi} |C_M(x)|\, dx = 1 \quad \text{for all } M.$$

Taking C to be 1 we see that (b) is true.

(c) Suppose $\epsilon > 0$ has been given. Then, for $\delta \leq |x| \leq \pi$ we have

$$0 \leq C_M(x) \leq \frac{1}{M} \frac{\sin^2(Mx/2)}{\sin^2(\delta/2)} \leq \frac{1}{M} \frac{1}{\sin^2(\delta/2)} \,. \tag{9.6}$$

If M is sufficiently large, then

$$\frac{1}{M} \frac{1}{\sin^2(\delta/2)} < \epsilon. \tag{9.7}$$

Combining (9.7) and (9.6) yields $0 \leq C_M(x) < \epsilon$, so $|C_M(x)| < \epsilon$ for M sufficiently large. Thus, (c) is true. ∎

Example 9.8
The kernel I_M defined by

$$I_M(x) = \begin{cases} M & \text{for } |x| \leq \frac{L}{M} \\ \\ 0 & \text{for } |x| > \frac{L}{M} \end{cases} \qquad (M = 1, 2, 3, \ldots)$$

is a summation kernel over $[-L, L]$. ∎

We leave the verification of Example 9.8 to the reader as an exercise. The kernel I_M is called the *point impulse kernel*.

The following theorem illustrates the importance of summation kernels.

THEOREM 9.9
Suppose that $\mathcal{P}_M$ is a summation kernel. Suppose also that f has period $2L$ and $\int_{-L}^{L} |f(x)|\, dx$ is finite (converges). Then, for each point x_0 of continuity of f,

$$\lim_{M \to \infty} (\mathcal{P}_M * f)(x_0) = f(x_0). \tag{9.10}$$

Moreover, if f is continuous, then

$$\lim_{M \to \infty} \left[\sup_{x \in \mathbf{R}} |(\mathcal{P}_M * f)(x) - f(x)| \right] = 0. \tag{9.11}$$

REMARK 9.12 Equation (9.11) says that $\mathcal{P}_M * f$ *converges uniformly* to f over the whole real line. Since, in this case, sup is just another way of saying *maximum*. ∎

PROOF OF THEOREM Suppose that x_0 is a point of continuity of f, and that $\epsilon > 0$ has been given. We begin by observing that property (a) of (9.1) yields

$$f(x_0) = f(x_0) \cdot 1 = f(x_0) \frac{1}{2L} \int_{-L}^{L} \mathcal{P}_M(u) \, du$$

$$= \frac{1}{2L} \int_{-L}^{L} f(x_0) \mathcal{P}_M(u) \, du. \qquad (9.13)$$

Consequently,

$$(\mathcal{P}_M * f)(x_0) - f(x_0) = \frac{1}{2L} \int_{-L}^{L} [f(x_0 - u) - f(x_0)] \mathcal{P}_M(u) \, du. \qquad (9.14)$$

Applying absolute values to (9.14) and bringing the absolute values inside the integral sign yields

$$|(\mathcal{P}_M * f)(x_0) - f(x_0)| \leq \frac{1}{2L} \int_{-L}^{L} |f(x_0 - u) - f(x_0)||\mathcal{P}_M(u)| \, du. \qquad (9.15)$$

The right-hand side of (9.15) can be expressed as

$$\frac{1}{2L} \int_{-L}^{L} |f(x_0 - u) - f(x_0)||\mathcal{P}_M(u)| \, du = I + J \qquad (9.16)$$

where

$$I = \frac{1}{2L} \int_{|u|<\delta} |f(x_0 - u) - f(x_0)||\mathcal{P}_M(u)| \, du \qquad (9.16a)$$

and

$$J = \frac{1}{2L} \int_{\delta \leq |u| \leq L} |f(x_0 - u) - f(x_0)||\mathcal{P}_M(u)| \, du. \qquad (9.16b)$$

[*Note*: The integral $\int_{|u|<\delta}$ in (9.16a) is shorthand for $\int_{-\delta}^{\delta}$, while $\int_{\delta \leq |u| \leq L}$ is shorthand for the sum of two integrals $\int_{-L}^{-\delta} + \int_{\delta}^{L}$.]

Now, we shall first show that the integral I is small when δ is small. Since x_0 is a point of continuity for f we will have

$$|f(x_0 - u) - f(x_0)| < \epsilon \quad \text{when} \quad |(x_0 - u) - x_0| < \delta \qquad (9.17)$$

provided δ is sufficiently small. Choosing such a δ, and noting that $|(x_0 - u) - x_0| = |u|$, we have

$$|f(x_0 - u) - f(x_0)| < \epsilon \quad \text{for} \quad |u| < \delta. \qquad (9.18)$$

Using (9.18), the integral I in (9.16a) is bounded as follows:

$$I \leq \frac{1}{2L} \int_{|u|<\delta} \epsilon |\mathcal{P}_M(u)|\, du \leq \epsilon C \tag{9.19}$$

where we used property (b) from (9.1) to get the last inequality.

We now turn to the integral J in (9.16b), where δ is the *fixed* value chosen to get (9.19). By the triangle inequality for absolute values

$$|f(x_0 - u) - f(x_0)| \leq |f(x_0 - u)| + |f(x_0)|. \tag{9.20}$$

Hence, the integral J in (9.16b) is bounded in the following way

$$J \leq \frac{1}{2L} \int_{\delta \leq |u| \leq L} |f(x_0 - u)||\mathcal{P}_M(u)|\, du$$
$$+ \frac{1}{2L} \int_{\delta \leq |u| \leq L} |f(x_0)||\mathcal{P}_M(u)|\, du. \tag{9.21}$$

Using property (c) from (9.1), we can further bound the terms on the right side of inequality (9.21). For M sufficiently large, we get

$$J \leq \frac{1}{2L} \int_{\delta \leq |u| \leq L} |f(x_0 - u)|\epsilon\, du + \frac{1}{2L} \int_{\delta \leq |u| \leq L} |f(x_0)|\epsilon\, du. \tag{9.22}$$

By enlarging the integration intervals to $[-L, L]$, we have

$$J \leq \epsilon \frac{1}{2L} \int_{-L}^{L} |f(x_0 - u)|\, du + \epsilon \frac{1}{2L} \int_{-L}^{L} |f(x_0)|\, du. \tag{9.23}$$

The first term on the right side of (9.23) equals

$$\epsilon \frac{1}{2L} \int_{-L}^{L} |f(v)|\, dv \quad \left(= \frac{\epsilon}{2L} \|f\|_1 \right)$$

because of the periodicity of f. The second term on the right side of (9.23) just equals $|f(x_0)|$. Therefore, (9.23) becomes

$$J \leq \epsilon \left[\frac{1}{2L} \|f\|_1 + |f(x_0)| \right]. \tag{9.24}$$

Using inequalities (9.19) and (9.24) in (9.16) we get, for all M sufficiently large,

$$|(\mathcal{P}_M * f)(x_0) - f(x_0)| \leq \epsilon \left[C + \frac{1}{2L} \|f\|_1 + |f(x_0)| \right]. \tag{9.25}$$

Since ϵ can be taken arbitrarily small, we have proved (9.10).

To complete the proof we must show that (9.11) holds. Here, we must use a result from advanced calculus. If the function f is continuous on the *closed* interval $[-L, L]$, then there exists a $\delta > 0$ such that

$$|f(x - u) - f(x)| < \epsilon \quad \text{when} \quad |(x - u) - x| < \delta \qquad (9.26)$$

for *all* x in $[-L, L]$. This is known as the *uniform continuity* of f, it is proved in every text on advanced calculus (see, for instance, [Ru,2] or [Ba]). Formula (9.26) says that

$$|f(x - u) - f(x)| < \epsilon \quad \text{when} \quad |u| < \delta. \qquad (9.27)$$

But, then, replacing x_0 by x in all of the calculations above from (9.13) through (9.25), we have for all x in $[-L, L]$

$$|(\mathcal{P}_M * f)(x) - f(x)| \leq \epsilon \left[C + \frac{1}{2L} \|f\|_1 + |f(x)| \right] \qquad (9.28)$$

provided M is chosen sufficiently large. Since f and $\mathcal{P}_M * f$ both have period $2L$, formula (9.28) holds for all x in **R**. If we define $\|f\|_\infty$ to be the maximum of $|f(x)|$ over **R**, then we have for all M sufficiently large

$$|(\mathcal{P}_M * f)(x) - f(x)| \leq \epsilon \left[C + \frac{1}{2L} \|f\|_1 + \|f\|_\infty \right]. \qquad (9.29)$$

Since the right side of (9.29) is a constant, and (9.29) holds for all x in **R**, we have for all M sufficiently large

$$\sup_{x \in \mathbf{R}} |(\mathcal{P}_M * f)(x) - f(x)| \leq \epsilon \left[C + \frac{1}{2L} \|f\|_1 + \|f\|_\infty \right]. \qquad (9.30)$$

Because ϵ can be taken arbitrarily small, we have proved (9.11). ∎

REMARK 9.31 Using the notation introduced in the proof above, we can rewrite (9.30) as $\lim_{M \to \infty} \|\mathcal{P}_M * f - f\|_\infty = 0$. It is said that $\mathcal{P}_M * f$ *converges to* f *in the supnorm over* **R**. ∎

References

For further discussion of applications of Fourier series to physical problems, see [Wa], [We], or [Str]. More discussion of filters can be found in [Ha], [Op-S], or [Ra-G].

Exercises

Section 1

4.1 Graph approximate solutions (using, say, 30 harmonics and 1024 points) to problem (1.1) using the following functions, times, and diffusion constants.

(a) $u(x,0) = 40(10x - x^2)$, $0 \le x \le 10$, $a^2 = 0.12$ (cast iron), $t = 0.5$, 1.0, 2.0, and 4.0 sec

(b) $u(x,0) = 80\exp(-(x-8) \wedge 20)$, $0 \le x \le 16$, $a^2 = 0.005$ (concrete), $t = 4.0$, 8.0, 16.0, and 32.0 sec

(c)
$$u(x,0) = \begin{cases} 100 & \text{for } 17 < x < 23 \\ 0 & \text{for } 0 < x < 17 \text{ and } 23 < x < 32 \end{cases}$$
$a^2 = 1.14$ (copper), $t = 0.1$, 0.5, 1.0, and 2.0 sec

(d)
$$u(x,0) = \begin{cases} 100 & \text{for } 9 < x < 11 \text{ and } 29 < x < 32 \\ 0 & \text{for } 0 < x < 9 \text{ and } 11 < x < 29 \text{ and } 32 < x < 40 \end{cases}$$
$a^2 = 0.0038$ (brick), $t = 4.0$, 8.0, 16.0, and 32.0 sec

4.2 For each of the examples in Exercise 4.1, determine estimates for the maximum errors involved in using a 40 harmonic partial sum approximation to $u(x,t)$. If an error estimate is more than 10^{-5}, then find some number of harmonics that will yield such accuracy.

4.3 Generalize the estimate (1.13) as follows. First, show that

$$|b_n| \le \frac{2}{L}\|f\|_1, \qquad \left(\|f\|_1 = \int_0^L |f(x)|\,dx\right) \qquad (a)$$

then show that

$$\left| u(x,t) - \sum_{n=1}^{M} b_n e^{-(n\pi a/L)^2 t} \sin\frac{n\pi x}{L} \right| \le \frac{2}{L}\|f\|_1 \frac{e^{-(M+1)^2(a\pi/L)^2 t}}{1 - e^{-(2M+1)^2(a\pi/L)^2 t}} \cdot \qquad (b)$$

4.4 Suppose $t \ge 0.01$ sec, $a^2 = 1.14$ (copper), and $L = 10$. How many harmonics M are needed to insure that

$$\left| u(x,t) - \sum_{n=1}^{M} b_n e^{-(n\pi a/L)^2 t} \sin\frac{n\pi x}{L} \right| \le 10^{-5}.$$

Hint: Use Exercise 4.3(b), assuming first that

$$1 - e^{-(2M+1)^2(a\pi/10)^2(0.01)} \le 0.5$$

and then checking this assumption after finding M.

4.5 Explain why as $t \to \infty$, we have

$$u(x,t) \approx b_1 e^{-(a\pi/L)^2 t} \sin\frac{n\pi x}{L}.$$

That is, for t sufficiently large, $u(x,t)$ is approximated by the first term in its series expansion. Using *FAS*, illustrate this result for each of the functions in Exercise 4.1.

4.6 Using Fourier cosine series expansions, derive a filtered cosine series solution to the following heat conduction problem:

$$\frac{\partial u}{\partial t} = a^2 \frac{\partial^2 u}{\partial x^2} \qquad \text{(heat equation)}$$
$$\frac{\partial u}{\partial x}(0,t) = 0, \quad \frac{\partial u}{\partial x}(L,t) = 0 \qquad \text{(boundary conditions)} \qquad (a)$$
$$u(x,0) = g(x) \qquad \text{(initial condition)}.$$

4.7 Graph approximate solutions to Exercise 4.6(a), using 40 harmonics and 1024 points, using the same functions, times, and diffusion constants as in Exercise 4.1.

Section 2

4.8 Using *FAS*, graph solutions to (2.1) for $c^2 = 10,000$, $L = 10$, $t = 0.01, 0.02, 0.03$, and 0.04, for the following initial conditions [*Hint*: for (d) and (e) use *Combine* on the display menu]:

(a) $f(x) = 0.1 \exp[-((x-1)/4)^{10}]$, $g(x) = 0$
(b) $f(x) = 0.1 \sin(0.3\pi x)$, $g(x) = 0$
(c) $f(x) = 0.1 \sin(0.4\pi x)$, $g(x) = 0$
(d) $f(x) = x(x-10)(x+40)$, $g(x) = 3x(x-10)$
(e) $f(x) = 0.2 \exp[-(x-3)^{10}]$, $g(x) = x(10-x)$
(f) $f(x) = 0$, $g(x) = \sin(0.3\pi x)$
(g) $f(x) = 0$, $g(x) = \sin(0.4\pi x)$

4.9 Suppose c^2 in Exercise 4.8 is changed to 40,000. Repeat the exercise and explain how (why) the new results differ from the previous ones.

4.10 Show that for $f(x) = \sin(n\pi x/L)$, $g(x) = 0$, a solution to (2.1) is $y(x,t) = \cos(nc\pi t/L)\sin(n\pi x/L)$. Describe what this solution looks like for $n = 1, 2, 3$, and 4.

4.11 Show that for $f(x) = 0$, $g(x) = \sin(n\pi x/L)$, a solution to (2.1) is $y(x,t) = (L/nc\pi)\sin(nc\pi t/L)\sin(n\pi x/L)$. Describe what this solution looks like for $n = 1, 2, 3$, and 4.

Section 3

4.12 Continue Example 3.15 for the times $t = 0.4$ and 0.5 sec. Notice how the basic form of the graph of $|S_{500}(x,t)|$ has *stabilized*.

4.13 Use *FAS* to calculate $|S_{500}(x,t)|$ for $t = 0.1, 0.2, 0.3, 0.4$, and 0.5 sec, given the following initial functions over the interval $[-32, 32]$. Use 4096 points, and use the mass of the electron for m.

(a)
$$\psi(x,0) = \begin{cases} 0 & \text{for } 1.5 < |x| < 32 \text{ and } |x| < 1 \\ 1 & \text{for } 1 < |x| < 1.5 \end{cases}$$

(b)
$$\psi(x,0) = \begin{cases} 0 & \text{for } 2.5 < |x| < 32 \text{ and } |x| < 2 \\ 1 & \text{for } 2 < |x| < 2.5 \end{cases}$$

(c)
$$\psi(x,0) = \begin{cases} 0 & \text{for } 3.5 < |x| < 32 \text{ and } |x| < 3 \\ 1 & \text{for } 3 < |x| < 3.5 \end{cases}$$

4.14 Analyze the results of Exercise 4.13, using concepts of quantum mechanics. What effects do you observe as the distance between the rectangles in the initial functions in Exercise 4.13 is increased?

4.15 Calculate $|S_{500}(x,t)|$ for $t = 0.1, 0.2$, and 0.3 sec, given the following initial functions (use 4096 points over the interval $[-32, 32]$ and use m equal to the mass of the electron).

(a) $\psi(x,0) = \exp[-(x/2)^{10}]$
(b) $\psi(x,0) = \exp[-(x/.5)^{10}]$
(c) $\psi(x,0) = \exp(-\pi x^2)$
(d) $\psi(x,0) = 2\exp(-4\pi x^2)$

4.16 Given that the mass of a neutron is many times greater than the mass of an electron (*it is part of this exercise for you to look up how many times greater it is*), calculate $|S_{500}(x,t)|$ for the functions and times t in Examples 3.10 and 3.15 using the mass of the neutron for m. How does neutron diffraction compare with electron diffraction?

4.17 Compute $|S_{500}(x,t)|$ for the functions given in Examples 3.10 and 3.15, for the times $t = 1$, 2, 4, and 8 sec (assume that m equals the mass of the electron). Give a physical interpretation for the increasingly erratic appearance of the wave function as t increases.

4.18 With regard to Exercise 4.16, for what time t_n will a neutron diffraction pattern be *identical* to an electron diffraction pattern at time t_e?

Section 4

4.19 For the following function, compute Cesàro, dlVP, hanning, and Hamming filtered Fourier series using 40 harmonics and 1024 points:
$$f(x) = \begin{cases} -1 & \text{for } -1 < x < 0 \\ 1 & \text{for } 0 < x < 1. \end{cases}$$
Compare all four of these filtered partial sums within the x-y window: $0 \le x \le 1$ and $0.98 \le y \le 1.02$.

4.20 For the function $f(x) = 1 - |x - 1|$ on the interval $0 \le x \le 2$, use *FAS* to estimate how many harmonics are needed to approximate the function to within ± 0.01 for each of the filters: Cesàro, dlVP, hanning, and Hamming.

Section 5

4.21 *Lanczos Filter.* For each of the following functions, graph filtered Fourier series partial sums using 10, 20, and 30 harmonics with $\text{sinc}(x)$ as the filter function (use $[-\pi, \pi]$ and 1024 points):

(a) $\begin{cases} 1 & \text{if } 0 < x < \pi \\ -1 & \text{if } -\pi < x < 0 \end{cases}$

(b) x^2

(c) x

(d) $\exp(-(x/2) \wedge 10)$

4.22 Suppose $\omega = \omega_3 = 30\pi$ in Example 5.9. Using *FAS*, draw approximate graphs of $y(x,t)$ in (5.24) for $\omega = 29.99\pi$ and the times $t = 1$, 3, and 12 sec. Then repeat this comparison using $\omega = 29.9999\pi$ and the same times t. Compare your graphs with those for Exercise 4.11.

4.23 Generalizing Exercise 4.22, what type of motion would you expect if $\omega = 39.9999\pi$ or $\omega = 49.9999\pi$? Confirm your answer using *FAS*.

4.24 Explain the results of the previous two exercises using (5.23).

4.25 Derive a series solution to (5.10) if the force function is $F(x,t) = \rho G(x) \cos \omega t$ and the initial conditions are $f(x) = 0$, $g(x) = 0$. You should obtain the following series solution (*assuming $\omega \neq \omega_n$ for all n*):
$$y(x,t) = \sum_{n=1}^{\infty} \left[\frac{\cos \omega t - \cos \omega_n t}{\omega_n^2 - \omega^2} \right] K_n \sin \frac{n\pi x}{L},$$
$$\left(K_n = \frac{2}{L} \int_0^L G(x) \sin \frac{n\pi x}{L} \, dx \right). \tag{a}$$

4.26 Using 200 harmonics, $L = 10$, $c^2 = 10,000$, and $\omega = 29.99\pi$ graph approximations to $y(x,t)$ in Exercise 4.25(a) [using the function in (5.19) for G] for times $t = 0.25, 0.5, 1$, and 2 sec.

4.27 Same problem as 4.26, but use $G(x) = 2\exp(-[((x-5)/4) \wedge 20])$.

4.28 Same problem as 4.26, but use $G(x) = 2\exp(-[((x-5)/4) \wedge 20])$ and $\omega = 99.9999\pi$.

Section 6

4.29 Check that (6.9) is correct.

4.30 Using $M = 10, 20$, and 30 harmonics, and 1024 points, graph the kernels for each of the filters below (use $[-\pi, \pi]$ as the interval).
   (a) Cesàro
   (b) dlVP
   (c) hanning
   (d) Hamming
   (e) Gauss (damping constant 0.001)
   (f) Riesz (damping power 3.8)

4.31 (a) Graph the 20 harmonic Cesàro filtered Fourier series partial sum for the function

$$f(x) = \begin{cases} -1 & \text{for } -\pi < x < 0 \\ 1 & \text{for } 0 < x < -\pi \end{cases}$$

over the interval $[-\pi, \pi]$ using 1024 points. (b) Graph the convolution over $[-\pi, \pi]$ of this function $f(x)$ with the *PSF* $\mathcal{P}_{20}$ for Cesàro filtering. Use 1024 points. (c) Check that the graphs found in (a) and (b) match, to a high degree of accuracy. (d) Repeat (a)–(c), but use 40 harmonics.

4.32 Repeat Exercise 4.31 using hanning, Hamming, and dlVP filtering.

4.33 For each of the functions given below, graph the convolution $f * \Delta_{1024}$, over the given interval, using 1024 points. Check that in each case $f * \Delta_{1024}$ is the same as f to a high degree of accuracy.
   (a) $f(x) = \exp(-x \wedge 2)$, interval $[-8,8]$
   (b) $f(x) = \begin{cases} 1 & \text{for } |x| < 1 \\ 0 & \text{for } |x| > 1 \end{cases}$, interval $[-3,3]$
   (c) $f(x) = x$, interval $[-4,4]$

4.34 Show that if f and g are functions of period $2L$, then $f * g = g * f$.

Section 7

4.35 Show that $(u * v)_j = (v * u)_j$ for any two sequences $\{u_j\}$ and $\{v_j\}$ of period N.

4.36 How many real multiplications are needed to calculate $\{(u*v)_j\}$ if $\{u_j\}$ and $\{v_j\}$ are real sequences? [*Hint*: Use the methods in Sections 5 and 9 of Chapter 3.]

Section 8

4.37 Verify (8.25).

4.38 Combine (8.9a) and (8.25) to draw graphs of the hanning kernel for $M = 10, 20$, and 30 harmonics. Compare these graphs with the ones obtained as solutions to Exercise 4.30(c).

4.39 Show that Hamming's kernel H_M satisfies

$$H_M(x) = 0.54 D_M(x) + 0.23 D_M\left(x - \frac{\pi}{M}\right) + 0.23 D_M\left(x + \frac{\pi}{M}\right).$$

Use this formula in combination with (8.9a) to draw graphs of Hamming's kernel for $M = 10, 20$, and 30 harmonics. Compare these graphs with the ones obtained as solutions to Exercise 4.30(d).

4.40 Using (8.20), graph Cesàro's kernel for $M = 10, 20$, and 30 harmonics. Compare these graphs with the ones obtained as solutions to Exercise 4.30(a).

4.41 Using *FAS*, graph the hanning and Hamming filtered Fourier series, using 40 harmonics, of

$$f(x) = \begin{cases} 1 & \text{for } 0 < x < 1 \\ 0 & \text{for } -1 < x < 0. \end{cases}$$

Change the x-interval to $[0, 1]$ and the y-interval to $[0.98, 1.02]$. Explain the appearance of the graphs in terms of the *PSFs* for the two filters.

4.42 Repeat Exercise 4.41, but now use the dlVP kernel and the Riesz kernel with damping power 3.8.

Section 9

4.43 Verify that I_M in Example 9.8 is a summation kernel.

4.44 Using *FAS*, calculate $f * I_M$ for $M = 16, 32$, and 64, where

$$f(x) = \exp(-(x/0.3) \wedge 10).$$

Use an interval of $[-.5, .5]$ and 1024 points.

4.45 Show that the dlVP kernel, using $2M$ harmonics, is a summation kernel. (*Remark*: The case of an odd number of harmonics is more difficult. The method for this case will be discussed in Section 10 of the next chapter.)

4.46 Show that Dirichlet's kernel is *not* a summation kernel.

5

Fourier Transforms

In this chapter, we will discuss the Fourier transform, which is one of the most important tools in pure and applied mathematics. We will describe its principal properties, including the concept of convolution, using *FAS* to provide many illustrations. Applications to mathematical physics will also be described. And, we will discuss two important tools of communication theory, Poisson summation and sampling theory.

1 Introduction

In this section we define the Fourier transform. Although it is possible to motivate the definition of the Fourier transform by way of Fourier series, it is logically simpler to just begin with the definition of the Fourier transform and then show its properties and applications.

Throughout this chapter we will use the following notation. Let $\|f\|_1$ be defined by

$$\|f\|_1 = \int_{-\infty}^{\infty} |f(x)| \, dx. \tag{1.1}$$

If the integral in (1.1) converges, then we will say that $\|f\|_1$ is finite and has the value to which the integral converges (if the integral diverges, then we will say that $\|f\|_1 = \infty$). Similarly, we define $\|f\|_2$ by

$$\|f\|_2 = \left[\int_{-\infty}^{\infty} |f(x)|^2 \, dx \right]^{\frac{1}{2}}. \tag{1.2}$$

Again, if the integral in (1.2) converges, then we will say that $\|f\|_2$ is finite and $\|f\|_2^2$ has the value to which the integral converges (if the integral diverges, then we will say that $\|f\|_2 = \infty$).

We can now give the definition of the Fourier transform.

DEFINITION 1.3   *Given a function f for which $\|f\|_1$ is finite, the **Fourier** **transform** of f is denoted by $\hat{f}$ and is defined as a function of u by*

$$\hat{f}(u) = \int_{-\infty}^{\infty} f(x)e^{-i2\pi ux}\, dx.$$

Here are some simple examples.

Example 1.4
Suppose that the function rec is defined by

$$\text{rec}\,(x) = \begin{cases} 1 & \text{if } |x| < .5 \\ .5 & \text{if } |x| = .5 \\ 0 & \text{if } |x| > .5 \end{cases} \tag{1.5}$$

then

$$\widehat{\text{rec}}\,(u) = \int_{-.5}^{.5} e^{-i2\pi ux}\, dx = \frac{e^{-i2\pi ux}}{-i2\pi u}\bigg|_{x=-.5}^{x=.5}$$

$$= \frac{e^{-i\pi u} - e^{i\pi u}}{-i2\pi u} = \frac{\sin \pi u}{\pi u}\ .$$

The transform of rec (x) is therefore sinc u. Using the notation $f(x) \xrightarrow{\ \mathcal{F}\ } \hat{f}(u)$ to denote the Fourier transform operation, we have

$$\text{rec}\,(x) \xrightarrow{\ \mathcal{F}\ } \text{sinc}\,(u). \tag{1.6}$$

See Figure 5.1.   ▯

Example 1.7
Suppose that f is defined by

$$f(x) = \begin{cases} e^{-2\pi x} & \text{for } x > 0 \\ 0 & \text{for } x < 0. \end{cases}$$

Then

$$\hat{f}(u) = \int_0^{\infty} e^{-2\pi x} e^{-i2\pi ux}\, dx = \frac{e^{-2\pi x(1+iu)}}{-2\pi(1+iu)}\bigg|_{x=0}^{x\to\infty}$$

$$= \frac{1}{2\pi(1+iu)}$$

since $e^{-2\pi x(1+iu)} \to 0$ as $x \to \infty$.

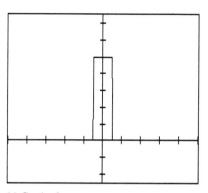

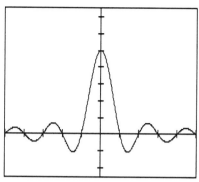

(a) Graph of rec

X interval: $[-5, 5]$ X increment = 1

Y interval: $[-.5, 1.5]$ Y increment = .2

(b) Graph of transform of rec

X interval: $[-5, 5]$ X increment = 1

Y interval: $[-.5, 1.5]$ Y increment = .2

FIGURE 5.1
Fourier transform of the rec function.

Similarly, the function

$$g(x) = \begin{cases} 0 & \text{for } x > 0 \\ e^{2\pi x} & \text{for } x < 0 \end{cases}$$

has Fourier transform

$$\hat{g}(u) = \frac{1}{2\pi(1 - iu)} \, . \quad \square$$

Example 1.8
This example will show that

$$e^{-2\pi|x|} \xrightarrow{\mathcal{F}} \frac{1}{\pi} \frac{1}{1 + u^2} \, . \tag{1.9}$$

SOLUTION   We need to evaluate the Fourier transform integral

$$\int_{-\infty}^{\infty} e^{-2\pi|x|} e^{-i2\pi ux} \, dx.$$

Splitting the integral into two integrals, we have

$$\int_{-\infty}^{\infty} e^{-2\pi ux} e^{-i2\pi ux} \, dx = \int_{0}^{\infty} e^{-2\pi x} e^{-i2\pi ux} \, dx + \int_{-\infty}^{0} e^{2\pi x} e^{-i2\pi ux} \, dx$$

$$= \hat{f}(u) + \hat{g}(u)$$

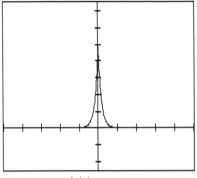

(a) Graph of $e^{-2\pi|x|}$

X interval: $[-5, 5]$ X increment = 1

Y interval: $[-.5, 1.5]$ Y increment = .2

(b) Graph of transform of $e^{-2\pi|x|}$

X interval: $[-5, 5]$ X increment = 1

Y interval: $[-.5, 1.5]$ Y increment = .2

FIGURE 5.2
Fourier transform of $e^{-2\pi|x|}$.

where f and g are the functions defined in Example 1.7. Using the results of Example 1.7 for $\hat{f}$ and $\hat{g}$, we have

$$\int_{-\infty}^{\infty} e^{-2\pi|x|} e^{-i2\pi ux} \, dx = \frac{1}{2\pi(1 + iu)} + \frac{1}{2\pi(1 - iu)} = \frac{1}{\pi} \frac{1}{1 + u^2}$$

and (1.9) is proved. See Figure 5.2. ∎

Example 1.10
Find the Fourier transform of

$$\Lambda(x) = \begin{cases} 1 - |x| & \text{for } |x| \leq 1 \\ 0 & \text{for } |x| > 1. \end{cases} \quad ☐$$

SOLUTION   Since $\Lambda = 0$ outside of the interval $[-1, 1]$ and Λ is an even function, we have

$$\hat{\Lambda}(u) = \int_{-1}^{1} (1 - |x|) \cos 2\pi ux \, dx - i \int_{-1}^{1} (1 - |x|) \sin 2\pi ux \, dx$$

$$= 2 \int_{0}^{1} (1 - x) \cos 2\pi ux \, dx.$$

Performing an integration by parts with this last integral, we obtain

$$\hat{\Lambda}(u) = \frac{1}{\pi u} \int_0^1 \sin 2\pi u x \, dx = \frac{1 - \cos 2\pi u}{2\pi^2 u^2}$$

$$= \frac{\sin^2 \pi u}{\pi^2 u^2} = \operatorname{sinc}^2 u.$$

Thus, $\Lambda(x) \xrightarrow{\mathcal{F}} \operatorname{sinc}^2 u.$ ∎

Example 1.11

We will calculate the Fourier transform of $f(x) = e^{-\pi x^2}$. The definition of Fourier transform yields

$$\hat{f}(u) = \int_{-\infty}^{\infty} e^{-\pi x^2} e^{-i2\pi u x} \, dx. \tag{1.12}$$

If we differentiate both sides of (1.12) with respect to u, then by differentiating under the integral sign, we get

$$\frac{d\hat{f}}{du} = \int_{-\infty}^{\infty} \frac{\partial}{\partial u} \left[e^{-\pi x^2} e^{-i2\pi u x} \right] dx$$

$$= \int_{-\infty}^{\infty} -i2\pi x e^{-\pi x^2} e^{-i2\pi u x} \, dx$$

$$= \int_{-\infty}^{\infty} i \frac{d}{dx} \left[e^{-\pi x^2} \right] e^{-i2\pi u x} \, dx. \tag{1.13}$$

(*Note*: The method of differentiating under the integral sign is a common technique. For further discussion, see [Wa, Chapter 6.2].)

Integrating by parts in the last integral in (1.13), we obtain

$$\frac{d\hat{f}}{du} = i e^{-i2\pi u x} e^{-\pi x^2} \Big|_{x \to -\infty}^{x \to \infty} - \int_{-\infty}^{\infty} 2\pi u e^{-\pi x^2} e^{-i2\pi u x} \, dx$$

$$= -2\pi u \int_{-\infty}^{\infty} e^{-\pi x^2} e^{-i2\pi u x} \, dx.$$

Thus,

$$\frac{d\hat{f}}{du} = -2\pi u \hat{f}(u). \tag{1.14}$$

Solving (1.14) for the unknown function $\hat{f}$, we get

$$\hat{f}(u) = \hat{f}(0) e^{-\pi u^2}. \tag{1.15}$$

It remains to find $\hat{f}(0)$, where

$$\hat{f}(0) = \int_{-\infty}^{\infty} e^{-\pi x^2} \, dx. \tag{1.16}$$

Squaring both sides of (1.16), and replacing the *dummy variable* x by y in one of the integrals, we have

$$\hat{f}(0)^2 = \int_{-\infty}^{\infty} e^{-\pi x^2} \, dx \int_{-\infty}^{\infty} e^{-\pi y^2} \, dy$$

$$= \int_{-\infty}^{\infty} \int_{-\infty}^{\infty} e^{-\pi(x^2+y^2)} \, dx \, dy. \tag{1.17}$$

Changing to polar coordinates, we get

$$\int_{-\infty}^{\infty} \int_{-\infty}^{\infty} e^{-\pi(x^2+y^2)} \, dx \, dy = \int_{0}^{2\pi} \int_{-\infty}^{\infty} e^{-\pi r^2} r \, dr \, d\theta$$

$$= \int_{0}^{2\pi} \left[\frac{-1}{2\pi} e^{-\pi r^2} \right] \Big|_{r=0}^{r \to \infty} \, d\theta = 1.$$

Thus, on returning to (1.17), we see that $\hat{f}(0)^2 = 1$. Since $\hat{f}(0) > 0$, as we can see from (1.16), we must have $\hat{f}(0) = 1$. Using this fact in (1.15) we have

$$e^{-\pi x^2} \xrightarrow{\mathcal{F}} e^{-\pi u^2}. \tag{1.18}$$

Formula (1.18) expresses the remarkable fact that $e^{-\pi x^2}$ Fourier transforms to itself (as a function of u). ☐

---

2   Properties of Fourier transforms

This section covers some important properties of Fourier transforms. For example, how they behave with respect to shifting, scaling, and differentiation.
We begin with the following theorem.

THEOREM 2.1
The Fourier transform operation $f \xrightarrow{\mathcal{F}} \hat{f}$ has the following properties:

(a) *[linearity] For all constants a and b,*

$$af + bg \xrightarrow{\mathcal{F}} a\hat{f} + b\hat{g}.$$

(b) *[scaling] For each positive constant ρ,*

$$f\left(\frac{x}{\rho}\right) \xrightarrow{\mathcal{F}} \rho\hat{f}(\rho u) \quad \text{and} \quad f(\rho x) \xrightarrow{\mathcal{F}} \frac{1}{\rho}\hat{f}\left(\frac{u}{\rho}\right)$$

*(c) [**shifting**] For each real constant c,*

$$f(x-c) \xrightarrow{\mathcal{F}} \hat{f}(u)e^{-i2\pi cu}.$$

*(d) [**modulation**] For each real constant c,*

$$f(x)e^{i2\pi cx} \xrightarrow{\mathcal{F}} \hat{f}(u-c).$$

PROOF The linearity property (a) is clear, so we leave the details to the reader. To prove (b), we make the change of variables $s = x/\rho$ in the following Fourier transform integral:

$$f\left(\frac{x}{\rho}\right) \xrightarrow{\mathcal{F}} \int_{-\infty}^{\infty} f\left(\frac{x}{\rho}\right) e^{-i2\pi ux}\,dx = \int_{-\infty}^{\infty} f(s)e^{-i2\pi u(\rho s)}\,d(\rho s)$$

$$= \rho \int_{-\infty}^{\infty} f(s)e^{-i2\pi(\rho u)s}\,ds = \rho \hat{f}(\rho u).$$

Thus, $f\left(\frac{x}{\rho}\right) \xrightarrow{\mathcal{F}} \rho\hat{f}(\rho u)$. Substituting $1/\rho$ in place of ρ, we see that $f(\rho x) \xrightarrow{\mathcal{F}}$ $(1/\rho)\hat{f}\left(u/\rho\right)$ and (b) is verified.

To prove (c) we make the change of variable $s = x - c$ in the following Fourier transform integral:

$$f(x-c) \xrightarrow{\mathcal{F}} \int_{-\infty}^{\infty} f(x-c)e^{-i2\pi ux}\,dx = \int_{-\infty}^{\infty} f(s)e^{-i2\pi u(s+c)}\,ds$$

$$= \int_{-\infty}^{\infty} f(s)e^{-i2\pi us}\,ds\, e^{-i2\pi cu}$$

$$= \hat{f}(u)e^{-i2\pi cu}.$$

Thus, (c) holds.

To prove (d), we note that $e^{i2\pi cx}e^{-i2\pi ux} = e^{-i2\pi(u-c)x}$, hence

$$f(x)e^{i2\pi cx} \xrightarrow{\mathcal{F}} \int_{-\infty}^{\infty} f(x)e^{-i2\pi(u-c)x}\,dx = \hat{f}(u-c)$$

and (d) holds. ∎

Here are some examples of Theorem 2.1.

Example 2.2
(a) Find the Fourier transform of $(1/\rho)e^{-\pi x^2/\rho^2}$ for any $\rho > 0$. ☐

SOLUTION Using scaling, linearity, and $e^{-\pi x^2} \xrightarrow{\mathcal{F}} e^{-\pi u^2}$ we have

$$\frac{1}{\rho}e^{-\pi x^2/\rho^2} \xrightarrow{\mathcal{F}} e^{-\pi\rho^2 u^2}, \qquad (\rho > 0). ∎$$

(b) Find the Fourier transform of $\mathrm{rec}\,(x-4) + \mathrm{rec}\,(x+4)$.   ⬚

SOLUTION   Using shifting and linearity, we have

$$\mathrm{rec}\,(x-4) + \mathrm{rec}\,(x+4) \xrightarrow{\;\mathcal{F}\;} \widehat{\mathrm{rec}}\,(u)e^{-i2\pi 4u} + \widehat{\mathrm{rec}}\,(u)e^{i2\pi 4u}.$$

Since $\widehat{\mathrm{rec}}\,(u) = \mathrm{sinc}\,u$ and $e^{-i2\pi 4u} + e^{i2\pi 4u} = 2\cos 8\pi u$, we have

$$\mathrm{rec}\,(x-4) + \mathrm{rec}\,(x+4) \xrightarrow{\;\mathcal{F}\;} 2(\mathrm{sinc}\,u)(\cos 8\pi u). \quad\blacksquare$$

(c) Find the Fourier transform of $e^{-2\pi|x|}\cos 6\pi x$.   ⬚

SOLUTION   Rewriting $\cos 6\pi x$ as $(1/2)e^{i2\pi 3x} + (1/2)e^{-i2\pi 3x}$ and then using linearity and modulation, we get

$$e^{-2\pi|x|}\cos 6\pi x \xrightarrow{\;\mathcal{F}\;} \frac{1}{2}(e^{-2\pi|x|})^{\wedge}(u-3) + \frac{1}{2}(e^{-2\pi|x|})^{\wedge}(u+3).$$

Hence,

$$e^{-2\pi|x|}\cos 6\pi x \xrightarrow{\;\mathcal{F}\;} \frac{1}{2\pi}\frac{1}{1+(u-3)^2} + \frac{1}{2\pi}\frac{1}{1+(u+3)^2}. \quad\blacksquare$$

(d) Find the Fourier transform of $e^{-2\pi\rho|x|}$ for all $\rho > 0$.   ⬚

SOLUTION   Using the scaling property, we have

$$e^{-2\pi\rho|x|} \xrightarrow{\;\mathcal{F}\;} \frac{1}{\rho}\left(\frac{1}{\pi}\frac{1}{1+(u/\rho)^2}\right).$$

Hence,

$$e^{-2\pi\rho|x|} \xrightarrow{\;\mathcal{F}\;} \frac{1}{\pi}\frac{\rho}{\rho^2+u^2}, \qquad (\rho > 0). \quad\blacksquare$$

Another useful property of Fourier transforms is the conversion of differentiation into multiplication by the transform variable. To be more precise, we have the following theorem.

THEOREM 2.3
(a) Suppose that $\|f\|_1$ and $\|f'\|_1$ are finite and that f' is continuous, then $f' \xrightarrow{\;\mathcal{F}\;} i2\pi u\hat{f}(u)$. (b) Suppose that $\|f\|_1$ and $\|xf(x)\|_1$ are finite, then $\hat{f}$ is differentiable and $xf(x) \xrightarrow{\;\mathcal{F}\;} \frac{i}{2\pi}\hat{f}'(u)$.

PROOF   (a) Using integration by parts, we have

$$f'(x) \xrightarrow{\;\mathcal{F}\;} \int_{-\infty}^{\infty} f'(x)e^{-i2\pi ux}\,dx = f(x)e^{-i2\pi ux}\Big|_{x\to-\infty}^{x\to\infty}$$

$$+ i2\pi u\int_{-\infty}^{\infty} f(x)e^{-i2\pi ux}\,dx.$$

By the Fundamental Theorem of Calculus, we have $f(x) = \int_0^x f'(s)\,ds + f(0)$. Consequently, the limits $\lim_{x\to\infty} f(x)$ and $\lim_{x\to-\infty} f(x)$ exist and are finite. If either of these limits were not 0, then

$$\int_{-\infty}^{\infty} f(x)\,dx = \int_0^{\infty} f(x)\,dx + \int_{\infty}^0 f(x)\,dx$$

would not converge (since at least one of the integrals on the right side would diverge). Therefore, both limits are 0 and

$$f'(x) \xrightarrow{\mathcal{F}} i2\pi u \int_{-\infty}^{\infty} f(x)e^{-i2\pi ux}\,dx = i2\pi u \hat{f}(u)$$

and (a) is proved.

To prove (b) we differentiate under the integral sign as follows:

$$\hat{f}'(u) = \frac{d}{du} \int_{-\infty}^{\infty} f(x)e^{-i2\pi ux}\,dx = \int_{-\infty}^{\infty} \frac{\partial}{\partial u}\left[f(x)e^{-i2\pi ux}\right]dx$$

$$= \int_{-\infty}^{\infty} -i2\pi x f(x)e^{-i2\pi ux}\,dx.$$

But, this shows that

$$-i2\pi x f(x) \xrightarrow{\mathcal{F}} \hat{f}'(u). \tag{2.4}$$

Multiplying both sides of (2.4) by $i/2\pi$ yields (b).   ∎

Example 2.5

(a) Using Theorem 2.3(b), we have

$$xe^{-\pi x^2} \xrightarrow{\mathcal{F}} \frac{i}{2\pi}\frac{d}{du}\left[e^{-\pi u^2}\right]$$

hence

$$xe^{-\pi x^2} \xrightarrow{\mathcal{F}} -iue^{-\pi u^2}.$$

Thus, $xe^{-\pi x^2}$ transforms into $-i$ times itself (as a function of u).

(b) Using the previous example, and Theorem 2.3(b),

$$x^2 e^{-\pi x^2} \xrightarrow{\mathcal{F}} \frac{i}{2\pi}\frac{d}{du}\left[-iue^{-\pi u^2}\right].$$

Hence,

$$x^2 e^{-\pi x^2} \xrightarrow{\mathcal{F}} \frac{1}{2\pi}e^{-\pi u^2} - u^2 e^{-\pi u^2}.$$

(c) From (b), it follows that

$$\left(x^2 - \frac{1}{4\pi}\right)e^{-\pi x^2} \xrightarrow{\mathcal{F}} -\left(u^2 - \frac{1}{4\pi}\right)e^{-\pi u^2}.$$

Thus, $(x^2 - (1/4\pi))e^{-\pi x^2}$ Fourier transforms into -1 times itself (as a function of u). □

REMARK 2.6   Bringing absolute values inside the integral sign and using

$$\left| e^{-i2\pi ux} \right| = 1$$

we have

$$|\hat{f}(u)| \le \|f\|_1 \tag{2.7}$$

which shows that $|\hat{f}|$ is bounded by the constant $\|f\|_1$.

Using Theorem 2.3(a), we can say a little more than (2.7). If $\|f\|_1$ and $\|f'\|_1$ are both finite and f' is continuous, then by applying (2.7) to f' in place of f and using $\widehat{(f')}(u) = i2\pi u\hat{f}(u)$, we have

$$|i2\pi u\hat{f}(u)| \le \|f'\|_1. \tag{2.8}$$

In other words, for all $u \ne 0$

$$|\hat{f}(u)| \le \frac{\|f'\|_1}{2\pi|u|}. \tag{2.9}$$

From (2.9) it follows that

$$\lim_{|u|\to\infty} \hat{f}(u) = 0 \tag{2.10}$$

when $\|f\|_1$ and $\|f'\|_1$ are both finite and f' is continuous. Actually, (2.10) is known to be true whenever $\|f\|_1$ is finite (this is known as the *Riemann–Lebesgue Lemma*, but we will not need this fact in the sequel). ∎

3   Inversion of Fourier transforms

In this section, we shall state some inversion theorems for Fourier transforms. The proofs of these theorems are fairly technical and are omitted here, although references will be given.

We begin with the conceptually simplest inversion theorem.

THEOREM 3.1
If f is continuous and $\|f\|_1$ and $\|\hat{f}\|_1$ are both finite, then

$$f(x) = \int_{-\infty}^{\infty} \hat{f}(u)e^{i2\pi ux}\, du$$

for all x in $\mathbf{R}^2$.

PROOF See [Wa, Chapter 6.4]. ∎

REMARK 3.2 The integral in Theorem 3.1 is also called a Fourier transform. Sometimes we might call it an *inverse Fourier transform*, or we might call it a *Fourier transform with positive exponent* (the Fourier transform in Definition 1.3 being called a Fourier transform with *negative* exponent). If $\|g\|_1$ is finite, then we define $\tilde{g}$ by

$$\tilde{g}(x) = \int_{-\infty}^{\infty} g(u)e^{i2\pi ux}\, du.$$

Hence, the result in Theorem 3.1 can be rewritten as $(\hat{f})^{\sim} = f$ or $\hat{f} \xrightarrow{\mathcal{F}^{-1}} f$. ∎

As an example of Theorem 3.1, consider the triangle function

$$\Lambda(x) = \begin{cases} 1 - |x| & \text{for } |x| \leq 1 \\ 0 & \text{for } |x| > 1. \end{cases}$$

We have, by Example 1.10,

$$\hat{\Lambda}(u) = \operatorname{sinc}^2 u = \frac{\sin^2 \pi u}{(\pi u)^2}\,.$$

Both Λ and $\hat{\Lambda}$ are continuous. That $\|\Lambda\|_1$ is finite is obvious. The finiteness of $\|\hat{\Lambda}\|_1$ follows from comparing it to the integrals of $1/(\pi u)^2$ for $|u| \geq 1$. It follows from Theorem 3.1 that

$$\Lambda(x) = \int_{-\infty}^{\infty} \operatorname{sinc}^2 u\, e^{i2\pi ux}\, du. \tag{3.3}$$

Let's examine (3.3) more closely. The integral on the right side of (3.3) converges absolutely, since for $|u| > 0$

$$\left| \frac{\sin^2 \pi u}{(\pi u)^2} e^{i2\pi ux} \right| \leq \frac{1}{(\pi u)^2}\,. \tag{3.4}$$

Inequality (3.4) can be used for estimating approximations to (3.3). For example, if we only integrate from $-L$ to L, for some finite positive L, then we have

$$\left| \Lambda(x) - \int_{-L}^{L} \operatorname{sinc}^2 u\, e^{i2\pi ux}\, du \right| = \left| \int_{|u|>L} \operatorname{sinc}^2 u\, e^{i2\pi ux}\, du \right|$$

$$\leq \int_{|u|>L} \frac{1}{(\pi u)^2}\, du = \frac{2}{\pi^2 L}\,.$$

For example, if we take $L = 32$, then $2/(\pi^2 L) \approx 0.006$. Hence, for *all* x,

$$\left| \Lambda(x) - \int_{-32}^{32} \operatorname{sinc}^2 u\, e^{i2\pi ux}\, du \right| \leq 0.006. \tag{3.5}$$

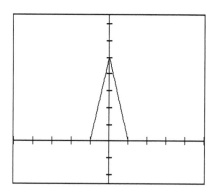

FIGURE 5.3
Inverse Fourier transform of sinc² over [−32, 32].

In Figure 5.3 we have shown a graph of the inverse Fourier transform of sinc²
over the interval [−32, 32], using 4096 points.

Here is a second inversion theorem. It is less restrictive in its hypotheses
than Theorem 3.1, in that it does not require f to be continuous nor $\|\hat{f}\|_1$ to be
finite.

THEOREM 3.6
If $\|f\|_1$ is finite, then

$$\lim_{L \to \infty} \int_{-L}^{L} \hat{f}(u)e^{i2\pi ux}\, du = \frac{1}{2}\left[f(x+) + f(x-)\right].$$

*provided f is Lipschitz from the left and right at x. Moreover, if f is also
continuous at x, then*

$$\lim_{L \to \infty} \int_{-L}^{L} \hat{f}(u)e^{i2\pi ux}\, du = f(x).$$

PROOF   See [Wa, Chapter 6.5].   ∎

The reader might compare this theorem to the Fourier series convergence
theorem (Theorem 5.4) in Chapter 1.

Here is an example of how this theorem works. The function

$$f(x) = \begin{cases} i\pi e^{-2\pi x} & \text{for } x > 0 \\ -i\pi e^{2\pi x} & \text{for } x < 0 \end{cases} \tag{3.7}$$

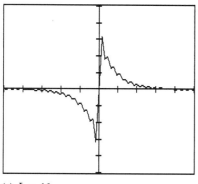
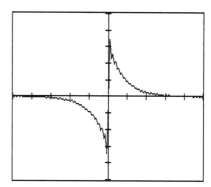

(a) $L = 16$
X interval: $[-1, 1]$ X increment $= .2$
Y interval: $[-5, 5]$ Y increment $= 1$

(b) $L = 32$
X interval: $[-1, 1]$ X increment $= .2$
Y interval: $[-5, 5]$ Y increment $= 1$

FIGURE 5.4
Calculation of (3.10) for $L = 16$ and $L = 32$.

has Fourier transform $u/(1 + u^2)$. But, the inversion integral

$$\int_{-\infty}^{\infty} \frac{u}{1 + u^2} e^{i2\pi u x} \, du \tag{3.8}$$

does not converge for $x = 0$. Since, for $x = 0$, the inversion integral in (3.8) is defined by

$$\int_{-\infty}^{\infty} \frac{u}{1 + u^2} \, du = \int_{0}^{\infty} \frac{u}{1 + u^2} \, du + \int_{-\infty}^{0} \frac{u}{1 + u^2} \, du \tag{3.9}$$

and both integrals on the right side of (3.9) diverge. However, the integrals

$$\int_{-L}^{L} \frac{u}{1 + u^2} \, du \quad (= 0)$$

converge to 0 as $L \to \infty$, which is precisely what Theorem 3.6 says should happen at $x = 0$ (in particular, $0 = (1/2)[f(0+) + f(0-)]$). In Figure 5.4 we show graphs obtained using *FAS* to approximate

$$\int_{-L}^{L} \frac{u}{1 + u^2} e^{i2\pi u x} \, du \tag{3.10}$$

for increasing values of L. As we can see from this figure, the convergence of the integral in (3.10) to $f(x)$ is very reminiscent of Fourier series convergence (there is even a Gibbs' phenomenon!). We take advantage of this similarity in our next definition.

DEFINITION 3.11 *If L is a positive real number, then $\mathcal{S}_L^f$ is defined by*

$$\mathcal{S}_L^f(x) = \int_{-L}^{L} \hat{f}(u) e^{i2\pi u x}\, du.$$

For example, Theorem 3.6 could be restated as

$$\mathcal{S}_L^f(x) \rightarrow \frac{1}{2}\left[f(x+) + f(x-)\right], \qquad \text{as } L \to \infty \tag{3.12}$$

provided f is Lipschitz from the left and right at x.

Here is our final convergence theorem. It is analagous to the completeness of Fourier series.

THEOREM 3.13
If $\|f\|_2 < \infty$, then $\lim_{L\to\infty} \|f - \mathcal{S}_L^f\|_2 = 0$. That is, $\mathcal{S}_L^f$ converges to f in the 2-Norm.

PROOF A proof can be found in [Ru, Chapter 9]. ∎

To understand Theorem 3.13 completely one must know how the Fourier transform $\hat{f}$ is defined when $\|f\|_2 < \infty$. There are many functions for which $\|f\|_2 < \infty$ but $\|f\|_1 = \infty$, in which case Definition 1.3 does not apply. There is a subtler definition of the Fourier transform $\hat{f}$ in this case. It is known that $\int_{-L}^{L} f(x) e^{-i2\pi u x}\, dx$ converges as $L \to \infty$, using the 2-Norm $\|\cdot\|_2$ to measure the magnitude of differences, to a function $g(u)$ for which $\|g\|_2 < \infty$. That is,

$$\lim_{L\to\infty} \left\| g(u) - \int_{-L}^{L} f(x) e^{-i2\pi u x}\, dx \right\|_2 = 0. \tag{3.14}$$

This function $g(u)$ is called the Fourier transform $\hat{f}(u)$. And it is known that *Parseval's equality*

$$\int_{-\infty}^{\infty} |f(x)|^2\, dx = \int_{-\infty}^{\infty} |\hat{f}(u)|^2\, du \tag{3.15}$$

holds. (See [Ru, Chapter 9], for proofs of these results.)

As an example of Theorem 3.13, suppose $f(x) = \text{rec}(x)$. Then, $\text{rec}(x) \xrightarrow{\mathcal{F}} \text{sinc}(u)$. If we use *FAS* to approximate the integral

$$S_L^{\text{rec}}(x) = \int_{-L}^{L} \text{sinc}(u) \, e^{i2\pi ux} \, du \tag{3.16}$$

(using, say, 1024 points) then the 2-Norms $\|\text{rec} - S_L^{\text{rec}}\|_2$ can be approximated. Here are some results for various values of L:

$$\|\text{rec} - S_8^{\text{rec}}\|_2 \approx 0.06 \qquad \|\text{rec} - S_{16}^{\text{rec}}\|_2 \approx 0.04$$

$$\|\text{rec} - S_{32}^{\text{rec}}\|_2 \approx 0.03 \qquad \|\text{rec} - S_{64}^{\text{rec}}\|_2 \approx 0.02 \tag{3.17}$$

4 The relation between Fourier transforms and DFTs

In this section we will explain how the Fourier transform of a function can be approximated by a DFT.

First, let's assume that the transform $\hat{f}$ can be approximated by an integral over the interval $[0, \Omega]$ for some large Ω (this would be the case, for example, if $f(x) = 0$ for $x < 0$). This case is a bit easier to handle than the more general case, which would be to assume that $\hat{f}$ can be approximated by an integral over the interval $[-(1/2)\Omega, (1/2)\Omega]$ for some large Ω. We will discuss this latter case by reduction to the first case.

Assuming that $\hat{f}$ is approximated by an integral over $[0, \Omega]$, we have

$$\hat{f}(u) \approx \int_0^{\Omega} f(x)e^{-i2\pi ux} \, dx. \tag{4.1}$$

The integral is then approximated by a uniform, left-endpoint sum

$$\int_0^{\Omega} f(x)e^{-i2\pi ux} \, dx \approx \sum_{j=0}^{N-1} f\left(j\frac{\Omega}{N}\right) e^{-i2\pi uj\Omega/N} \frac{\Omega}{N}.$$

Thus,

$$\hat{f}(u) \approx \frac{\Omega}{N} \sum_{j=0}^{N-1} f\left(j\frac{\Omega}{N}\right) e^{-i2\pi uj\Omega/N}. \tag{4.2}$$

To convert the right side of (4.2) into a DFT we replace u by k/Ω for $k = 0$, $\pm 1, \pm 2, \ldots$

$$\hat{f}\left(\frac{k}{\Omega}\right) \approx \frac{\Omega}{N} \sum_{j=0}^{N-1} f\left(j\frac{\Omega}{N}\right) e^{-i2\pi jk/N}. \tag{4.3}$$

The right side of (4.3) is equal to Ω/N times the N-point DFT of the sequence

$$\left\{ f\left(j\frac{\Omega}{N} \right) \right\}_{j=0}^{N-1}.$$

The computer, of course, is programmed to compute an N-point FFT of $\{f(j(\Omega/N))\}_{j=0}^{N-1}$, call it $\{F_k\}$. By periodicity, we can define F_{-k} by

$$F_{-k} = F_{N-k} \tag{4.4}$$

and, hence

$$\hat{f}\left(\frac{k}{\Omega} \right) \approx \frac{\Omega}{N}F_k \quad \text{for } k = 0, 1, \ldots, \tfrac{1}{2}N$$

$$\hat{f}\left(\frac{-k}{\Omega} \right) \approx \frac{\Omega}{N}F_{N-k} \quad \text{for } k = 1, \ldots, \tfrac{1}{2}N - 1. \tag{4.5}$$

Because the right side of (4.3) is a multiple of a DFT, it has period N in k. Therefore, it is hopeless to expect that the approximation in (4.3) will be valid for *all* integers k (since $\hat{f}$ is typically *not* periodic). As we did for Fourier series, we shall assume that $|k| \le N/8$. This assumption will serve to remove the grossest of aliasing errors.

Now, suppose that we wish to approximate $\hat{f}$ by an integral over $[-(1/2)\Omega, (1/2)\Omega]$. Then we will have

$$\hat{f}(u) \approx \int_{-\frac{1}{2}\Omega}^{\frac{1}{2}\Omega} f(x)e^{-i2\pi ux}\, dx. \tag{4.6}$$

We will show that $\{\hat{f}(k/\Omega)\}$ can be approximated by a constant multiple of a DFT. We have

$$\hat{f}\left(\frac{k}{\Omega} \right) \approx \int_{-\frac{1}{2}\Omega}^{\frac{1}{2}\Omega} f(x)e^{-i2\pi kx/\Omega}\, dx. \tag{4.7}$$

Splitting the integral in (4.7) into two integrals, we can change variables so that we are integrating over $[0, \Omega]$

$$\int_{-\frac{1}{2}\Omega}^{\frac{1}{2}\Omega} f(x)e^{-i2\pi kx/\Omega}\, dx = \int_{-\frac{1}{2}\Omega}^{0} f(x)e^{-i2\pi kx/\Omega}\, dx$$

$$+ \int_{0}^{\frac{1}{2}\Omega} f(x)e^{-i2\pi kx/\Omega}\, dx. \tag{4.8}$$

Substituting $t = x + \Omega$ into the first integral on the right side of (4.8) yields

$$\int_{-\frac{1}{2}\Omega}^{\frac{1}{2}\Omega} f(x)e^{-i2\pi kx/\Omega}\, dx = \int_{\frac{1}{2}\Omega}^{\Omega} f(t - \Omega)e^{-i2\pi kt/\Omega}\, dt$$

$$+ \int_{0}^{\frac{1}{2}\Omega} f(x)e^{-i2\pi kx/\Omega}\, dx$$

$$= \int_{\frac{1}{2}\Omega}^{\Omega} f(x - \Omega)e^{-i2\pi kx/\Omega}\, dx$$

$$+ \int_{0}^{\frac{1}{2}\Omega} f(x)e^{-i2\pi kx/\Omega}\, dx.$$

Thus, if we define g by

$$g(x) = \begin{cases} f(x) & \text{for } 0 \le x < \frac{1}{2}\Omega \\ \frac{1}{2}f(\frac{1}{2}\Omega) + \frac{1}{2}f(-\frac{1}{2}\Omega) & \text{for } x = \frac{1}{2}\Omega \\ f(x - \Omega) & \text{for } \frac{1}{2}\Omega < x \le \Omega \end{cases} \tag{4.9}$$

we have

$$\int_{-\frac{1}{2}\Omega}^{\frac{1}{2}\Omega} f(x)e^{-i2\pi kx/\Omega}\, dx = \int_{0}^{\Omega} g(x)e^{-i2\pi kx/\Omega}\, dx. \tag{4.10}$$

Using (4.10) we have from (4.7) that

$$\hat{f}\left(\frac{k}{\Omega}\right) \approx \int_{0}^{\Omega} g(x)e^{-i2\pi kx/\Omega}\, dx. \tag{4.11}$$

Then, as we did above for (4.1), we can approximate the right side of (4.11) by Ω/N times an N-point DFT

$$\hat{f}\left(\frac{k}{\Omega}\right) \approx \frac{\Omega}{N} \sum_{j=0}^{N-1} g\left(j\frac{\Omega}{N}\right) e^{-i2\pi jk/N}. \tag{4.12}$$

In the discussion above, we mentioned the restriction: $|k| \le N/8$. *FAS* actually computes the entire FFT of $\{g(j\Omega/N)\}$ (or $\{f(j\Omega/N)\}$) when working over the interval $[0, \Omega)$ and connects these FFT values by line segments. An example should make this point clearer.

Example 4.13
Compare the computer calculation of the Fourier transform of rec and its exact transform, sinc. ☐

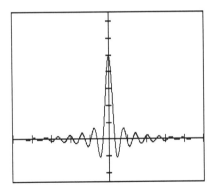

X interval: [−16, 16]   X increment = 3.2
Y interval: [−.5, 1.5]   Y increment = .2

FIGURE 5.5
FAS **calculated transform of rec using the interval** [−16, 16] **and 1024 points.**

SOLUTION   If rec (x) is Fourier transformed using *FAS*, by choosing the interval
[−16, 16] and 1024 points, then the graph in Figure 5.5 results. If this graph is
saved and then compared with the graph of sinc (x), we get the graphs shown
in Figure 5.6. We see in Figure 5.6(a) that the graphs are a reasonably close
match over the interval [−4, 4] where $|k| \leq N/8$. But, in Figure 5.6(b) we see
the difference between the two graphs over [6, 16]. The graph of the computed
transform seems to be much more severely damped toward 0 as we approach
the end of the interval.

To expand the range of the computed transform *and improve its accuracy*,
we increase the number of points used *and* increase the size of the interval over
which we transform. If rec is transformed over [−32, 32] using 4096 points,
then the computed transform will be graphed over the interval [−T, T] where

$$T = \frac{(\text{number of points})/2}{\text{length of } [-32, 32]} = \frac{2048}{64} = 32.$$

Also, the number of functional values of rec used per unit length is $4096/64 =$
64 values/unit length, which is greater than $1024/32 = 32$ values/unit length
used in the previous computation. In Figure 5.7 we show the graph of the computed transform of rec and the graph of its exact transform, sinc. In particular,
notice the improved match of the computed transform to sinc over the interval
[6, 16].   ∎

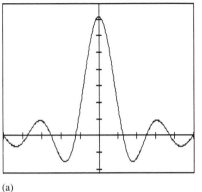
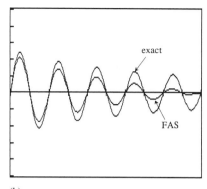

(a)

X interval: $[-4, 4]$ X increment = .8
Y interval: $[-.3, 1.1]$ Y increment = .14

(b)

X interval: $[6, 16]$ X increment = 1
Y interval: $[-.1, .1]$ Y increment = .02

FIGURE 5.6
Comparison of *FAS* calculated transform of rec (using interval $[-16, 16]$ and 1024 points) and the exact transform sinc.

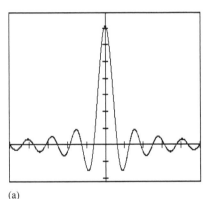
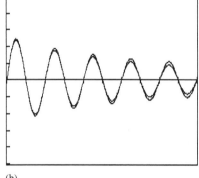

(a)

X interval: $[-8, 8]$ X increment = 1.6
Y interval: $[-.3, 1.1]$ Y increment = .14

(b)

X interval: $[6, 16]$ X increment = 1
Y interval: $[-.1, .1]$ Y increment = .02

FIGURE 5.7
Comparison of *FAS* calculated transform of rec (using interval $[-32, 32]$ and 4096 points) and the exact transform sinc.

REMARK 4.14 In formulas (4.3) and (4.12) the transform $\hat{f}$ is evaluated at integer multiples of $1/\Omega$ where Ω is the length of the interval over which f was transformed. The quantity $1/\Omega$ is called the *frequency resolution* for the computed transform. The frequency resolution is improved, made *finer*, by increasing the length of the interval over which the function is transformed.

On the other hand, in order to capture more rapid oscillation in a function, that function must be evaluated more frequently. The number of values per unit length is given by N/Ω. To increase N/Ω, which is called the *sampling frequency*, one must either increase N or decrease Ω. The frequency resolution $1/\Omega$ and sampling frequency N/Ω can both be improved, but they are at odds with one another. To create twice as fine a frequency resolution one can double Ω, but then, in order to sample twice as frequently, one must increase N by a factor of 4. This is what we did in the previous example. ∎

When comparing graphs of a computed transform with the original function, *FAS* requires the same number of points *and* the same interval (so the graphs can be overlaid simultaneously in the *Graph* procedure). This is possible, *using integer values for the interval*, for the following choices of numbers of points and intervals: 256 points with interval $[-8, 8]$, 1024 points with interval $[-16, 16]$, or 4096 points with interval $[-32, 32]$. We will always restrict ourselves to these choices when it is desirable to compare a function with its computed transform.

5   Convolution — An introduction

In this section we shall introduce the concept of *convolution*, which is one of the most important ideas in Fourier analysis. One of the great advantages of FFTs is that they provide a very fast way of computing approximations to convolutions.

To motivate the need for convolutions we will look at an example. We will show how convolution arises in describing the solution to the following problem. This problem is known as *Dirichlet's problem for the upper half plane*.

Example 5.1
Find a function $H(x, y)$ that satisfies

$$\frac{\partial^2 H}{\partial x^2} + \frac{\partial^2 H}{\partial y^2} = 0 , \quad (-\infty < x < \infty , \ y > 0) \qquad (5.1\text{a})$$

$$H(x, 0) = f(x) , \qquad (-\infty < x < \infty). \qquad (5.1\text{b})$$

SOLUTION   We solve (5.1a) and (5.1b) by Fourier transforming the problem. Define $\hat{H}(u, y)$ by

$$\hat{H}(u, y) = \int_{-\infty}^{\infty} H(x, y) e^{-i2\pi u x} \, dx. \qquad (5.2)$$

Or, by Fourier inversion,

$$H(x, y) = \int_{-\infty}^{\infty} \hat{H}(u, y) e^{i2\pi ux} \, du. \tag{5.3}$$

By differentiating the right side of (5.3) twice under the integral sign, we get

$$\frac{\partial^2 H}{\partial x^2} = \int_{-\infty}^{\infty} \hat{H}(u, y) \frac{\partial^2}{\partial x^2} \left[e^{i2\pi ux} \right] \, du$$

$$= \int_{-\infty}^{\infty} -(2\pi u)^2 \hat{H}(u, y) e^{i2\pi ux} \, du. \tag{5.4}$$

Similarly,

$$\frac{\partial^2 H}{\partial y^2} = \int_{-\infty}^{\infty} \frac{\partial^2 \hat{H}(u, y)}{\partial y^2} e^{i2\pi ux} \, du. \tag{5.5}$$

By combining (5.4) and (5.5) we obtain from (5.1a) that

$$0 = \frac{\partial^2 H}{\partial x^2} + \frac{\partial^2 H}{\partial y^2} = \int_{-\infty}^{\infty} \left[\frac{\partial^2 \hat{H}(u, y)}{\partial y^2} - (2\pi u)^2 \hat{H}(u, y) \right] e^{i2\pi ux} \, du. \tag{5.6}$$

Since the function in brackets in (5.6) is being inverse Fourier transformed to the 0-function, we conclude that it must also be the 0-function. That is,

$$\frac{\partial^2 \hat{H}(u, y)}{\partial y^2} - (2\pi u)^2 \hat{H}(u, y) = 0. \tag{5.7}$$

Now, by putting $y = 0$ into (5.2), we obtain from $H(x, 0) = f(x)$ in (5.1b)

$$\hat{H}(u, 0) = \int_{-\infty}^{\infty} f(x) e^{-i2\pi ux} \, dx = \hat{f}(u). \tag{5.8}$$

Formulas (5.7) and (5.8) result in the Fourier transformed problem of finding $\hat{H}(u, y)$ which satisfies

$$\frac{\partial^2 \hat{H}(u, y)}{\partial y^2} = (2\pi u)^2 \hat{H}(u, y), \qquad (-\infty < u < \infty, \ y > 0) \tag{5.9a}$$

$$\hat{H}(u, 0) = \hat{f}(u), \qquad (-\infty < u < \infty). \tag{5.9b}$$

To solve (5.9a) and (5.9b) we *fix* some value of u, then (5.9a) is an *ordinary* differential equation

$$\frac{d^2 \hat{H}(u, y)}{dy^2} = (2\pi u)^2 \hat{H}(u, y)$$

which has solution

$$\hat{H}(u, y) = a \, e^{-2\pi |u| y} + b \, e^{2\pi |u| y} \tag{5.10}$$

where a and b are constants (actually, a and b depend on u).

In order to recover $H(x, y)$ from formula (5.3) we *reject the part of $\hat{H}$ in* (5.10) *which has an infinite integral over* **R**, namely, $b\,e^{2\pi|u|y}$. We do this by putting $b = 0$, so that

$$\hat{H}(u, y) = a\,e^{-2\pi|u|y}. \tag{5.11}$$

Comparing (5.11) with (5.9b) we see that

$$\hat{H}(u, y) = \hat{f}(u)\,e^{-2\pi|u|y}. \tag{5.12}$$

Now, to find $H(x, y)$ we substitute the expression for $\hat{H}$ in (5.12) into the integral in (5.3) obtaining

$$H(x, y) = \int_{-\infty}^{\infty} \hat{f}(u)e^{-2\pi|u|y}e^{i2\pi ux}\,du. \tag{5.13}$$

Formula (5.13) says that a solution $H(x, y)$ to (5.1a) and (5.1b) can be obtained by Fourier transforming f, multiplying by $e^{-2\pi|u|y}$, and then inverse Fourier transforming back to get $H(x, y)$.

We will now show how to describe this solution procedure in terms of convolution. Replacing $\hat{f}(u)$ in (5.13) by the integral which defines it, we get

$$H(x, y) = \int_{-\infty}^{\infty} \left[\int_{-\infty}^{\infty} f(s)e^{-i2\pi us}\,ds \right] e^{-2\pi|u|y}e^{i2\pi ux}\,du. \tag{5.14}$$

Interchanging integrals and combining the exponentials, yields

$$H(x, y) = \int_{-\infty}^{\infty} f(s) \left[\int_{-\infty}^{\infty} e^{-2\pi|u|y}e^{i2\pi u(x-s)}\,du \right] ds. \tag{5.15}$$

Changing variables ($u = -v$) and using Example 2.2(d) with y in place of ρ *and $x - s$ as the transform variable*, we obtain

$$\int_{-\infty}^{\infty} e^{-2\pi|u|y}e^{i2\pi u(x-s)}\,du = \int_{-\infty}^{\infty} e^{-2\pi|v|y}e^{-i2\pi v(x-s)}\,dv$$

$$= \frac{1}{\pi}\frac{y}{y^2 + (x-s)^2}. \tag{5.16}$$

Using this last result back in (5.15) we get

$$H(x, y) = \int_{-\infty}^{\infty} f(s) \left[\frac{1}{\pi}\frac{y}{y^2 + (x-s)^2} \right] ds. \tag{5.17}$$

Formula (5.17) expresses our solution H to (5.1a) and (5.1b) as a *convolution* (integral) of $f(x)$ with the function $(1/\pi)(y/(y^2 + x^2))$. This can be seen from the formal definition of convolution given in Definition 5.18 below. ∎

DEFINITION 5.18 *If f and g are two functions, with $\|f\|_1$ and $\|g\|_1$ finite,
then the **convolution** of f and g is denoted by f ∗ g and is defined by*

$$(f * g)(x) = \int_{-\infty}^{\infty} f(s)g(x - s)\, ds.$$

Notice that the previous definition of convolution given in Theorem 6.3, Chapter 4, referred to convolution *over an interval* $[-L, L]$. *When no reference is given to any interval*, then we are referring to the convolution defined in (5.18).
If we define $_yP$ by

$$_yP(x) = \frac{1}{\pi}\frac{y}{y^2 + x^2}, \qquad (y > 0) \tag{5.19}$$

then we can express (5.17) as

$$H(x, y) = (f * {_yP})(x). \tag{5.20}$$

It is an easy exercise to show that

$$f * g = g * f, \qquad (commutativity). \tag{5.21}$$

Hence we also have

$$H(x, y) = ({_yP} * f)(x). \tag{5.22}$$

In the next section we shall generalize what we have done here and give further examples of how convolutions can be used to solve some important problems. We close this section by describing how *FAS* can be used to approximate convolutions.

FAS will compute as an approximation to $f * g$ a discrete version of

$$\int_{-L}^{L} f(s)g(x - s)\, ds. \tag{5.23}$$

[How this is done was explained in Section 7 of Chapter 4.] For suitably chosen (large enough) values of L, the integral in (5.23) will provide good approximations to $(f * g)(x)$. For instance, we can approximate $H(x, y) = (f * {_yP})(x)$ for $y = 0.5$, 1, and 2. For each value of y, the function $_yP(x)$ is a specific function of x. The results are shown in Figure 5.8(b)–(d) for the function $f(x) = \text{rec}\,(x)$ and in Figure 5.9(b)–(d) for $f(x) = \text{rec}\,(x + 1) + \text{rec}\,(x - 1)$. In both cases, the interval used was $[-L, L] = [-8, 8]$, and 1024 points were used. These graphs were obtained by following the procedure for graphing convolutions described in the user's manual for *FAS* (see Appendix A).
Here is another example.

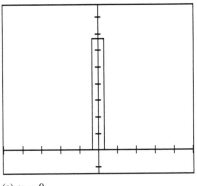

(a) $y = 0$
X interval: $[-8, 8]$ X increment = 1.6
Y interval: $[-.2, 1.3]$ Y increment = .15

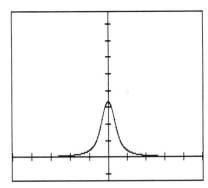

(b) $y = 0.5$
X interval: $[-8, 8]$ X increment = 1.6
Y interval: $[-.2, 1.3]$ Y increment = .15

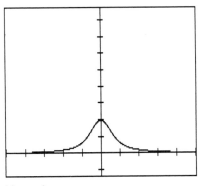

(c) $y = 1$
X interval: $[-8, 8]$ X increment = 1.6
Y interval: $[-.2, 1.3]$ Y increment = .15

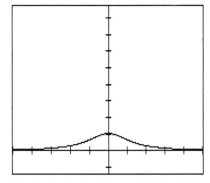

(d) $y = 2$
X interval: $[-8, 8]$ X increment = 1.6
Y interval: $[-.2, 1.3]$ Y increment = .15

FIGURE 5.8
Graph of a solution $H(x, y)$ to (5.1a) and (5.1b) for $y = 0$, 0.5, 1, and 2.

Example 5.24
Suppose that $f(x) = \text{rec}\,(x)$ and $g(x) = \text{rec}\,(x - 1)$. Then

$$(f * g)(x) = (g * f)(x) = \int_{-\infty}^{\infty} \text{rec}\,(x - s)\text{rec}\,(s - 1)\,ds$$

$$= \int_{.5}^{1.5} \text{rec}\,(x - s)\,ds.$$

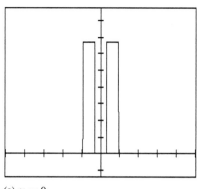

(a) $y = 0$
X interval: $[-8, 8]$ X increment $= 1.6$
Y interval: $[-.2, 1.3]$ Y increment $= .15$

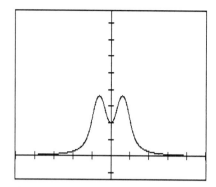

(b) $y = 0.5$
X interval: $[-8, 8]$ X increment $= 1.6$
Y interval: $[-.2, 1.3]$ Y increment $= .15$

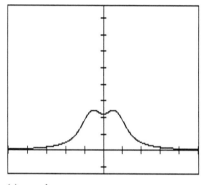

(c) $y = 1$
X interval: $[-8, 8]$ X increment $= 1.6$
Y interval: $[-.2, 1.3]$ Y increment $= .15$

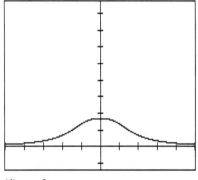

(d) $y = 2$
X interval: $[-8, 8]$ X increment $= 1.6$
Y interval: $[-.2, 1.3]$ Y increment $= .15$

FIGURE 5.9
Graph of a solution $H(x, y)$ to (5.1a) and (5.1b) for $y = 0$, 0.5, 1, and 2.

By substituting $v = x - s$ we obtain

$$(f * g)(x) = \int_{x-1.5}^{x-.5} \text{rec}\,(v)\, dv.$$

We leave it as an exercise for the reader to finish the calculation, obtaining

$$(f * g)(x) = \begin{cases} 0 & \text{if } x < 0 \text{ or } x > 2 \\ x & \text{if } 0 \le x \le 1 \\ 2 - x & \text{if } 1 \le x \le 2. \end{cases} \qquad (5.25)$$

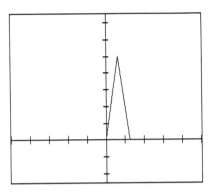

X interval: $[-8, 8]$ X increment = 1.6
Y interval: $[-.5, 1.5]$ Y increment = .2

FIGURE 5.10
Convolution of rec(x) and rec($x - 1$).

We show graphs of this exact expression for $f * g$ and the *FAS* calculated version in Figure 5.10. The difference is hardly noticeable when using the interval $[-8, 8]$ and 1024 points. We have calculated the maximum difference between the two graphs, and it is about 0.0078. ▯

---

6   The convolution theorem

In this section we shall discuss the *convolution theorem*, which generalizes the work done in the previous section. To motivate this theorem, look again at (5.13) and (5.20). The function $H(x, y) = (f *_y P)(x)$ is a convolution, and it was obtained by inverse Fourier transforming the product $\hat{f}(u)e^{-2\pi|u|y}$. In general, whenever a product is Fourier transformed, a convolution results. More precisely, we have the following theorem.

THEOREM 6.1   CONVOLUTION
The inverse Fourier transform of $\hat{f}\hat{g}$ is the convolution $f * g$.

PROOF   By Fourier inversion, we have

$$\mathcal{F}^{-1}(\hat{f}\hat{g}) = \int_{-\infty}^{\infty} \hat{f}(u)\hat{g}(u)e^{i2\pi ux}\, du$$

$$= \int_{-\infty}^{\infty} \left[\int_{-\infty}^{\infty} f(s)e^{-i2\pi us}\, ds \right] \hat{g}(u)e^{i2\pi ux}\, du.$$

Interchanging integrals and combining exponentials, we obtain

$$\mathcal{F}^{-1}(\hat{f}\hat{g})(x) = \int_{-\infty}^{\infty} f(s) \left[\int_{-\infty}^{\infty} \hat{g}(u) e^{i2\pi u(x-s)} \, du \right] ds. \qquad (6.2)$$

The integral in brackets in (6.2) is a Fourier inversion integral of $\hat{g}(u)$, with variable $x - s$ instead of x, so we will get $g(x - s)$ for that integral. Thus, we have shown that

$$\mathcal{F}^{-1}(\hat{f}\hat{g}) = f * g. \quad \blacksquare \qquad (6.3)$$

By applying the Fourier transform $\mathcal{F}$ to both sides of (6.3) we get the following theorem, which is also called the convolution theorem.

THEOREM 6.4 CONVOLUTION
The Fourier transform of $f * g$ is $\hat{f}\hat{g}$.

There are many important applications of these convolution theorems. First, we will describe another application to a partial differential equation problem, similar to the one solved in Section 5.

Example 6.5

Find a function $F(x, y)$ that satisfies the following heat equation problem ($-\infty < x < \infty,\ t > 0$)

$$\frac{\partial F}{\partial t} = a^2 \frac{\partial^2 F}{\partial x^2} \qquad \text{(heat equation, diffusion constant } a^2) \qquad (6.5a)$$

$$F(x, 0) = f(x) \qquad \text{(initial condition).} \qquad (6.5b)$$

SOLUTION We proceed similarly to the solution of (5.1a) and (5.1b). Define $\hat{F}(u, t)$ by

$$\hat{F}(u, t) = \int_{-\infty}^{\infty} F(x, t) e^{-i2\pi u x} \, dx. \qquad (6.6)$$

Hence, by Fourier inversion,

$$F(x, t) = \int_{-\infty}^{\infty} \hat{F}(u, t) e^{i2\pi u x} \, du. \qquad (6.7)$$

By differentiating under the integral sign, we have

$$\frac{\partial F(x, t)}{\partial t} = \int_{-\infty}^{\infty} \frac{\partial \hat{F}(u, t)}{\partial t} e^{i2\pi u x} \, du$$

$$a^2 \frac{\partial^2 F(x, t)}{\partial x^2} = \int_{-\infty}^{\infty} \hat{F}(u, t) \left[-(2\pi a)^2 u^2 \right] e^{i2\pi u x} \, du. \qquad (6.8)$$

Substituting the integrals from (6.8) back into (6.5a) and rearranging terms into one integral we have

$$\int_{-\infty}^{\infty} \left[\frac{\partial \hat{F}(x,t)}{\partial t} + (2\pi a)^2 u^2 \hat{F}(u,t) \right] e^{i2\pi ux} \, du = 0. \qquad (6.9)$$

Since the Fourier inverse of the function in brackets in (6.9) is the 0-function, we conclude that the function in brackets is also the 0-function. Therefore,

$$\frac{\partial \hat{F}(u,t)}{\partial t} + (2\pi a)^2 u^2 \hat{F}(u,t) = 0. \qquad (6.10)$$

Now, by putting $t = 0$ in (6.6), we conclude from (6.5b) that

$$\hat{F}(u,0) = \hat{f}(u). \qquad (6.11)$$

Hence, our transformed problem is

$$\frac{\partial \hat{F}(u,t)}{\partial t} = -(2\pi a)^2 u^2 \hat{F}(u,t) \qquad (6.12a)$$

$$\hat{F}(u,0) = \hat{f}(u). \qquad (6.12b)$$

Fixing a value of u turns (6.12a) into an *ordinary* differential equation

$$\frac{d\hat{F}}{dt} = -(2\pi a)^2 u^2 \hat{F}$$

which has solution

$$\hat{F}(u,t) = A \, e^{-(2\pi a)^2 u^2 t}. \qquad (6.13)$$

Putting $t = 0$ in (6.13) and using (6.12b) yields $A = \hat{f}(u)$. Hence, (6.7) becomes

$$F(x,t) = \int_{-\infty}^{\infty} \hat{f}(u) e^{-(2\pi a)^2 u^2 t} e^{i2\pi ux} \, du. \qquad (6.14)$$

Now we can apply the convolution theorem (Theorem 6.1) to express $F(x,t)$ as a convolution. In (6.14) we have a Fourier inversion of a product of two transforms. These transforms are $f \xrightarrow{\mathcal{F}} \hat{f}$ and

$$\frac{1}{(4\pi a^2 t)^{\frac{1}{2}}} e^{-\pi x^2/(4\pi a^2 t)} \xrightarrow{\mathcal{F}} e^{-(2\pi a)^2 u^2 t}.$$

Thus, by the convolution theorem (Theorem 6.1), our solution $F(x,t)$ to (6.5a) and (6.5b) is

$$F(x,t) = (f * {}_t H)(x) \qquad (6.15)$$

where

$$ {}_t H(x) = \frac{1}{(4\pi a^2 t)^{\frac{1}{2}}} e^{-\pi x^2/(4\pi a^2 t)}. \quad \blacksquare \qquad (6.16)$$

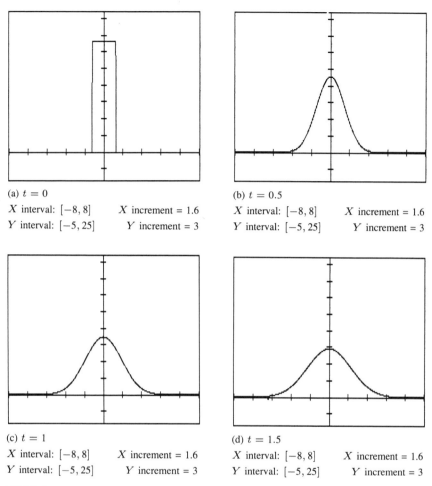

(a) $t = 0$
X interval: $[-8, 8]$ X increment = 1.6
Y interval: $[-5, 25]$ Y increment = 3

(b) $t = 0.5$
X interval: $[-8, 8]$ X increment = 1.6
Y interval: $[-5, 25]$ Y increment = 3

(c) $t = 1$
X interval: $[-8, 8]$ X increment = 1.6
Y interval: $[-5, 25]$ Y increment = 3

(d) $t = 1.5$
X interval: $[-8, 8]$ X increment = 1.6
Y interval: $[-5, 25]$ Y increment = 3

FIGURE 5.11
Graph of a solution $F(x, t)$ to (6.5a) and (6.5b) for $t = 0$, 0.5, 1, and 1.5.

It is easy to calculate computer approximations to (6.15) using *FAS*. For example, suppose that $f(x) = 20 \operatorname{rec}(x/2)$ and $a^2 = 1$. We will use for our second function

$$\frac{1}{(4\pi t)^{\frac{1}{2}}} e^{-x^2/(4t)}$$

for $t = 0.5$, 1.0, and 1.5. If we use 1024 points and the interval $[-8, 8]$, then we obtain the graphs shown in Figure 5.11. These graphs can be interpreted as the evolution of temperature, from the initial temperature $f(x) = 20 \operatorname{rec}(x/2)$, through a homogeneous insulated rod.

Here is another application of convolution.

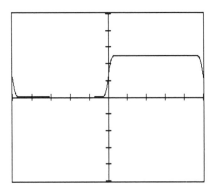

X interval: $[-16, 16]$ X increment = 3.2
Y interval: $[-2, 2]$ Y increment = .4

FIGURE 5.12
Convolution of $e^{-\pi x^2}$ and Heaviside step function over $[-16, 16]$ using 1024 points.

Example 6.17

In probability theory, a *probability density function* (p.d.f.) is a nonnegative function f satisfying $\int_{-\infty}^{\infty} f(x)\,dx = 1$. For example, $f(x) = e^{-\pi x^2}$ is a p.d.f. For a p.d.f. there is associated a *cumulative distribution function* (*distribution,* for short) defined by $F(x) = \int_{-\infty}^{x} f(s)\,ds$. Graph the distribution $F(x)$ for p.d.f. $f(x) = e^{-\pi x^2}$.   □

SOLUTION   We leave it as an exercise for the reader to show that $F(x) = (f * \mathcal{H})(x)$ where $\mathcal{H}$ is the *Heaviside step function* defined by

$$\mathcal{H}(x) = \begin{cases} 0 & \text{if } x < 0 \\ 1 & \text{if } x > 0. \end{cases} \tag{6.18}$$

By doing a convolution of $f(x) = e^{-\pi x^2}$ with $\mathcal{H}(x)$, over $[-16, 16]$ using 1024 points, one obtains the graph shown in Figure 5.12. For reasons that we will explain below, there is an undesirable behavior at the ends of the interval $[-16, 16]$ (e.g., the graph obviously decreases near ± 16, but $F(x)$ *never* decreases, because its derivative is $e^{-\pi x^2}$ which is always positive). Nevertheless, if we change the x-interval to, say, $[-10, 10]$, then we get the graph shown in Figure 5.13 and this graph is a good approximation to the distribution $F(x)$ for the p.d.f. $e^{-\pi x^2}$ over the interval $[-10, 10]$.   ∎

REMARK 6.19   As we saw in Example 6.17, the computer approximation of the convolution of $e^{-\pi x^2}$ with $\mathcal{H}(x)$ has some undesirable behavior at the ends of the interval $[-16, 16]$. The reason for this is that, as we showed in Section 7 of

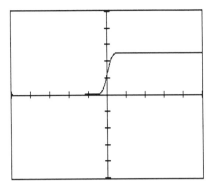

X interval: $[-10, 10]$ X increment $= 2$

Y interval: $[-2, 2]$ Y increment $= .4$

FIGURE 5.13
Approximate distribution for the p.d.f. $e^{-\pi x^2}$.

Chapter 4, an FFT algorithm is used for approximating the *periodic* (or *cyclic*) convolution

$$(f * g)(x) = \frac{1}{2L} \int_{-L}^{L} f(s)g(x - s)\, ds \tag{6.20}$$

where f and g are periodic functions with period $2L$ (in the example above, $L = 16$). What is graphed in Figure 5.12 is $2L = 32$ times the *periodic* convolution in (6.20), where f and g are the *periodic extensions* of the restrictions to $[-16, 16]$ of $e^{-\pi x^2}$ and $\mathcal{H}(x)$. To see why the troublesome behavior near the endpoints ± 16 occurs, we show in Figure 5.14(a) what the components of (6.20) look like for $x = 15.75$. From Figure 5.14(a) it is clear that there will only be a *partial overlap* of $e^{-\pi x^2}$ and *not* the more complete overlap that the nonperiodic convolution would have [as shown in Figure 5.14(b)]. However, as we show in Figure 5.15, the periodic and nonperiodic overlaps will be much closer when $x = 9.75$ (due to $e^{-\pi x^2}$ being so near 0 when $|x| \geq 10$). ∎

7 Filtering

In this section, we introduce the concept of *filtering* a function before Fourier transforming it. This technique is widely used in signal processing. Let's begin with an example.

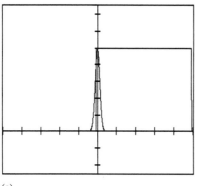

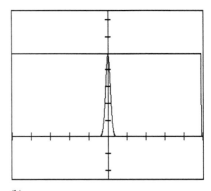

(a)

X interval: $[-16, 16]$	X increment $= 3.2$
Y interval: $[-.5, 1.5]$	Y increment $= .2$

(b)

X interval: $[-16, 16]$	X increment $= 3.2$
Y interval: $[-.5, 1.5]$	Y increment $= .2$

FIGURE 5.14
(a) Graph of components of cyclic (periodic) convolution of $e^{-\pi x^2}$ and Heaviside step function over $[-16, 16]$ for $x = 15.75$. (b) Graph of the same components for the convolution over **R** for $x = 15.75$.

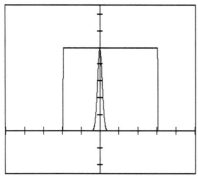

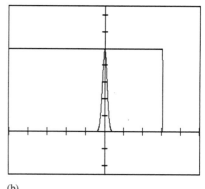

(a)

X interval: $[-16, 16]$	X increment $= 3.2$
Y interval: $[-.5, 1.5]$	Y increment $= .2$

(b)

X interval: $[-16, 16]$	X increment $= 3.2$
Y interval: $[-.5, 1.5]$	Y increment $= .2$

FIGURE 5.15
(a) Graph of components of cyclic (periodic) convolution of $e^{-\pi x^2}$ and Heaviside step function over $[-16, 16]$ for $x = 9.75$. (b) Graph of the same components for the convolution over **R** for $x = 9.75$.

Example 7.1

Suppose the function $f(x) = \cos(.8\pi x)$ is Fourier transformed over the interval $[-32, 32]$ using 1024 points. The result is shown in Figure 5.16. Notice that the frequency 0.4 that characterizes $f(x)$ is clearly marked by a *peak* at the point 0.4. However, there is a very noticeable oscillation about the x-axis away from this peak. This phenomenon is known as *leakage*. Leakage occurs because

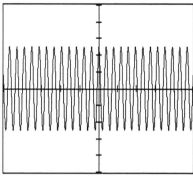

(a) cos(.8πx)

X interval: [−32, 32]   X increment = 6.4
Y interval: [−2, 2]   Y increment = .4

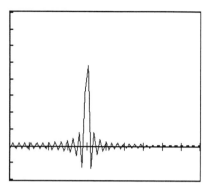

(b) transform of cos(.8πx)

X interval: [0, 1]   X increment = .1
Y interval: [−10, 40]   Y increment = 5

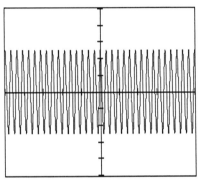

(c) cos(πx)

X interval: [−32, 32]   X increment = 6.4
Y interval: [−2, 2]   Y increment = .4

(d) transform of cos(πx)

X interval: [0, 1]   X increment = .1
Y interval: [−10, 40]   Y increment = 5

FIGURE 5.16
Fourier transforms of two cosine functions over the interval $[-32, 32]$ **using 1024 points.**

the FFT method yields approximations of

$$\left\{ \hat{f}\left(\frac{k}{\Omega}\right) \right\}_{k=-\frac{1}{2}N}^{k=\frac{1}{2}N} = \left\{ \hat{f}\left(\frac{k}{64}\right) \right\}_{k=-512}^{k=512}.$$

However, the frequency 0.4 does not equal $k/64$ for any integer k. This failure to realize a precise frequency location results in leakage. That's the explanation on the frequency side. On the other hand, from the spatial (x-variable) side, if we look at the graph of $f(x)$ in Figure 5.16(a) we see that it does not fit

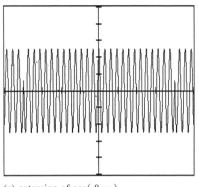

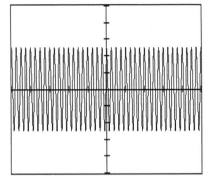

(a) extension of $\cos(.8\pi x)$ (b) extension of $\cos(\pi x)$

X interval: $[-40, 40]$ X increment = 8 X interval: $[-40, 40]$ X increment = 8

Y interval: $[-2, 2]$ Y increment = .4 Y interval: $[-2, 2]$ Y increment = .4

FIGURE 5.17
Periodic extensions, period 64, of two cosine functions (defined initially on
$[-32, 32]$**).**

very well in the window determined by $-32 \leq x \leq 32$. In fact, its periodic extension beyond $[-32, 32]$ *no longer represents the actual function* $f(x)$. See Figure 5.17(a). When this occurs, it is said that the *effect of the window becomes visible*. In fact, we can see that the pattern of the transform in Figure 5.16(b) looks a lot like the transform of rec $(x/64)$.

To see this last point more clearly, it helps to look at the function $g(x) = \cos \pi x$ [see Figure 5.16(c)] for which leakage does not occur. As we can see in Figure 5.16(d), the graph of $\hat{g}(u)$ has a peak centered on the frequency 0.5 and *no leakage*. For this example, $k/64 = 0.5$ has the solution $k = 32$ so the frequency 0.5 belongs to the set $\{k/64\}_{k=-512}^{k=512}$. On the other hand, as we show in Figure 5.17(b), the periodic extension of $g(x)$ beyond $[-32, 32]$ still represents the function $g(x)$. And, in this case, no window effects are visible.

Based on these considerations, a standard method of reducing leakage is to multiply the original function by a function that damps down to 0 at the edges of the x-interval *and whose transform does not exhibit the oscillations that the transform of* rec $(x/64)$ *does*. For example, if we transform the function $h(x) = \cos(.8\pi x)[.5 + .5\cos(\pi x/32)]$, then the transform $\hat{h}$ shown in Figure 5.18(b) results. Or, if we transform the function $G(x) = \cos(.8\pi x)e^{-.005\pi x^2}$, then the transform $\hat{G}$ shown in Figure 5.18(d) results. We say that $\hat{h}(u)$ is a *hanning filtered Fourier transform* of $f(x)$ and that $\hat{G}(u)$ is a *Gauss filtered Fourier transform of* $f(x)$.   ◻

A filtered Fourier transform is defined as follows.

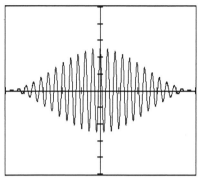

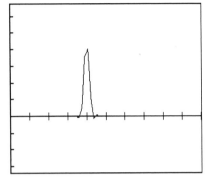

(a) cos(.8πx), hanning filtered

X interval: [−32, 32]   X increment = 6.4
Y interval: [−2, 2]   Y increment = .4

(b) hanning filtered transform

X interval: [0,1]   X increment = .1
Y interval: [−12, 24]   Y increment = 3.6

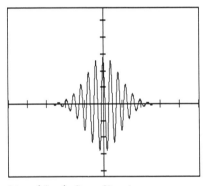

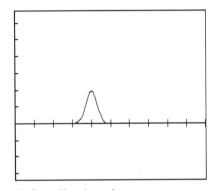

(c) cos(.8πx), Gauss filtered

X interval: [−32, 32]   X increment = 6.4
Y interval: [−2, 2]   Y increment = .4

(d) Gauss filtered transform

X interval: [0,1]   X increment = .1
Y interval: [−12, 24]   Y increment = 3.6

FIGURE 5.18
Some filtered Fourier transforms.

DEFINITION 7.2   *If $\|f\|_1$ is finite and F is a bounded, continuous function, then the **filtered Fourier transform** of f (using the filter function F) is*

$$\int_{-\infty}^{\infty} F(s)f(s)e^{-i2\pi us}\,ds.$$

When we use *FAS* to graph filtered Fourier transforms, we employ the following approximation (if $[-L, L]$ is the chosen interval)

$$\int_{-\infty}^{\infty} F(s)f(s)e^{-i2\pi us}\,ds \approx \int_{-L}^{L} F(s)f(s)e^{-i2\pi us}\,ds. \tag{7.3}$$

When using the interval $[-L, L]$, the most common filters are

Filter name	Filter function, $F(s)$
Cesàro	$1 - |s/L|$
hanning	$.5 + .5\cos(\pi s/L)$
Hamming	$.54 + .46\cos(\pi s/L)$
Gauss	$\exp[-T(\pi s/L)^2]$
Parzen	$1 - (N/(N+1))|s/L|$
Welch	$1 - (N/(N+1))^2 |s/L|^2$

(7.4)

(*Note*: The parameter T in the Gauss filter denotes the *damping constant*. The parameter N in the Parzen and Welch filters denotes the number of points used by *FAS*.)

It is also possible to use *FAS* to design your own filter function. To see how to do this, we first have to rewrite the filters in (7.4) in a more normalized form. If we substitute x in place of s/L, then each of the functions in (7.4) are even functions of x over the interval $[-1, 1]$, in which case they need only be specified over the interval $[0, 1]$. Employing this procedure, we express (7.4) as

Filter name	Filter function
Cesàro	$1 - x$
hanning	$.5 + .5\cos(\pi x)$
Hamming	$.54 + .46\cos(\pi x)$
Gauss	$\exp[-T(\pi x)^2]$
Parzen	$1 - (N/(N+1))x$
Welch	$1 - (N/(N+1))^2 x^2$

(7.5)

FAS also has the feature that you can *design your own filters*. To design your own filter you must specify some function, say $g(x)$, over the interval $[0, 1]$; *FAS* will then convert automatically to $F(s) = g(s/L)$ and use an FFT to approximate the right-hand side of (7.3). To start this process, you just choose *user* on the filter menu.

We will close this section with another illustration of filtering.

Example 7.6 NOISE SUPPRESSION

If you compute the Fourier transform of $\sin(\exp(x^2))$ over $[-4, 4]$, using 1024 points, then the graph shown in Figure 5.19(a) results. This graph is a model for a *noisy signal*. Many signals have the kind of background noise apparent in Figure 5.19(a). If we apply a Gauss filter, damping constant = 2, to $\sin(\exp x^2)$, then we obtain the graph shown in Figure 5.19(b). *The noisy background has been removed.*

In practice this procedure would run as follows:

Step 1. Measure noisy signal.

Step 2. Perform FFT.

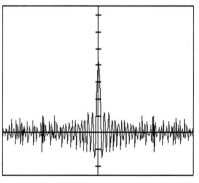

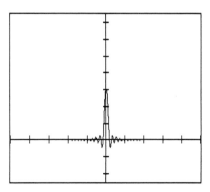

(a) Noisy signal
X interval: $[-20, 20]$ X increment = 4
Y interval: $[-1, 3]$ Y increment = .4

(b) Filtered signal
X interval: $[-20, 20]$ X increment = 4
Y interval: $[-1, 3]$ Y increment = .4

FIGURE 5.19
Filtering of a noisy signal.

Step 3. Multiply by filter.

Step 4. Perform inverse FFT.

If the signal is susceptible, then the reconstructed signal obtained at the end of this procedure will have a much less noisy background.   □

8 Poisson summation

In this section, we will explore the relationship between Fourier series and Fourier transforms. This relationship is governed by a process known as *Poisson summation*.

Suppose first that f is a function that is 0 outside of the interval $[-L, L]$. Then the Fourier transform of f reduces to

$$\hat{f}(u) = \int_{-L}^{L} f(x)e^{-i2\pi ux}\, dx. \tag{8.1}$$

Substituting $n/2L$ for u in (8.1) and multiplying by $1/2L$ we get

$$\frac{1}{2L}\hat{f}\left(\frac{n}{2L}\right) = \frac{1}{2L}\int_{-L}^{L} f(x)e^{-in\pi x/L}\, dx. \tag{8.2}$$

Formula (8.2) says that *the Fourier series coefficients of $f(x)$ over the interval* $[-L, L]$ *are given by* $\left\{(1/2L)\hat{f}(n/2L)\right\}_{n=-\infty}^{\infty}$ *where $\hat{f}$ is the Fourier transform* *of* f. Or, expressed another way,

$$f \sim \sum_{n=-\infty}^{\infty} \left[\frac{1}{2L} \hat{f} \left(\frac{n}{2L} \right) \right] e^{in\pi x/L} \tag{8.3}$$

is the Fourier series expansion, period $2L$, of a function f that is 0 outside the interval $[-L, L]$.

Formula (8.3) can be generalized beyond the case of a function that is 0 outside of a finite interval. One version of this is the following theorem.

THEOREM 8.4   POISSON SUMMATION
Suppose that $\|f\|_1$ is finite. If $f_\mathbf{P}$ is defined by

$$f_\mathbf{P}(x) = \sum_{n=-\infty}^{\infty} f(x - 2nL)$$

then $f_\mathbf{P}$ is a periodic function, period $2L$, and has Fourier series

$$f_\mathbf{P} \sim \sum_{n=-\infty}^{\infty} \frac{1}{2L} \hat{f} \left(\frac{n}{2L} \right) e^{in\pi x/L}. \tag{8.4a}$$

REMARK 8.5   (a) The function $f_\mathbf{P}$ is called the *periodization* of f. (b) If f is 0 outside of $[-L, L]$, then $f_\mathbf{P}$ is just the periodic extension of f and (8.4a) reduces to (8.3).   ∎

PROOF OF THEOREM   Clearly, $f_\mathbf{P}$ has period $2L$. To prove that the Fourier series expansion in (8.4a) is correct, let c_n stand for the nth Fourier series coefficient of $f_\mathbf{P}$, then

$$c_n = \frac{1}{2L} \int_{-L}^{L} f_\mathbf{P}(x) e^{-in\pi x/L} \, dx$$

$$= \frac{1}{2L} \int_{-L}^{L} \left[\sum_{k=-\infty}^{\infty} f(x - 2kL) e^{-in\pi x/L} \right] dx$$

$$= \sum_{k=-\infty}^{\infty} \left[\frac{1}{2L} \int_{-L}^{L} f(x - 2kL) e^{-in\pi x/L} \, dx \right].$$

Now, substituting $x + 2kL$ in place of x, we get

$$c_n = \sum_{k=-\infty}^{\infty} \frac{1}{2L} \int_{2kL-L}^{2kL+L} f(x) e^{-in\pi x/L} \, dx$$

$$= \frac{1}{2L} \int_{-\infty}^{\infty} f(x) e^{-i2\pi(n/2L)x} \, dx = \frac{1}{2L} \hat{f} \left(\frac{n}{2L} \right).$$

Thus, the Fourier series coefficient c_n of $f_{\mathbf{P}}$ equals $(1/2L)\hat{f}(n/2L)$. This says that (8.4a) holds and our theorem is proved. ∎

There is an important corollary to this theorem, which we will also call *Poisson summation*.

THEOREM 8.6   POISSON SUMMATION
Suppose that f is continuous and $\|f\|_1$ is finite. If

$$\sum_{n=-\infty}^{\infty} f(x - 2nL)$$

converges uniformly for $|x| \leq L$ and

$$\sum_{n=-\infty}^{\infty} \frac{1}{2L}\left|\hat{f}\left(\frac{n}{2L}\right)\right|$$

converges, then

$$\sum_{n=-\infty}^{\infty} f(x - 2nL) = \sum_{n=-\infty}^{\infty} \frac{1}{2L}\hat{f}\left(\frac{n}{2L}\right) e^{in\pi x/L}. \qquad (8.6a)$$

PROOF   By Theorem 8.4, $f_{\mathbf{P}}(x) = \sum_{n=-\infty}^{\infty} f(x - 2nL)$ has the right side of (8.6a) as its Fourier series. Since the sum defining $f_{\mathbf{P}}(x)$ converges uniformly and each term is a continuous function, a well-known theorem of advanced calculus says that $f_{\mathbf{P}}(x)$ is a continuous function. Also, since

$$\sum_{n=-\infty}^{\infty} \frac{1}{2L}\left|\hat{f}\left(\frac{n}{2L}\right)\right|$$

converges, the Weierstrass M-test says that the Fourier series

$$\sum_{n=-\infty}^{\infty} \frac{1}{2L}\hat{f}\left(\frac{n}{2L}\right) e^{in\pi x/L}$$

converges uniformly to a continuous function. Moreover, we know that this continuous function has the same Fourier coefficients as $f_{\mathbf{P}}(x)$. These two functions must therefore be the same. Thus, (8.6a) holds. ∎

We close this section with a simple example of Poisson summation. In the next two sections we will examine some further applications.

Example 8.7

Since $(1/\pi)\rho/(\rho^2 + x^2) \xrightarrow{\mathcal{F}} e^{-2\pi\rho|u|}$, we have by Theorem 8.6

$$\sum_{n=-\infty}^{\infty} \frac{\rho}{\rho^2 + (x-n)^2} = \sum_{n=-\infty}^{\infty} \pi e^{-2\pi\rho|n|} e^{i2n\pi x}. \tag{8.8}$$

If, for example, we put $x = 0$ and $\rho = 1$ in (8.8) we obtain

$$\sum_{n=-\infty}^{\infty} \frac{1}{1+n^2} = \sum_{n=-\infty}^{\infty} \pi e^{-2\pi|n|}. \tag{8.9}$$

It is interesting that by using geometric series, one can determine the precise sums of the right-hand sides of (8.8) and (8.9). We leave it as an exercise for the reader to show that

$$\sum_{n=-\infty}^{\infty} \frac{\rho}{\rho^2 + (x-n)^2} = \frac{\pi(1 - e^{-4\pi\rho})}{2 - 2e^{-2\pi\rho}\cos 2\pi x} \tag{8.10}$$

$$\sum_{n=-\infty}^{\infty} \frac{1}{1+n^2} = \frac{\pi}{2}(1 + e^{-2\pi}). \quad \square \tag{8.11}$$

9 An example of Poisson summation in quantum mechanics

In this section we will discuss an example from quantum mechanics that makes use of Poisson summation.

Example 9.1

Draw graphs that approximate ψ, where ψ satisfies

$$i\frac{\partial\psi}{\partial t} = -\frac{\hbar}{2m}\frac{\partial^2\psi}{\partial t^2} \qquad \text{(Schrödinger's equation)} \tag{9.1a}$$

$$\psi(x,0) = f(x) \qquad \text{(initial condition)} \tag{9.1b}$$

for the function $f(x) = 0.5\,\text{rec}\,(x/4)$ and times $t = 0.1$ and 0.2 sec. Assume that m is the mass of an electron. $\quad \square$

SOLUTION   We will make use of the solution found for Example 6.5. If we multiply both sides of Schrödinger's equation in (9.1a) by $-i$ then we get

$$\frac{\partial \psi}{\partial t} = a^2 \frac{\partial^2 \psi}{\partial x^2}, \qquad \left(a^2 = \frac{i\hbar}{2m} \right) \tag{9.2a}$$

$$\psi(x, 0) = f(x). \tag{9.2b}$$

Therefore, using the solution (6.14) found for (6.5a) and (6.5b), *but now replacing a^2 by $i\hbar/2m$*, we get

$$\psi(x, t) = \int_{-\infty}^{\infty} \hat{f}(u) e^{-(2\pi u)^2 (i\hbar/2m)t} e^{i2\pi u x} \, du. \tag{9.3}$$

And, instead of (6.15) and (6.16), we have

$$\psi(x, t) = (f *_t F)(x) \tag{9.4}$$

where

$$_t F(x) = \frac{1}{[4\pi(i\hbar/2m)t]^{\frac{1}{2}}} e^{-\pi x^2/(4\pi ti\hbar/2m)}. \tag{9.5}$$

We will now show how to approximate the integral in (9.3) by a *filtered Fourier series partial sum*.

If we take Ω sufficiently large, then we can approximate the integral $\int_{-\infty}^{\infty}$ in (9.3) by $\int_{-\Omega}^{\Omega}$, obtaining

$$\psi(x, t) \approx \int_{-\Omega}^{\Omega} \hat{f}(u) e^{-(2\pi u)^2 (i\hbar/2m)t} e^{i2\pi u x} \, du. \tag{9.6}$$

Now, if we approximate the integral in (9.6) by a sum (using the points $u = n\Omega/M$ for $n = -M, \ldots, 0, \ldots, M$), then we obtain

$$\psi(x, t) \approx \sum_{n=-M}^{M} \hat{f}\left(n\frac{\Omega}{M} \right) e^{-(2\pi n\Omega/M)^2 (i\hbar/2m)t} e^{i2\pi n\Omega x/M} \frac{\Omega}{M}. \tag{9.7}$$

If we *define* L by $L = M/2\Omega$, then we can rewrite (9.7) as

$$\psi(x, t) \approx \sum_{n=-M}^{M} \left[\frac{1}{2L} \hat{f}\left(\frac{n}{2L} \right) \right] e^{-i(\hbar/2m)(n\pi/L)^2 t} e^{i\pi n x/L}. \tag{9.8}$$

By comparing the coefficients $\{(1/2L)\hat{f}(n/2L)\}$ with the coefficients in the Fourier series in (8.3), we see that $\{(1/2L)\hat{f}(n/2L)\}$ are *Fourier series* coefficients for f *provided $f = 0$ outside $[-L, L]$.* Therefore, the series on the right side of (9.8) is a *filtered Fourier series partial sum*, using M harmonics. In fact, we have already discussed how to graph it in Section 3 of Chapter 4. If we examine formula (3.9a) of Chapter 4, we see that we are dealing with exactly the same series as in (9.8) since the Fourier coefficient c_n is $(1/2L)\hat{f}(n/2L)$ for each n. Using *FresCos* and *FresSin* filters with wave constant $\hbar/2m = 0.578$

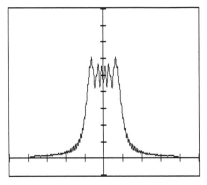

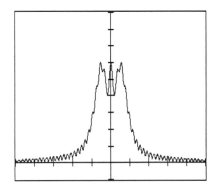

(a) $t = 0.1$ sec
X interval: $[-10, 10]$ X increment = 4
Y interval: $[-.1, .9]$ Y increment = .1

(b) $t = 0.2$ sec
X interval: $[-10, 10]$ X increment = 4
Y interval: $[-.1, .9]$ Y increment = .1

FIGURE 5.20
Graphs of $|\psi(x,t)|$ for Example 9.1.

and time constants $t = 0.1$ and $t = 0.2$ sec, we obtain the graphs for $|\psi|$ shown in Figure 5.20. For those graphs we used $M = 512$ harmonics, 4096 points, and $\Omega = 8$. (*Note*: The interval used for the Fourier series calculations was $[-L, L] = [-32, 32]$ *not* $[-\Omega, \Omega] = [-8, 8]$.) ∎

REMARK 9.9   By rewriting (9.4) and (9.5) we have

$$\psi(x,t) = \frac{1}{[4\pi(i\hbar/2m)t]^{\frac{1}{2}}} \int_{-\infty}^{\infty} f(s)e^{-\pi(x-s)^2/(4\pi ti\hbar/2m)} \, ds \qquad (9.10)$$

which we can use to explain why $|\psi(x,t)|$ has a *stability of form* (asymptotic behavior) as $t \to \infty$ (see Exercise 3.12, Chapter 4). To be precise, we will explain why

$$|\psi(x,t)| \approx \frac{1}{[4\pi(\hbar/2m)t]^{\frac{1}{2}}} \left| \hat{f}\left(\frac{x}{4\pi(\hbar/2m)t} \right) \right| \qquad (9.11)$$

as $t \to \infty$. To simplify notation, we define $q(t)$ by

$$q(t) = 4\pi i(\hbar/2m)t \qquad (9.12)$$

and observe that $1/q(t) \to 0$ as $t \to \infty$. If we expand the exponential inside the integral in (9.10), we get

$$e^{-\pi(x-s)^2/q(t)} = e^{-\pi x^2/q(t)}e^{-\pi s^2/q(t)}e^{2\pi xs/q(t)}. \qquad (9.13)$$

Now, suppose that for some positive number P,

$$f(s) = 0, \quad \text{for } |s| \geq \tfrac{1}{2}P \tag{9.14}$$

which is the assumption that we used in our example above (and it was also satisfied in the examples discussed in Chapter 4, Section 3, too). If we now let $t \to \infty$, we will have for all $|s| \leq \tfrac{1}{2}P$ *as soon as t is sufficiently large*

$$e^{-\pi s^2/q(t)} \approx e^0 = 1 \tag{9.15}$$

hence

$$e^{-\pi(x-s)^2/q(t)} \approx e^{-\pi x^2/q(t)} e^{2\pi x s/q(t)}. \tag{9.16}$$

Using (9.16) and (9.14) back in (9.10), and also using the definition of $q(t)$, we have

$$\psi(x,t) = \frac{1}{q(t)^{\frac{1}{2}}} \int_{-\frac{1}{2}P}^{\frac{1}{2}P} f(s) e^{-\pi(x-s)^2/q(t)} \, ds$$

$$\approx \frac{1}{q(t)^{\frac{1}{2}}} \int_{-\frac{1}{2}P}^{\frac{1}{2}P} f(s) e^{-\pi x^2/q(t)} e^{2\pi x s/q(t)} \, ds. \tag{9.17}$$

Factoring $e^{-\pi x^2/q(t)}$ outside of this last integral yields

$$\psi(x,t) \approx \frac{e^{-\pi x^2/q(t)}}{q(t)^{\frac{1}{2}}} \int_{-\frac{1}{2}P}^{\frac{1}{2}P} f(s) e^{2\pi x s/q(t)} \, ds. \tag{9.18}$$

And, using (9.14) again, we have

$$\psi(x,t) \approx \frac{e^{-\pi x^2/q(t)}}{q(t)^{\frac{1}{2}}} \int_{-\infty}^{\infty} f(s) e^{2\pi x s/q(t)} \, ds. \tag{9.19}$$

Replacing $q(t)$ by $4\pi i(\hbar/2m)t$ inside the integral yields

$$\psi(x,t) \approx \frac{e^{-\pi x^2/q(t)}}{q(t)^{\frac{1}{2}}} \int_{-\infty}^{\infty} f(s) e^{-i2\pi s[x/(4\pi(\hbar/2m)t)]} \, ds. \tag{9.20}$$

Thus,

$$\psi(x,t) \approx \frac{e^{-\pi x^2/q(t)}}{q(t)^{\frac{1}{2}}} \hat{f}\left(\frac{x}{4\pi(\hbar/2m)t}\right). \tag{9.21}$$

And, since $|e^{i\phi}| = 1$ for *all* real ϕ, we have

$$\left| e^{-\pi x^2/q(t)} \right| = \left| e^{i\pi x^2/[4\pi(\hbar/2m)t]} \right| = 1.$$

Therefore, we obtain

$$|\psi(x,t)| \approx \frac{1}{|q(t)|^{\frac{1}{2}}} \left| \hat{f}\left(\frac{x}{4\pi(\hbar/2m)t}\right) \right|. \tag{9.22}$$

And replacing $|q(t)|^{1/2}$ by $[4\pi(\hbar/2m)t]^{1/2}$ yields

$$|\psi(x,t)| \approx \frac{1}{[4\pi(\hbar/2m)t]^{\frac{1}{2}}} \left| \hat{f}\left(\frac{x}{4\pi(\hbar/2m)t}\right) \right| \qquad (9.23)$$

for sufficiently large t. Formula (9.23) is the *asymptotic form* for $|\psi(x,t)|$ as $t \to \infty$. As $t \to \infty$, the *magnitude* of ψ approaches a scaled version of $|\hat{f}|$. ∎

10 Summation kernels arising from Poisson summation

In this section we will show how a summation kernel (see Section 9, Chapter 4) can be analyzed using Poisson summation. As an example, we will show that the hanning kernel is a summation kernel.

For simplicity of notation we will assume throughout this section that we are dealing with Fourier series with period 2π.

Using the functional notation for filter coefficients introduced in Section 5 of Chapter 4, let's suppose that the *PSF* $\mathcal{P}_M$ is defined by

$$\mathcal{P}_M(x) = \sum_{n=-M}^{M} F\left(\frac{n}{M}\right) e^{inx} \qquad (10.1)$$

where F is an even, continuous function on $[-1,1]$. Extending F to be 0 outside of the interval $[-1,1]$, we can rewrite (10.1) as

$$\mathcal{P}_M(x) = \sum_{n=-\infty}^{\infty} F\left(\frac{n}{M}\right) e^{inx}. \qquad (10.2)$$

Notice that F is still an even function.

We will now show that Poisson summation can be applied to (10.2) so that

$$\mathcal{P}_M(x) = \sum_{n=-\infty}^{\infty} M\hat{F}\left(\frac{M(x-2n\pi)}{2\pi}\right). \qquad (10.3)$$

Moreover, we can relax the requirement that F be 0 outside of $[-1,1]$. Here is the theorem.

THEOREM 10.4
Suppose that F is even and continuous and that F and its Fourier transform $\hat{F}$ satisfy

$$|F(x)| \leq A\left(1+|x|\right)^{-1-\alpha} \qquad (10.5)$$

$$|\hat{F}(u)| \leq B\left(1+|u|\right)^{-1-\beta} \qquad (10.6)$$

for some positive constants A, α, B, β. Then, for all x in **R**,

$$\sum_{n=-\infty}^{\infty} F\left(\frac{n}{M}\right) e^{inx} = \sum_{n=-\infty}^{\infty} M\hat{F}\left(\frac{M(x-2n\pi)}{2\pi}\right). \qquad (10.7)$$

REMARK 10.8 Notice that when F is continuous and is 0 outside of $[-1,1]$, then inequality (10.5) is definitely satisfied for some A and α. ∎

PROOF OF THEOREM Since F is even, it follows that $\hat{F}$ is also even. Moreover, by Fourier inversion,

$$\hat{\hat{F}}(x) = F(-x) = F(x),$$

so F is the Fourier transform of $\hat{F}$. Inequality (10.5) insures that $\sum_{n=-\infty}^{\infty} |F(n/M)|$ converges. In fact, by (10.5), we have

$$\sum_{n=-\infty}^{\infty} \left| F\left(\frac{n}{M}\right) \right| \leq \sum_{n=-\infty}^{\infty} \frac{A}{(1+|n/M|)^{1+\alpha}}$$

and the sum on the right side converges *for each fixed M*, by the integral test from calculus. Also, inequality (10.6) insures that

$$\sum_{n=-\infty}^{\infty} M\hat{F}\left(\frac{M(x-2n\pi)}{2\pi}\right)$$

converges uniformly for all x in $[-\pi, \pi]$. In fact, for $|n| \geq 1$

$$M\left|\hat{F}\left(\frac{M(x-2n\pi)}{2\pi}\right)\right| \leq BM\frac{(2\pi)^{1+\beta}}{(2\pi+M|x-2n\pi|)^{1+\beta}}. \qquad (10.9)$$

Since $|n| \geq 1$ and x is in $[-\pi, \pi]$, it follows that

$$|x - 2n\pi| \geq |n|\pi \qquad (10.10)$$

by a simple consideration of distance on the real line. Thus, for $|n| \geq 1$

$$M\left|\hat{F}\left(\frac{M(x-2n\pi)}{2\pi}\right)\right| \leq BM\frac{(2\pi)^{1+\beta}}{(2\pi+M\pi|n|)^{1+\beta}}. \qquad (10.11)$$

Since

$$\sum_{|n|\geq 1} BM\frac{(2\pi)^{1+\beta}}{(2\pi+M\pi|n|)^{1+\beta}} = \sum_{n=1}^{\infty} 2BM\frac{(2\pi)^{1+\beta}}{(2\pi+M\pi n)^{1+\beta}} \qquad (10.12)$$

converges (by the integral test), it follows from the Weierstrass M-test that

$$\sum_{n=-\infty}^{\infty} M\hat{F}\left(\frac{M(x-2n\pi)}{2\pi}\right)$$

converges uniformly for all x in $[-\pi, \pi]$.

We have now verified all the conditions for Poisson summation to be applied to the series $\sum_{n=-\infty}^{\infty} F(n/M)e^{inx}$ which defines $\mathcal{P}_M(x)$. Since we showed above that $\hat{\hat{F}} = F$, we have by change of scale

$$M\hat{F}\left(\frac{Mx}{2\pi}\right) \xrightarrow{\mathcal{F}} 2\pi F\left(\frac{2\pi u}{M}\right).$$

Consequently, Theorem 8.6 yields (for $2L = 2\pi$)

$$\sum_{n=-\infty}^{\infty} M\hat{F}\left(\frac{M(x-2n\pi)}{2\pi}\right) = \sum_{n=-\infty}^{\infty} F\left(\frac{n}{M}\right)e^{inx}$$

which is the same as (10.7). ∎

Based on Theorem 10.4 we can prove the following theorem, which tells us when the *PSF* $\mathcal{P}_M$ in (10.2) will be a summation kernel.

THEOREM 10.13
Suppose that F is even and continuous, that $F(0) = 1$, and that (10.5) and (10.6) are valid. Then, the kernel $\mathcal{P}_M$ defined by

$$\mathcal{P}_M(x) = \sum_{n=-\infty}^{\infty} F\left(\frac{n}{M}\right)e^{inx}$$

is a summation kernel.

PROOF We need to check the three properties (a)–(c) in (9.1) from Chapter 4 (for $L = 2\pi$).

(a) By integrating term by term

$$\frac{1}{2\pi}\int_{-\pi}^{\pi} \mathcal{P}_M(x)\,dx = \sum_{n=-\infty}^{\infty} F\left(\frac{n}{M}\right)\frac{1}{2\pi}\int_{-\pi}^{\pi} e^{inx}\,dx$$

$$= F\left(\frac{0}{M}\right) \cdot 1 = 1$$

so (a) holds.

(b) We know from Theorem 10.4 that (10.7) is true. Making use of (10.6), and also (10.11) and (10.12), we have

$$
\begin{aligned}
|\mathcal{P}_M(x)| &\leq \sum_{n=-\infty}^{\infty} \left| M\hat{F}\left(\frac{M(x-2n\pi)}{2\pi}\right) \right| \\
&= \left| M\hat{F}\left(\frac{Mx}{2\pi}\right) \right| + \sum_{|n|\geq 1} \left| M\hat{F}\left(\frac{M(x-2n\pi)}{2\pi}\right) \right| \\
&\leq \left| M\hat{F}\left(\frac{Mx}{2\pi}\right) \right| + 2\sum_{n=1}^{\infty} BM\frac{(2\pi)^{1+\beta}}{(2\pi+M\pi n)^{1+\beta}} \cdot
\end{aligned}
\tag{10.14}
$$

Since the numerical series

$$
2\sum_{n=1}^{\infty} BM\frac{(2\pi)^{1+\beta}}{(2\pi+M\pi n)^{1+\beta}}
$$

converges, let's denote its sum by the constant S. Then, by (10.6) applied to $|M\hat{F}(Mx/2\pi)|$ we can rewrite (10.14) as

$$
|\mathcal{P}_M(x)| \leq B\frac{M(2\pi)^{1+\beta}}{(2\pi+M|x|)^{1+\beta}} + S.
$$

Consequently,

$$
\begin{aligned}
\frac{1}{2\pi}\int_{-\pi}^{\pi} |\mathcal{P}_M(x)|\, dx &\leq \frac{B}{2\pi}\int_{-\pi}^{\pi} \frac{M(2\pi)^{1+\beta}}{(2\pi+M|x|)^{1+\beta}}\, dx + \frac{1}{2\pi}\int_{-\pi}^{\pi} S\, dx \\
&= \frac{B}{2\pi}\int_{-M\pi}^{M\pi} \frac{(2\pi)^{1+\beta}}{(2\pi+|t|)^{1+\beta}}\, dt + S \\
&\leq \frac{B}{2\pi}\int_{-\infty}^{\infty} \frac{(2\pi)^{1+\beta}}{(2\pi+|t|)^{1+\beta}}\, dt + S = \frac{2B}{\beta} + S.
\end{aligned}
$$

Putting C equal $2B/\beta + S$, we see that (b) holds.

(c) We now assume that some $\delta > 0$, has been given and $\delta \leq |x| \leq \pi$. By (10.14) and (10.6) applied to $|M\hat{F}(Mx/2\pi)|$ we have

$$
\begin{aligned}
|\mathcal{P}_M(x)| &\leq B\frac{M(2\pi)^{1+\beta}}{(2\pi+M|x|)^{1+\beta}} + 2\sum_{n=1}^{\infty} \frac{BM(2\pi)^{1+\beta}}{(2\pi+M\pi n)^{1+\beta}} \\
&\leq B\frac{M(2\pi)^{1+\beta}}{(M\delta)^{1+\beta}} + 2\sum_{n=1}^{\infty} \frac{BM(2\pi)^{1+\beta}}{(M\pi n)^{1+\beta}} \\
&= \frac{1}{M^{\beta}}\left[\frac{B(2\pi)^{1+\beta}}{\delta^{1+\beta}} + 2\sum_{n=1}^{\infty} \frac{B2^{1+\beta}}{n^{1+\beta}}\right].
\end{aligned}
\tag{10.15}
$$

Since the series

$$2\sum_{n=1}^{\infty} \frac{B2^{1+\beta}}{n^{1+\beta}}$$

converges, we can represent the quantity in brackets in (10.15) by a constant, let's call it D. Then (10.15) becomes

$$|\mathcal{P}_M(x)| \leq \frac{D}{M^\beta}, \qquad (\delta \leq |x| \leq \pi). \tag{10.16}$$

Letting $M \to \infty$ in (10.16) we see that the right side of the inequality tends to 0. Hence, for any given $\epsilon > 0$, if M is taken sufficiently large we will have $D/M^\beta < \epsilon$ and then

$$|\mathcal{P}_M(x)| < \epsilon, \qquad (\delta \leq |x| \leq \pi) \tag{10.17}$$

so (c) is true. ∎

As an application of this theory we will now show that the hanning kernel is a summation kernel.

Example 10.18
The hanning kernel, defined by

$$\mathcal{P}_M(x) = \sum_{|n| \leq M} \left[0.5 + 0.5 \cos\left(\frac{\pi n}{M}\right) \right] e^{inx}, \tag{10.19}$$

is a summation kernel over $[-\pi, \pi]$. □

SOLUTION For the kernel $\mathcal{P}_M$ defined by (10.19) we have that $\mathcal{P}_M(x) = \sum_{n=-\infty}^{\infty} F(n/M)e^{inx}$ where the function F is defined by

$$F(x) = \begin{cases} 0.5 + 0.5\cos(\pi x) & \text{for } |x| \leq 1 \\ 0 & \text{for } |x| > 1. \end{cases} \tag{10.20}$$

It is easy to see that this function F is continuous and even, and that $F(0) = 1$. Since F is 0 outside of $[-1, 1]$ and continuous, we also have (10.5). In particular, if we take $A = 4$ and $\alpha = 1$ it is easy to check that (10.5) holds (just draw graphs of $F(x)$ and $4[1 + |x|]^{-2}$). We also have that

$$F(x) = 0.5 \operatorname{rec}(x/2) + 0.25 \operatorname{rec}(x/2)\left[e^{i\pi x} + e^{-i\pi x}\right] \tag{10.21}$$

so, by scaling and modulation properties,

$$\hat{F}(u) = \operatorname{sinc}(2u) + 0.5 \operatorname{sinc}\left[2\left(u - \frac{1}{2}\right)\right] + 0.5 \operatorname{sinc}\left[2\left(u + \frac{1}{2}\right)\right]. \tag{10.22}$$

Replacing sinc v by $(\sin \pi v)/(\pi v)$ and simplifying yields

$$\hat{F}(u) = \frac{-\pi \sin(2\pi u)}{(2\pi u)(4\pi^2 u^2 - \pi^2)} = \left[\frac{-\text{sinc}\,(2u)}{\pi}\right]\left[\frac{1}{4u^2 - 1}\right]. \tag{10.23}$$

Since $|\text{sinc}\,v| \leq 1$ for all v in $\mathbf{R}$, it follows from (10.23) that (10.6) holds when $\beta = 1$ and B is chosen sufficiently large (since $\lim_{|u|\to\infty}(1+|u|)^2/(4u^2-1) = 1/4$).

Since (10.5) and (10.6) are valid, it follows from Theorem 10.13 that the hanning kernel is a summation kernel. ∎

Here is another example.

Example 10.24
The kernel $\mathcal{P}_M$ defined by

$$\mathcal{P}_M(x) = \sum_{n=-\infty}^{\infty} e^{-(n\pi/M)^2} e^{inx}$$

is a summation kernel over $[-\pi, \pi]$. ☐

SOLUTION The function $F(x) = e^{-\pi x^2}$ satisfies all the requirements of Theorem 10.13. ∎

---

11 The sampling theorem

This section describes a very important concept in signal processing, known as the *sampling theorem*. The theory of Poisson summation and the theory of sampling are closely related. We begin with the definition of those functions to which the sampling theory applies.

DEFINITION 11.1 *A function f is called **band limited** if its Fourier transform $\hat{f}$ is 0 outside of a finite interval $[-L, L]$.*

The sampling theorem says that a band-limited function can be recovered from its *samples*

$$\left\{f\left(\frac{n}{2L}\right)\right\}_{n=-\infty}^{n=\infty}$$

provided $\hat{f}$ is 0 outside of the interval $[-L, L]$. This theorem has many important applications. For instance, the construction of compact disc players uses sampling theory (see [Mo]). In this case, the frequencies that the human ear is sensitive to lie in a finite range (20–20,000 Hz) and, consequently, the recorded music can be digitized (sampled) effectively. In fact, there is software available now that allows a PC to be used for digitally analyzing and synthesizing music. Another application of sampling theory is to the sending of multiple telephone messages along a *single* cable. By digitizing the messages (sampling at points separated by $1/2L$) gaps are created within which other digitized messages can be sent. Using a fiber optic cable, upwards of 8000 simultaneous messages can be transmitted. Besides illustrating the wonder of fiber optics, this result also illustrates the power of the sampling theorem.

Here is the sampling theorem.

THEOREM 11.2
Suppose that f is band limited. If $\hat{f}$ is 0 outside of $[-L, L]$, then

$$f(x) = \sum_{n=-\infty}^{\infty} f\left(\frac{n}{2L}\right) \operatorname{sinc}(2Lx - n). \tag{11.2a}$$

PROOF   We will prove a very general version of the sampling theorem which yields (11.2a) as a special case.

Since $\hat{f}$ is 0 outside of $[-L, L]$, we can periodically extend it. In fact, if we use $\hat{f}_{\mathbf{P}}$ to denote the periodic extension of $\hat{f}$ with period $2L$, then

$$\hat{f}_{\mathbf{P}}(u) = \sum_{n=-\infty}^{\infty} \hat{f}(u - 2nL). \tag{11.3}$$

By Poisson summation, we know that the Fourier series for $\hat{f}_{\mathbf{P}}$ is [use $\hat{f}$ in place of f in (8.4a)]

$$\hat{f}_{\mathbf{P}} \sim \sum_{n=-\infty}^{\infty} \frac{1}{2L} \hat{\hat{f}}\left(\frac{n}{2L}\right) e^{in\pi u/L}.$$

Since $\hat{\hat{f}}(x) = f(-x)$, we have

$$\hat{f}_{\mathbf{P}}(u) \sim \sum_{n=-\infty}^{\infty} \frac{1}{2L} f\left(\frac{-n}{2L}\right) e^{in\pi u/L}. \tag{11.4}$$

Now, suppose we multiply (11.4) by a function $W(u)$, known in sampling theory as a *window function*, which satisfies

$$W(u) = \begin{cases} 1 & \text{when } \hat{f}(u) \neq 0 \\ 0 & \text{when } |u| > L \end{cases} \tag{11.5}$$

and is bounded for all other u-values. For such a window function, we have

$$\hat{f}_P(u)W(u) = \hat{f}(u)W(u) = \hat{f}(u) \tag{11.6}$$

because $\hat{f}(u) = 0$ for $|u| > L$.

For example, if we take

$$W(u) = \begin{cases} 1 & \text{when } |u| \leq L \\ 0 & \text{when } |u| > L \end{cases} \tag{11.7}$$

then (11.5) and (11.6) hold.

Multiplying both sides of (11.4) by $W(u)$, and using (11.6), we get

$$\hat{f}(u) \sim \sum_{n=-\infty}^{\infty} \frac{1}{2L} f\left(\frac{-n}{2L}\right) W(u)e^{in\pi u/L}. \tag{11.8}$$

Multiplying both sides of (11.8) by $e^{i2\pi ux}$ and integrating with respect to u from $-L$ to L, we obtain the following equality:

$$\int_{-L}^{L} \hat{f}(u)e^{i2\pi ux}\, du = \sum_{n=-\infty}^{\infty} f\left(\frac{-n}{2L}\right) \frac{1}{2L} \int_{-L}^{L} W(u)e^{i2\pi(x+n/2L)u}\, du. \tag{11.9}$$

We will explain why there is equality in (11.9) in Remark 11.15 below. But first, let's just assume that the equality is valid.

If we define $\mathbf{S}(x)$ by

$$\mathbf{S}(x) = \frac{1}{2L} \int_{-L}^{L} W(u)e^{i2\pi ux}\, du \tag{11.10}$$

then (11.9) becomes

$$\int_{-L}^{L} \hat{f}(u)e^{i2\pi ux}\, du = \sum_{n=-\infty}^{\infty} f\left(\frac{-n}{2L}\right) \mathbf{S}\left(x + \frac{n}{2L}\right). \tag{11.11}$$

Since $\hat{f}$ is 0 outside of $[-L, L]$, we have

$$\int_{-L}^{L} \hat{f}(u)e^{i2\pi ux}\, du = \int_{-\infty}^{\infty} \hat{f}(u)e^{i2\pi ux}\, du = f(x) \tag{11.12}$$

by Fourier inversion. Consequently, (11.11) becomes

$$f(x) = \sum_{n=-\infty}^{\infty} f\left(\frac{-n}{2L}\right) \mathbf{S}\left(x + \frac{n}{2L}\right). \tag{11.13}$$

Substituting $-n$ in place of n, and noting that the sum in (11.13) is over *all* integers, we obtain

$$f(x) = \sum_{n=-\infty}^{\infty} f\left(\frac{n}{2L}\right) \mathbf{S}\left(x - \frac{n}{2L}\right). \tag{11.14}$$

Formula (11.14) describes a very general sampling theorem. The data specific to the function f is the set of sample values

$$\left\{ f\left(\frac{n}{2L}\right)\right\}_{n=-\infty}^{\infty}$$

The *reconstruction function* $\mathbf{S}$ is defined by formula (11.10).
   If we take for W the function in (11.7), then we have

$$\mathbf{S}(x) = \frac{1}{2L}\int_{-L}^{L} e^{i2\pi ux}\, du = \operatorname{sinc}(2Lx)$$

and (11.14) reduces to (11.2a).   ∎

REMARK 11.15   In the proof above we did not explain why (11.9) holds. We will now prove the following more precise version of that equality:

$$\int_{-L}^{L}\hat{f}(u)e^{i2\pi ux}\, du$$

$$= \lim_{M\to\infty}\sum_{n=-M}^{M} f\left(\frac{-n}{2L}\right)\frac{1}{2L}\int_{-L}^{L} W(u)e^{i2\pi(x+n/2L)u}\, du \quad (11.16)$$

To prove (11.16) we use $\hat{f}(u) = \hat{f}(u)W(u)$ from (11.6), and we have

$$\left|\int_{-L}^{L}\hat{f}(u)e^{i2\pi ux}\, du - \sum_{n=-M}^{M} f\left(\frac{-n}{2L}\right)\frac{1}{2L}\int_{-L}^{L} W(u)e^{i2\pi(x+n/2L)u}\, du\right|$$

$$= \left|\int_{-L}^{L}\left[\hat{f}(u) - \sum_{n=-M}^{M}\frac{1}{2L}f\left(\frac{-n}{2L}\right)e^{in\pi u/L}\right]W(u)e^{i2\pi ux}\, du\right|$$

$$\le \int_{-L}^{L}\left|\hat{f}(u) - \sum_{n=-M}^{M}\frac{1}{2L}f\left(\frac{-n}{2L}\right)e^{in\pi u/L}\right||W(u)|\, du. \quad (11.17)$$

Now, using the symbol $S_M^{\hat{f}}(u)$ in place of the M-harmonic Fourier series partial sum for $\hat{f}$, we obtain from (11.17)

$$\left|\int_{-L}^{L}\hat{f}(u)e^{i2\pi ux}\, du - \sum_{n=-M}^{M} f\left(\frac{-n}{2L}\right)\frac{1}{2L}\int_{-L}^{L} W(u)e^{i2\pi(x+n/2L)u}\, du\right|$$

$$\le \int_{-L}^{L}\left|\hat{f}(u) - S_M^{\hat{f}}(u)\right||W(u)|\, du \le \left\|\hat{f} - S_M^{\hat{f}}\right\|_2 \|W\|_2, \quad (11.18)$$

the last inequality being the well-known *Schwarz inequality* (see [Wa, Chapter 2.4]) where the 2-Norms are taken over the interval $[-L, L]$.

If we now use the completeness relation from Theorem 3.17 in Chapter 4, we have (since $\hat{f}$ is bounded by $\|f\|_1$ its 2-Norm over $[-L, L]$ is finite)

$$\lim_{M \to \infty} \left\| \hat{f} - S_M^{\hat{f}} \right\|_2 = 0. \tag{11.19}$$

Comparing (11.19) with the last inequality in (11.18) we see that (11.16) holds.

Formula (11.16) justifies using (11.9) in the proof of the sampling theorem. Moreover, using the definition of **S** in (11.10), and formula (11.12), in inequality (11.18), we have also shown that for *all* x in **R**

$$\left| f(x) - \sum_{n=-M}^{M} f\left(\frac{n}{2L}\right) \mathbf{S}\left(x - \frac{n}{2L}\right) \right| \le \left\| \hat{f} - S_M^{\hat{f}} \right\|_2 \|W\|_2. \tag{11.20}$$

Inequality (11.20) is a useful estimate for the magnitude of the difference between f and a *partial sum* of the reconstruction series for f given in (11.14).

∎

REMARK 11.21 For a band-limited function f, a frequently used value of L is the *smallest possible* value for which $\hat{f}(u) = 0$ when $|u| \ge L$. In this case, one must use the window defined in (11.7), and the sampling series must be the one in (11.2a). The sampling rate, $2L$ samples/unit length, is called the *Nyquist rate* when this smallest value of L is used. ∎

The following example shows why it is sometimes advantageous to use a different window W than the one given in (11.7).

Example 11.22

Suppose $\hat{f}$ is 0 outside of $[-2, 2]$. Compare the reconstruction function **S** obtained from using (11.7) for $L = 2$, and from the window

$$W(u) = \begin{cases} 1 & \text{if } -2 \le u \le 2 \\ 0.5 + 0.5\cos\pi(u-2) & \text{if } 2 \le u \le 3 \\ 0.5 + 0.5\cos\pi(u+2) & \text{if } -3 \le u \le -2 \\ 0 & \text{if } |u| \ge 3. \end{cases} \tag{11.23}$$

SOLUTION If we use *FAS* to approximate **S** in formula (11.10), using the interval $[-16, 16]$, 1024 points, and the function in (11.23) for the window function W, then we obtain the graph shown in Figure 5.21(a). Meanwhile, the function $\text{sinc}(4x)$, which results from the window function in (11.7) for $L = 2$, is shown in Figure 5.21(b). The advantage of using the window function W in (11.23) is that its reconstruction function *damps down* (decays) more quickly away from 0.

∎

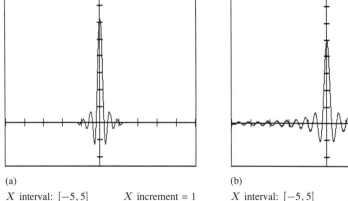

(a)

X interval: $[-5,5]$ X increment $= 1$

Y interval: $[-.5, 1.5]$ Y increment $= .2$

(b)

X interval: $[-5,5]$ X increment $= 1$

Y interval: $[-.5, 1.5]$ Y increment $= .2$

FIGURE 5.21
Graphs of two reconstruction functions.

12   Aliasing

The sampling theorem proved in the previous section applies to *band-limited* functions only. Not all functions are band limited, however. In this section we will discuss what happens if f is not band limited. It is still possible to obtain an approximate sampling theorem, if some error is allowed to exist between the original function f and its sampling series.

To begin, we will call a function *almost band limited* if there are positive constants A and α for which

$$|\hat{f}(u)| \leq A \left[1 + |u|\right]^{-1-\alpha} \tag{12.1}$$

holds for all u in **R**.

Now, suppose that f is almost band limited but *not* band limited. If we look again at (11.3), we see that $\hat{f}_{\mathbf{P}}$ is no longer the periodic extension of $\hat{f}$. For any given L, it is the *periodization* of $\hat{f}$ having period $2L$, that we defined in Remark 8.5(a). Nevertheless, by the Poisson summation theorem (Theorem 8.4), we still have formula (11.4). If we use the window function

$$W(u) = \begin{cases} 1 & \text{if } |u| \leq L \\ 0 & \text{if } |u| > L \end{cases} \tag{12.2}$$

then, multiplying (11.4) by $W(u)$ and integrating over $-L \leq u \leq L$, we obtain a new form of (11.9). Namely,

$$\int_{-L}^{L} \hat{f}_{\mathbf{P}}(u)e^{i2\pi ux} \, du$$

$$= \sum_{n=-\infty}^{\infty} f\left(\frac{-n}{2L}\right) \frac{1}{2L} \int_{-L}^{L} W(u)e^{i2\pi(x+n/2L)u} \, du. \qquad (12.3)$$

Since the window W is the same one as in (11.7), we obtain from (12.3)

$$\int_{-L}^{L} \hat{f}_{\mathbf{P}}(u)e^{i2\pi ux} \, du = \sum_{n=-\infty}^{\infty} f\left(\frac{n}{2L}\right) \operatorname{sinc}(2Lx - n). \qquad (12.4)$$

We now define the *alias* of f, denoting it by $\mathcal{A}_f$, to be

$$\mathcal{A}_f(x) = \int_{-L}^{L} \hat{f}_{\mathbf{P}}(u)e^{i2\pi ux} \, du \qquad (12.5)$$

where $\hat{f}_{\mathbf{P}}$ is defined by

$$\hat{f}_{\mathbf{P}}(u) = \sum_{n=-\infty}^{\infty} \hat{f}(u - 2nL). \qquad (12.6)$$

[*Note*: The condition (12.1) insures that the series in (12.6) converges, as shown in the proof of Theorem 10.4.]

Thus, we have from (12.5) and (12.4)

$$\mathcal{A}_f(x) = \sum_{n=-\infty}^{\infty} f\left(\frac{n}{2L}\right) \operatorname{sinc}(2Lx - n). \qquad (12.7)$$

Formula (12.7) shows that *the sampling series for f reconstructs the alias for f and not f itself.*

We will now obtain an estimate of the difference between f and its alias $\mathcal{A}_f$. We have

$$|f(x) - \mathcal{A}_f(x)| = \left| \int_{-\infty}^{\infty} \hat{f}(u)e^{i2\pi ux} \, du - \int_{-L}^{L} \hat{f}_{\mathbf{P}}(u)e^{i2\pi ux} \, du \right|. \qquad (12.8)$$

Using the definition of $\hat{f}_{\mathbf{P}}$ in (12.6), we get

$$\int_{-L}^{L} \hat{f}_{\mathbf{P}}(u)e^{i2\pi ux} \, du = \int_{-L}^{L} \sum_{n=-\infty}^{\infty} \hat{f}(u - 2nL)e^{i2\pi ux} \, du$$

$$= \sum_{n=-\infty}^{\infty} \int_{-L}^{L} \hat{f}(u - 2nL)e^{i2\pi ux} \, du.$$

Substituting $u + 2nL$ in place of u in the integral in the last sum above yields

$$\int_{-L}^{L} \hat{f}_{\mathbf{P}}(u) e^{i2\pi ux} \, du = \sum_{n=-\infty}^{\infty} \int_{-L+2nL}^{L+2nL} \hat{f}(u) e^{i2\pi ux} e^{i4\pi nLx} \, du. \tag{12.9}$$

Also, we can rewrite $\int_{-\infty}^{\infty} \hat{f}(u) e^{i2\pi ux} \, du$ in a similar way:

$$\int_{-\infty}^{\infty} \hat{f}(u) e^{i2\pi ux} \, du = \sum_{n=-\infty}^{\infty} \int_{-L+2nL}^{L+2nL} \hat{f}(u) e^{i2\pi ux} \, du. \tag{12.10}$$

Substituting from (12.9) and (12.10) into the right-hand side of (12.8) *and noting that the $n = 0$ term cancels*, we get (*Note*: the primed sums mean that the $n = 0$ term is *omitted*.)

$$|f(x) - A_f(x)| = \left| \sum_{n=-\infty}^{\infty}{}' \int_{-L+2nL}^{L+2nL} \hat{f}(u) e^{i2\pi ux} \left[1 - e^{i4\pi nLx} \right] du \right|$$

$$\leq \sum_{n=-\infty}^{\infty}{}' \int_{-L+2nL}^{L+2nL} |\hat{f}(u)| \left| 1 - e^{i4\pi nLx} \right| du$$

$$\leq \sum_{n=-\infty}^{\infty}{}' 2 \int_{-L+2nL}^{L+2nL} |\hat{f}(u)| \, du = 2 \int_{|u|>L} |\hat{f}(u)| \, du. \tag{12.11}$$

Thus, we have for *all* x in $\mathbf{R}$,

$$|f(x) - A_f(x)| \leq 2 \int_{|u|>L} |\hat{f}(u)| \, du. \tag{12.12}$$

Furthermore, if we use (12.1) we obtain

$$|f(x) - A_f(x)| \leq \frac{4A}{\alpha} \frac{1}{L+1} \tag{12.13}$$

Since $(4A/\alpha)(L+1)^{-1} \to 0$ as $L \to \infty$, inequality (12.13) shows that *for an almost band-limited function f we can always replace f with its alias A_f and have negligible error when L is taken sufficiently large.* Or, put another way, *when the sampling rate $2L$ samples/unit length is large enough, there will be negligible error between f and its sampling series.* This last result holds because (12.7) and (12.13) give

$$\left| f(x) - \sum_{n=-\infty}^{\infty} f\left(\frac{n}{2L} \right) \operatorname{sinc}(2Lx - n) \right| \leq \frac{4A}{\alpha} \frac{1}{L+1}. \tag{12.14}$$

13   Sine and cosine transforms

Besides Fourier transforms, there are two other transforms commonly used in
Fourier analysis: the Fourier sine and Fourier cosine transforms. Since the
theory of these transforms is so closely related to that of Fourier transforms, we
shall be very brief. Fourier sine and Fourier cosine transforms bear the same
relationship to the Fourier transform as Fourier sine and cosine series have to
Fourier series. In particular, if f is an odd function over **R**, then

$$\hat{f}(u) = \int_{-\infty}^{\infty} f(x) \cos 2\pi ux \, dx - i \int_{-\infty}^{\infty} f(x) \sin 2\pi ux \, dx$$

$$= -2i \int_{0}^{\infty} f(x) \sin 2\pi ux \, dx. \tag{13.1}$$

Similarly, if f is even, then

$$\hat{f}(u) = 2 \int_{0}^{\infty} f(x) \cos 2\pi ux \, dx. \tag{13.2}$$

Based on (13.1) and (13.2) we make the following definition.

DEFINITION 13.3   *If $\int_{0}^{\infty} |f(x)| \, dx$ is finite, then the **Fourier sine transform**
of f is denoted by $\hat{f}_S$ and is defined by*

$$\hat{f}_S(u) = 2 \int_{0}^{\infty} f(x) \sin 2\pi ux \, dx.$$

*The **Fourier cosine transform** of f is denoted by $\hat{f}_C$ and is defined by*

$$\hat{f}_C(u) = 2 \int_{0}^{\infty} f(x) \cos 2\pi ux \, dx.$$

Example 13.4
Calculate the Fourier cosine and Fourier sine transform of

$$f(x) = \begin{cases} 1 & \text{if } 0 < x < 1 \\ 0 & \text{if } x > 1 \end{cases} \tag{13.5}$$

and compare with graphs generated by *FAS* using 1024 points and interval $[0, 16]$.
∏

SOLUTION   For the Fourier sine transform $\hat{f}_S$ we have

$$\hat{f}_S(u) = 2 \int_{0}^{1} \sin 2\pi ux \, dx = \frac{1 - \cos 2\pi u}{\pi u} = \frac{2 \sin^2 \pi u}{\pi u}$$

$$= (\operatorname{sinc} u)(2 \sin \pi u). \tag{13.6}$$

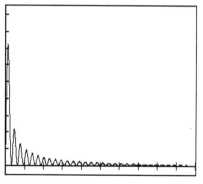

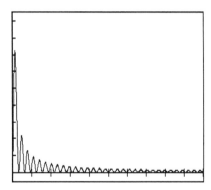

(a) *FAS* calculated sine transform

X interval: $[0, 32]$ X increment = 3.2

Y interval: $[-.1, 1.9]$ Y increment = .2

(b) exact sine transform

X interval: $[0, 32]$ X increment = 3.2

Y interval: $[-.1, 1.9]$ Y increment = .2

FIGURE 5.22
Graphs of Fourier sine transforms. In (a), the graph of *FAS* calculated transform of the function in (13.5). In (b), the graph of the exact transform of that function.

On the other hand, using *FAS*, we choose *Sine T*, an interval of $[0, 16]$, 1024 points, and then enter the function in (13.5). The graph produced by *FAS* is shown in Figure 5.22(a). For comparison, we have graphed the exact transform given in (13.6) in Figure 5.22(b).

For the cosine transform $\hat{f}_C$ we have

$$\hat{f}_C(u) = 2 \int_0^1 \cos 2\pi u x \, dx = \frac{\sin 2\pi u}{\pi u}$$

$$= 2\mathrm{sinc}(2u). \tag{13.7}$$

Using *FAS* (choosing *Cos T* and then proceeding as above for *Sine T*), we can approximate $\hat{f}_C$. The approximate graph is shown in Figure 5.23(a), while the exact transform shown in (13.7) is graphed in Figure 5.23(b).

As the reader can see, for both the Fourier sine and Fourier cosine transforms, there is a good match (at the level of resolution of the graphs shown) between the exact transforms and the computer approximations. ∎

As the reader might expect, *FAS* uses a *fast sine transform* to approximate a Fourier sine transform and a *fast cosine transform* to approximate a Fourier cosine transform. Since the theory is essentially the same as we described in Section 4 for Fourier transforms, we will omit any further discussion of the methods used.

There are also inversion theorems for Fourier sine and Fourier cosine transforms that are similar to those described for the Fourier transform. In fact,

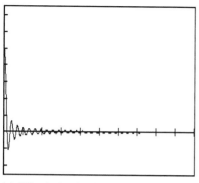

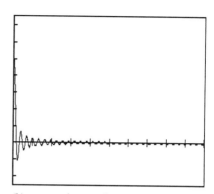

(a) *FAS* calculated cosine transform

X interval: $[0, 32]$ X increment $= 3.2$

Y interval: $[-1, 3]$ Y increment $= .4$

(b) exact cosine transform

X interval: $[0, 32]$ X increment $= 3.2$

Y interval: $[-1, 3]$ Y increment $= .4$

FIGURE 5.23
Graphs of Fourier cosine transforms. In (a), the graph of *FAS* calculated transform of the function in (13.5). In (b), the graph of the exact transform of that function.

these inversion theorems are just corollaries of the ones for the Fourier transform. The key facts are contained in formulas (13.1) and (13.2). If f_o is the *odd extension* of f, then the Fourier transform of f_o, denoted by $\hat{f}_o$, satisfies $\hat{f}_o(u) = -i\hat{f}_S(u)$. Consequently, since $\hat{f}_o(u)$ is an odd function of u, inversion of the Fourier transform $\hat{f}_o$ yields

$$f_o(x) = \int_{-\infty}^{\infty} \hat{f}_o(u)e^{i2\pi ux}\, du = 2i \int_0^{\infty} \hat{f}_o(u) \sin 2\pi ux\, dx$$

$$= 2 \int_0^{\infty} \hat{f}_S(u) \sin 2\pi ux\, dx.$$

Therefore, for $x > 0$, $f(x)$ *can be obtained by performing a Fourier sine transform of $\hat{f}_S(u)$*. Similarly, $f(x)$ *can be obtained by performing a Fourier cosine transform of $\hat{f}_C(u)$*. [*Note:* just like for the Fourier transform, sometimes $(1/2)f(x+) + (1/2)f(x-)$ is obtained by inversion, rather than $f(x)$.] We have shown that the Fourier sine and Fourier cosine transforms *are their own inverses*.

References

For further discussion of Fourier transforms, see [Sn], [Br,2], and [Wa]. More discussion of filtering can be found in [Ha], [Op-S], or [Ra-G]. For more on sampling theory, see [St] and [Je].

Exercises

Section 1
5.1 Using Definition 1.3, calculate the Fourier transforms of the following functions
  (a) $3 \operatorname{rec}(2x)$
  (b) $\operatorname{rec}(x-4) + \operatorname{rec}(x+4)$
  (c) $e^{-|x|}$
  (d) $\operatorname{rec}(x/8) \cos 2\pi x$ [*Hint*: $\cos 2\pi x = (1/2)e^{i2\pi x} + (1/2)e^{-i2\pi x}$.]
  (e) $xe^{-\pi x^2}$ [*Hint*: Integrate by parts.]

5.2 Use *FAS* to draw the graphs of the Fourier transforms (negative exponent) of the following functions. [Consult the user's manual for details of the *Four T* procedure.]
  (a) $xe^{-\pi x^2}$
  (b) $\exp(-[(x-2)/2]^{20})$
  (c) $f(x) = \begin{cases} \operatorname{sqr}(1-x^2) & \text{for } |x| < 1 \\ 0 & \text{for } |x| < 1 \end{cases}$
  (d) $f(x) = \begin{cases} 1 - |x| & \text{for } |x| \le 1 \\ 0 & \text{for } |x| > 1 \end{cases}$

5.3 Compute the transforms of the following functions using *FAS* and compare them to the exact transforms using Definition 1.3.
  (a) $f(x) = \begin{cases} 1 - 3|x| & \text{for } |x| < \frac{1}{3} \\ 0 & \text{for } |x| > \frac{1}{3} \end{cases}$
  [For $\hat{f}(u)$ you should get $(1/3)\operatorname{sinc}^2(u/3)$.]
  (b) $f(x) = \operatorname{rec}(3x)$ [For $\hat{f}(u)$ you should get $(1/3)\operatorname{sinc}(u/3)$.]
  (c) $f(x) = xe^{-2|x|}$ [For $\hat{f}(u)$ you should get $i\pi u/(1 + \pi^2 u^2)^2$.]
  (d) $(1/2)\operatorname{rec}(x-2) - (1/2)\operatorname{rec}(x+2)$ [For $\hat{f}(u)$ you should get $-i(\sin 4\pi u)(\operatorname{sinc} u)$.]

Section 2
5.4 Repeat Exercise 5.1, but now use the properties discussed in Section 2.

5.5 Suppose that f, f', and f'' are all continuous and $\|f\|_1$, $\|f'\|_1$, and $\|f''\|_1$ are all finite. Show that
$$|\hat{f}(u)| \le \frac{\|f''\|_1}{4\pi^2 u^2}, \qquad (u \ne 0). \tag{5.6}$$

Also, show that for some positive constant A (dependent on f),
$$|\hat{f}(u)| \le A\left[1 + |u|\right]^{-2}, \qquad \text{all } u \in \mathbf{R}.$$

5.7 Find the exact Fourier transforms of the following functions (check your answers with *FAS*):
  (a) $x^3 e^{-\pi x^2}$
  (b) $xe^{-2\pi |x|}$
  (c) $\begin{cases} \sin 8\pi x & \text{if } |x| < 3 \\ 0 & \text{if } |x| > 3 \end{cases}$
  (d) $e^{-|x|} \cos 6\pi x$
  (e) $e^{-|x-1|} + e^{|x+1|}$

Section 3

5.8 Check the estimates in (3.17). (See the User's Manual in Appendix A for details on how to calculate *Norms* in *FAS*.)

5.9 Repeat the estimates in (3.17), but use the function defined in (3.7) in place of rec.

5.10 Examine S_8^{rec}, S_{16}^{rec}, S_{32}^{rec}, and S_{64}^{rec} near the point $x = 0.5$ in order to observe *Gibbs' phenomenon*.

5.11 Repeat Exercise 5.10, but use the function defined in (3.7) and examine the graphs near $x = 0$.

5.12 If $e^{-2\pi|x|}$ is used instead of Λ, what number should replace 0.006 in (3.5)? (Again, use 32 for L.)

Section 4

5.13 Compare the computer calculation of Λ and its exact transform $sinc^2$.

5.14 Why is there less error involved in Exercise 5.13 than in Example 4.13?

Section 5

5.15 Verify (5.25).

5.16 Check (5.21). Also, show that $f * (g * h) = (f * g) * h$ *(associativity)*.

5.17 Using *FAS*, compute $f * f$, $f * f * f$, and $f * f * f * f$ for $f(x) = rec(x)$.

5.18 Repeat Exercise 5.17 for $f(x) = rec(x - 1)$, $f(x) = \Lambda(x - 2)$, and for $f(x) = 1/(1 + x^2)$.

5.19 Using *FAS*, compute $(f *_y P)(x)$ for $y = 0.01, 0.05, 0.25, 1.25, 6$ and $f(x) = e^{-|x|}$.

5.20 Repeat Exercise 5.19, but use $f(x) = \Lambda(x)$, and $f(x) = 1/(1 + x^2)$.

5.21 Using *FAS*, illustrate for the functions given in Exercises 5.19 and 5.20 that

$$\lim_{y \to 0+} (f *_y P)(x) = f(x).$$

(A proof of this limit can be found in [Wa, Chapter 7.1].)

5.22 Use Fourier transforms to solve the following for W (given f)

$$\frac{\partial^2 W}{\partial x^2} = \frac{\partial^2 W}{\partial t^2}, \quad W(x,0) = f(x), \quad \frac{\partial W}{\partial t}(x,0) = 0.$$

Graph this solution for $f(x) = e^{-.5|x|}$ and $t = 0.1, 0.2, 0.3, 0.4$, and 0.5.

5.23 Use Fourier transforms to solve the following for W (given g)

$$\frac{\partial^2 W}{\partial x^2} = \frac{\partial^2 W}{\partial t^2}, \quad W(x,0) = 0, \quad \frac{\partial W}{\partial t}(x,0) = g(x).$$

Graph this solution for $f(x) = e^{-.5|x|}$ and $t = 0.1, 0.2, 0.3, 0.4$, and 0.5.

Section 6

5.24 Using *FAS*, graph approximations to $(f *_t H)(x)$ for the following functions and values of a^2. Use $t = 0.1, 0.5, 1, 2$.

(a) $40 \, rec(x)$, $a^2 = 0.5$

(b) $20 \, rec(x - 2) + 20 \, rec(x + 2)$, $a^2 = 1$

(c) $20 \, rec(x/2)$, $a^2 = 4$

(d) $40/(1 + x^2)$, $a^2 = 1$

(e) $e^{-|x|}$, $a^2 = 2$

(f) $\exp(-(x/3) \wedge 10)$, $a^2 = 2$

5.25 Using *FAS*, check that $\lim_{t \to 0+}(f *_t H)(x) = f(x)$ for each of the functions in Exercise 5.24(d)–(f), check also that the limit holds for points of continuity of the functions in 5.24(a)–(c). (A proof of the limit can be found in [Wa, Chapter 7.2].)

5.26 Using the convolution theorem, prove that $_tH * _\tau H = _{(t+\tau)}H$ for each $t > 0$ and $\tau > 0$. Also, show that $_yP * _\Upsilon P = _{(y+\Upsilon)}P$ holds for each $y > 0$ and $\Upsilon > 0$. (See (5.19) for the definition of $_yP$.) Using *FAS*, check these identities (don't use huge values of t, τ, y, or Υ).

5.27 Using *FAS*, graph the distributions for the following p.d.f.s over the interval $[-10, 10]$.
   (a) $e^{-\pi x^2}$
   (b) $\text{rec}(x)$
   (c) $\Lambda(x)$
   (d) $\frac{1}{\pi}\frac{1}{1+x^2}$
   (e) $\text{rec}(x - 1)$
   (f) $\Lambda(x + 1)$

Section 7

5.28 Express the four-step process in Example 7.6 in terms of convolution. Can you give a physical interpretation of why noise suppression works? [*Hint*: Think of convolution as a kind of averaging and noise as a sort of random interference.]

5.29 Given the function $\cos 16\pi x + 0.8\cos 15\pi x - 0.6\cos 18\pi x$, perform a filtered Fourier transform on the interval $[-16, 16]$ using 1024 points. Use Parzen, hanning, Hamming, and Welch filters.

5.30 Repeat Exercise 5.29, but use $\sin 16\pi x + 0.8\cos 15\pi x - 0.3\sin 19\pi x$.

5.31 Repeat Exercise 5.29, but use the constant function $f(x) = 1$. Compare the shapes of the peaks of the filtered transforms in Exercise 5.29 with the shape of each peak in this exercise. Explain the similarity.

Section 8

5.32 Show that (8.10) and (8.11) are true. Also, find the exact sums of

$$\sum_{n=0}^{\infty} \frac{1}{1+n^2} \quad \text{and} \quad \sum_{n=1}^{\infty} \frac{1}{1+n^2}.$$

5.33 Using Poisson summation, show that for $t > 0$

$$\sum_{n=-\infty}^{\infty} e^{-2n^2\pi^2 t} e^{i2\pi nx} = \frac{1}{(2\pi t)^{\frac{1}{2}}} \sum_{n=-\infty}^{\infty} e^{-(x-n)^2/2t}. \qquad (a)$$

The identity (a) is known as *Jacobi's identity*.

5.34 Approximate $\sum_{n=-\infty}^{\infty} e^{-2n^2\pi^2 t} e^{i2\pi nx}$ for $t = 1$, using a filtered Fourier series partial sum. Using *FAS*, graph this approximation over $[-1/2, 1/2]$ with 2048 points.

5.35 Show that $\sum_{n=-\infty}^{\infty} e^{-2n^2\pi^2 t}\cos 2\pi nx > 0$ for all $x \in \mathbf{R}$ and all $t > 0$.

5.36 Find the exact value of $\sum_{n=1}^{\infty} 1/(a^2 + n^2)$ for all $a > 0$.

Section 9

5.37 Using for $f(x)$ the function $2\text{rec}(4x)$, and using 4096 points and an interval of $[-32, 32]$, check formula (9.23) on *FAS* for times $t = 0.2$, 0.4, and 0.5 (assume that m = mass of an electron, so that $\hbar/2m \approx 0.578$).

5.38 Using the convolution theorem, show that $_tF * _\tau F = _{(t+\tau)}F$ for all $t > 0$ and $\tau > 0$.

5.39 Graph approximations to $|\psi|$ for $t = 0.1$, 0.2, 0.3, and 0.4, $\hbar/2m = 0.578$, and $f(x) = e^{-\pi x^2}$. Check formula (9.23) for $t = 0.3$ and 0.4.

5.40 Show that if $f(x) = e^{-\pi x^2}$, then

$$(f * {}_tF)(x) = \frac{1}{[1 + 4\pi i(\hbar/2m)t]^{\frac{1}{2}}} e^{-\pi x^2/[1+4\pi i(\hbar/2m)t]}. \tag{a}$$

Using (a), give an alternate derivation of (9.23) for sufficiently large t.

Section 10

5.41 Use Theorem 10.13 to give another proof that the Cesàro kernel is a summation kernel over $[-\pi, \pi]$.

5.42 Show that the dlVP kernel V_M, see (4.9) and (4.10) in Chapter 4, is a summation kernel over $[-\pi, \pi]$.

5.43 Suppose that F is continuous and even, $F(0) = 1$, $F = 0$ outside of $[-1, 1]$ *and* the derivatives F' and F'' are both continuous. Show that (10.5) and (10.6) hold for this function F, hence $\mathcal{P}_M(x) = \sum_{n=-M}^{M} F(n/M)e^{inx}$ is a summation kernel. [*Hint:* Use estimates like (2.9), and (5.6) on p. 194, to bound $|\hat{F}(u)|$.]

5.44 Given a positive constant $\rho > 0$, the *Riesz kernel* is defined by

$$\mathcal{P}_M(x) = \sum_{n=-M}^{M} \left[1 - \left(\frac{n}{M}\right)^2\right]^\rho e^{inx}.$$

Use the result of Exercise 5.43 to show that the Riesz kernel is a summation kernel when $\rho \geq 2$.

REMARK It is known that the Riesz kernel is a summation kernel when $\rho > 0$. See [St-W, Chapter 7]. ∎

Section 11

5.45 For the function $f(x) = \text{sinc}^2(x)$, show that the Nyquist rate is 2. Using *FAS*, graph the approximation to $\text{sinc}^2(x)$ given by

$$\sum_{n=-9}^{9} \text{sinc}^2(n/2)\text{sinc}(2x - n)$$

over $[-8, 8]$, using 1024 points. [*Note:* Simplify the approximating sum by excluding the terms for which $\text{sinc}^2(n/2) = 0$, and replace $\text{sinc}^2(n/2)$ by $(4/\pi^2)/n^2$ when n is odd.]

5.46 Repeat Exercise 5.45 but use

$$\sum_{n=-15}^{15} \text{sinc}^2(n/2)\text{sinc}(2x - n).$$

[*Note:* $\text{sinc}(2x - n)$ may be written as \$2x - n)$ and *FAS* will accept it.]

5.47 Estimate the accuracy of the approximations in Exercises 5.45 and 5.46 using Sup-Norm and 2-Norm in *FAS*.

5.48 If $\hat{f}(u) = 0$ for $|u| > L$, explain why $f * 2L\text{sinc}(2Lx) = f$. Using *FAS*, check this result for $\text{sinc}^2(x)$ over the interval $[-8, 8]$. Use 1024 points.

Section 12

5.49 Show that for a function f satisfying (12.1)

$$\left| \mathcal{A}_f(x) - \sum_{n=-M}^{M} f\left(\frac{n}{2L}\right)\text{sinc}(2Lx - n) \right| \leq \left\| \hat{f}_\mathbf{P} - S_M^\mathbf{P} \right\|_2 \|W\|_2$$

where

$$\|W\|_2 = \left[\int_{-L}^{L} |W(u)|^2 \, du \right]^{1/2}$$

and S_M^P is the M-harmonic partial sum of the Fourier series for $\hat{f}_P$. Using this result, show that

$$\left| f(x) - \sum_{n=-M}^{M} f\left(\frac{n}{2L}\right) \text{sinc}(2Lx - n) \right| \leq \frac{4A}{\alpha} \frac{1}{L+1} + (2L)^{\frac{1}{2}} \|\hat{f}_P - S_M^P\|_2. \quad \text{(a)}$$

5.50 Using *FAS*, graph the approximation to $f(x) = 0.5 \exp(-0.25\pi x \wedge 2)$ given by

$$\sum_{n=-9}^{9} f(n)\text{sinc}(x - n).$$

Use the interval $[-8, 8]$ and 1024 points. [Also, use $x - n)$ in place of $\text{sinc}(x - n)$ and group like terms.] Use Exercise 5.49(a) to estimate the approximation error.

5.51 Repeat Exercise 5.50, but use $f(x) = (1/\pi)(1 + x \wedge 2)$.

5.52 Explain why $A_f = f * 2L\text{sinc}(2Lx)$.

5.53 Use the identity in Exercise 5.52 to draw the alias A_f of $f(x) = 0.5 \exp(-0.25\pi x \wedge 2)$. Compare this graph of the alias to the result for 5.50. (*Note:* To avoid endpoint errors, the alias graph should be computed over $[-16, 16]$ using 1024 points, then the x-interval reduced to $[-8, 8]$.)

Section 13

5.54 Using *FAS*, graph the Fourier sine and cosine transforms of the following functions (use the interval $[0, 16]$ and 512 points).

(a) $\text{rec}\,(x - 2)$
(b) $1/(4 + (x - 6)^2)$
(c) $e^{-0.4x} \cos 12\pi x$
(d) $e^{-0.4x} \sin 12\pi x$

5.55 Explain why a Fourier sine transform and a Fourier cosine transform can be approximated by a fast sine transform and a fast cosine transform, respectively.

6

Fourier Optics

In this chapter, we will discuss some applications of Fourier analysis to optics. We will show how *FAS* can be used to analyze some important problems in optics, such as the behavior of diffraction gratings and imaging with lenses.

1 Introduction, diffraction and coherency of light

In this section we will derive the fundamental formula of diffraction theory. Although we do not have space for a completely rigorous discussion (see [Go] or [Bo-W]), our treatment should capture the main ideas. If the presentation below is found too difficult, some readers might wish *to take formula (1.20) for granted* (it is a classical formula in diffraction theory); however, in that case, do read Notation 1.2, where the notation used throughout this chapter is given.

We shall assume that unpolarized light of wavelength λ traveling from a point P to a point Q along a curve $\mathbf{C}$ can be described as follows. The light amplitude $\psi(Q, t)$ at the point Q is given by

$$\psi(Q,t) = \frac{\psi(P,t)}{\lambda|\mathbf{C}|} e^{i\frac{2\pi}{\lambda}\int_{\mathbf{C}} \eta(s)\,ds} \tag{1.1}$$

where $\psi(P, t)$ is the light amplitude at P, $|\mathbf{C}|$ is the length of the curve $\mathbf{C}$, and $\int_{\mathbf{C}} \eta(s)\,ds$ is the line integral of the index of refraction $\eta(s)$ along the curve $\mathbf{C}$. [*Note*: we are omitting a *time delay* $t_{\mathbf{C}}$ in (1.1). Namely, $\psi(P, t)$ should read $\psi(P, t - t_{\mathbf{C}})$. We do this for simplicity. At the distances we shall consider, the speed of light is so great that $t_{\mathbf{C}} \approx 0$.]

Although formula (1.1) may appear strange, it will allow us to quickly derive the fundamental formulas of optical diffraction and imaging. Formula (1.1) is related to Fermat's principle and to the Feynman path integral method in optics (which is a more rigorous development of the approach described here).

A couple of examples might help to clarify the meaning of formula (1.1).

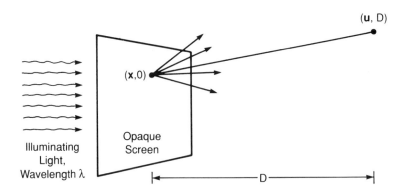

FIGURE 6.1
Diffraction from a single point.

First, though, we need to adopt some convenient notation.

NOTATION 1.2
To enable us to write our formulas compactly, we will adopt the following vector notation. A vector **x** *will have two components* $\mathbf{x} = (x, y)$ *and the point* (x, y, D) *will be denoted by* $(\mathbf{x}, D)$. *When we want to express integrals like*

$$\int_{-\infty}^{\infty} \int_{-\infty}^{\infty} f(x, y)\, dx\, dy \qquad \text{or} \qquad \int_{-\infty}^{\infty} \int_{-\infty}^{\infty} g(u, v)\, du\, dv$$

then we will write

$$\int_{\mathbf{R}^2} f(\mathbf{x})\, d\mathbf{x} \qquad \text{or} \qquad \int_{\mathbf{R}^2} g(\mathbf{u})\, d\mathbf{u}.$$

Finally, we will write $|\mathbf{u} - \mathbf{x}|$ *for the distance*

$$\left[(u - x)^2 + (v - y)^2 \right]^{\frac{1}{2}}$$

between **u** *and* **x**.

Example 1.3
Suppose that light of wavelength λ is shined onto an opaque screen with a single point $P = (\mathbf{x}, 0)$ cut out of it. See Figure 6.1. Assume that the index of refraction η is a *constant*, say $\eta = 1$ for simplicity. Then the light at a point $Q = (\mathbf{u}, D)$, on a parallel plane D units in front of the screen, will be caused by the light traveling along the straight line segment $\mathbf{C}$ from $(\mathbf{x}, 0)$ to $(\mathbf{u}, D)$. (See, however, Remark 1.7 below.) Since $\int_{\mathbf{C}} 1\, ds = |\mathbf{C}|$, we obtain from (1.1) that

$$\psi(\mathbf{u}, D, t) = \frac{\psi(\mathbf{x}, 0, t) e^{i \frac{2\pi}{\lambda} |\mathbf{C}|}}{\lambda |\mathbf{C}|}. \tag{1.4}$$

Now, we *define* the *light intensity* $I(\mathbf{u}, D)$ of the light at $(\mathbf{u}, D)$ by

$$I(\mathbf{u}, D) = \frac{1}{T} \int_0^T |\psi(\mathbf{u}, D, t)|^2 dt \qquad (1.5)$$

for some *huge* value of T. (For example, T might be the time involved in taking a photograph, which is huge relative to the fluctuations of visible light.)

Using (1.4) in (1.5), we obtain (since $|e^{i\phi}| = 1$ for all real ϕ)

$$I(\mathbf{u}, D) = \frac{I(\mathbf{x}, 0)}{\lambda^2 |\mathbf{C}|^2} . \qquad (1.6)$$

Formula (1.6) says that *the light intensity decreases in inverse proportion to the square of the distance,* $|\mathbf{C}|$, *from* $(\mathbf{x}, 0)$ *to* $(\mathbf{u}, D)$. This result is consistent with light being an electromagnetic phenomenon, and it shows the purpose of dividing by $|\mathbf{C}|$ in formula (1.1). [The reason for also dividing by λ will only be clear after our work in the next section.] ꠶

REMARK 1.7 In the example above we made the assumption that only light traveling along the *straight line segment* $\mathbf{C}$ from $(\mathbf{x}, 0)$ to $(\mathbf{u}, D)$ contributes to the light $\psi(\mathbf{u}, D, t)$ at $(\mathbf{u}, D)$. This assumption is consistent with classical geometrical optics, since the light rays are straight in a medium with constant index of refraction. However, a more rigorous approach like Feynman's path integral method would assume that the photons of light can travel *all* possible paths $\mathbf{C}$ from $(\mathbf{x}, 0)$ to $(\mathbf{u}, D)$. Then, an integral of quantities similar to those in (1.4) would be formed over *all* paths $\mathbf{C}$ from $(\mathbf{x}, 0)$ to $(\mathbf{u}, D)$. We will not pursue this approach here; however, since *when a long time average is taken* (as we did above) the results of this more rigorous method end up reducing to formulas like the ones described in this section. For a fascinating nontechnical discussion of Feynman's method, see [Fe]. ▮

This last example showed why the factor $1/|\mathbf{C}|$ is in formula (1.1). Our next example will show why the *phase factor*

$$e^{i\frac{2\pi}{\lambda} \int_{\mathbf{C}} \eta(s) \, ds}$$

is in formula (1.1).

Example 1.8

Suppose that light of wavelength λ is shined onto an opaque screen with two points cut out of it. Let's say these points are at $(\mathbf{x}, 0)$ and $(\mathbf{z}, 0)$. Assuming again that we have a constant index of refraction $\eta = 1$, we now must sum the contributions of the light from $(\mathbf{x}, 0)$ and $(\mathbf{z}, 0)$. If we let $\mathbf{C_x}$ be the line segment from $(\mathbf{x}, 0)$ to $(\mathbf{u}, D)$, then (1.4) generalizes to

$$\psi(\mathbf{u}, D, t) = \frac{\psi(\mathbf{x}, 0, t)}{\lambda |\mathbf{C_x}|} e^{i\frac{2\pi}{\lambda}|\mathbf{C_x}|} + \frac{\psi(\mathbf{z}, 0, t)}{\lambda |\mathbf{C_z}|} e^{i\frac{2\pi}{\lambda}|\mathbf{C_z}|} . \tag{1.9}$$

We will now calculate the intensity $I(\mathbf{u}, D)$ using formula (1.5). First, we observe that

$$|\psi(\mathbf{u}, D, t)|^2 = \frac{|\psi(\mathbf{x}, 0, t)|^2}{\lambda^2 |\mathbf{C_x}|^2} + \frac{|\psi(\mathbf{z}, 0, t)|^2}{\lambda^2 |\mathbf{C_z}|^2}$$
$$+ 2Re\left[\frac{\psi(\mathbf{x}, 0, t)\psi^*(\mathbf{z}, 0, t)}{\lambda^2 |\mathbf{C_x}| \cdot |\mathbf{C_z}|} e^{i\frac{2\pi}{\lambda}(|\mathbf{C_x}| - |\mathbf{C_z}|)}\right] . \tag{1.10}$$

Since our next step will be to compute $1/T$ times the integral $\int_0^T dt$ of the right side of (1.10), the following definition will be useful. The *cross-correlation function* $\Gamma(\mathbf{x}, \mathbf{z})$ is defined by

$$\Gamma(\mathbf{x}, \mathbf{z}) = \frac{1}{T} \int_0^T \psi(\mathbf{x}, 0, t)\psi^*(\mathbf{z}, 0, t)\, dt. \tag{1.11}$$

With this definition in mind, we now use (1.5) to calculate the intensity $I(\mathbf{u}, D)$ from (1.10). We obtain

$$I(\mathbf{u}, D) = \frac{I(\mathbf{x}, 0)}{\lambda^2 |\mathbf{C_x}|^2} + \frac{I(\mathbf{z}, 0)}{\lambda^2 |\mathbf{C_z}|^2} + 2Re\left[\frac{\Gamma(\mathbf{x}, \mathbf{z})e^{i\frac{2\pi}{\lambda}(|\mathbf{C_x}| - |\mathbf{C_z}|)}}{\lambda^2 |\mathbf{C_x}| \cdot |\mathbf{C_z}|}\right] . \tag{1.12}$$

Formula (1.12) is very complicated, reflecting the fact that we placed no constraints upon the nature of the light radiating onto the opaque screen (other than it is of one wavelength λ). There are two fundamental constraints that can be used to simplify (1.12). The first is called *incoherency*. The light will be called *incoherent* if the cross-correlation function Γ satisfies

$$\Gamma(\mathbf{x}, \mathbf{z}) = 0 \quad \text{for } \mathbf{x} \neq \mathbf{z} \qquad [\textit{Incoherency}]. \tag{1.13}$$

In this case, (1.12) simplifies to

$$\frac{I(\mathbf{x}, 0)}{\lambda^2 |\mathbf{C_x}|^2} + \frac{I(\mathbf{z}, 0)}{\lambda^2 |\mathbf{C_z}|^2} . \tag{1.14}$$

Thus, *in the case of incoherent illumination, the intensity at $(\mathbf{u}, D)$ is the sum of the intensities that would result from the points $(\mathbf{x}, 0)$ and $(\mathbf{z}, 0)$ individually.*

The second form of constraint is called *spatial coherency* (*coherency*, for short). It is coherency that we will deal with most frequently in this chapter. A simple condition specifying coherency is the following. The light will be said to be *coherent* if the cross-correlation function $\Gamma(\mathbf{x}, \mathbf{z})$ satisfies

$$\Gamma(\mathbf{x}, \mathbf{z}) = 1 \qquad [\textit{Coherency}]. \tag{1.15}$$

If the light is coherent, then using (1.15) in (1.12) yields

$$I(\mathbf{u}, D) = \frac{I(\mathbf{x}, 0)}{\lambda^2 |\mathbf{C_x}|^2} + \frac{I(\mathbf{z}, 0)}{\lambda^2 |\mathbf{C_z}|^2} + \frac{2 \cos \left[\frac{2\pi}{\lambda} (|\mathbf{C_x}|^2 - |\mathbf{C_z}|^2) \right]}{\lambda^2 |\mathbf{C_x}| \cdot |\mathbf{C_z}|}. \tag{1.16}$$

Thus, in the case of coherent illumination, we see from formula (1.16) that the intensity $I(\mathbf{u}, D)$ contains an *oscillatory term*

$$\frac{2 \cos \left[\frac{2\pi}{\lambda} (|\mathbf{C_x}| - |\mathbf{C_z}|) \right]}{\lambda^2 |\mathbf{C_x}| \cdot |\mathbf{C_z}|}.$$

As the point $(\mathbf{u}, D)$ varies this term will vary between

$$\frac{2}{\lambda^2 |\mathbf{C_x}| \cdot |\mathbf{C_z}|} \quad \text{and} \quad \frac{-2}{\lambda^2 |\mathbf{C_x}| \cdot |\mathbf{C_z}|}$$

thus creating *interference fringes* superimposed upon the background intensity described by

$$\frac{I(\mathbf{x}, 0)}{\lambda^2 |\mathbf{C_x}|^2} + \frac{I(\mathbf{z}, 0)}{\lambda^2 |\mathbf{C_z}|^2}.$$

Much experimental work has verified that *all* types of illumination will satisfy the coherency condition (at least to a good approximation) *provided* the points $(\mathbf{x}, 0)$ and $(\mathbf{z}, 0)$ are sufficiently close (how close is called the *coherency interval* of the light). Some light has such a small coherency interval that it can be assumed incoherent, while other kinds of light satisfy the coherency condition (1.15) over a wide range. For example, in Figure 6.2 we show a diagram illustrating how coherent light is created. If light from a strong source is channeled through a point drilled in an opaque screen (pinhole filter), and then collimated by a lens sitting at a focal distance from the point in the screen, then the light immediately behind the collimating lens is coherent. Often a laser is used as the initial source, since the light from the laser has a long coherency interval to begin with. □

In the following sections we will mostly examine coherent light, so (1.15) will be in effect.

We close this section by determining the intensity $I(\mathbf{u}, D)$ generated from an *aperture* in an opaque screen that is illuminated by coherent light. See Figure 6.3. In this case, we generalize (1.4) and (1.9), by forming an *integral* that consists of *superpositioning all contributions* along light rays $\mathbf{C_x}$ from $(\mathbf{x}, 0)$ to $(\mathbf{u}, D)$. Thus, we have

$$\psi(\mathbf{u}, D, t) = \int_{\mathbf{R}^2} A(\mathbf{x}) \frac{\psi(\mathbf{x}, 0, t)}{\lambda |\mathbf{C_x}|} e^{i \frac{2\pi}{\lambda} |\mathbf{C_x}|} d\mathbf{x}. \tag{1.17}$$

In (1.17) we have introduced the *aperture function* $A(\mathbf{x})$ to account for the finite

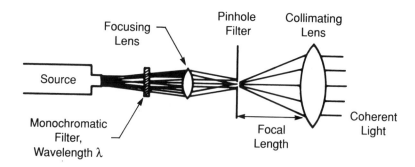

FIGURE 6.2
Creation of coherent, monochromatic light.

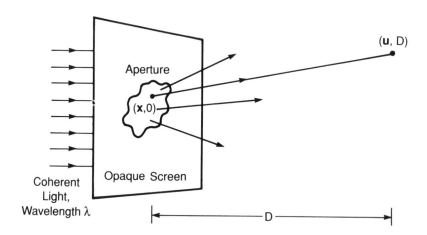

FIGURE 6.3
Diffraction from an aperture.

extent of the aperture. In particular, we assume that

$$A(\mathbf{x}) = 0, \qquad \text{for } |\mathbf{x}| > R. \tag{1.18}$$

where R is a finite positive number that marks off the extent of the aperture. The type of aperture function that we will typically look at has the form

$$A(\mathbf{x}) = \begin{cases} 1 & \text{if } (\mathbf{x}, 0) \text{ is in the aperture} \\ 0 & \text{if } (\mathbf{x}, 0) \text{ is not in the aperture.} \end{cases} \tag{1.19}$$

However, other types of aperture functions could be used to take into account varying levels of light transmitted through the aperture as well as other effects. Now, we calculate the intensity $I(\mathbf{u}, D)$ using (1.5) and (1.17):

$$I(\mathbf{u}, D) = \frac{1}{T} \int_0^T \psi(\mathbf{u}, D, t) \psi^*(\mathbf{u}, D, t)\, dt$$

$$= \frac{1}{T} \int_0^T \left[\int_{\mathbf{R}^2} A(\mathbf{x}) \frac{\psi(\mathbf{x}, 0, t)}{\lambda |\mathbf{C_x}|} e^{i\frac{2\pi}{\lambda}|\mathbf{C_x}|}\, d\mathbf{x} \right]$$

$$\left[\int_{\mathbf{R}^2} A^*(\mathbf{z}) \frac{\psi^*(\mathbf{z}, 0, t)}{\lambda |\mathbf{C_z}|} e^{-i\frac{2\pi}{\lambda}|\mathbf{C_z}|}\, d\mathbf{z} \right] dt$$

$$= \int_{\mathbf{R}^2} \int_{\mathbf{R}^2} \frac{A(\mathbf{x})}{\lambda |\mathbf{C_x}|} e^{i\frac{2\pi}{\lambda}|\mathbf{C_x}|} \frac{A^*(\mathbf{z})}{\lambda |\mathbf{C_z}|} e^{-i\frac{2\pi}{\lambda}|\mathbf{C_z}|}$$

$$\left[\frac{1}{T} \int_0^T \psi(\mathbf{x}, 0, t) \psi^*(\mathbf{z}, 0, t)\, dt \right] d\mathbf{x}\, d\mathbf{z}$$

$$= \int_{\mathbf{R}^2} \int_{\mathbf{R}^2} \frac{A(\mathbf{x})}{\lambda |\mathbf{C_x}|} e^{i\frac{2\pi}{\lambda}|\mathbf{C_x}|} \frac{A^*(\mathbf{z})}{\lambda |\mathbf{C_z}|} e^{-i\frac{2\pi}{\lambda}|\mathbf{C_z}|} \Gamma(\mathbf{x}, \mathbf{z})\, d\mathbf{x}\, d\mathbf{z}.$$

Using the coherency assumption, that $\Gamma(\mathbf{x}, \mathbf{z}) = 1$ *over the extent of the aperture*, it follows that the integrals above *separate out* and

$$I(\mathbf{u}, D) = \left[\int_{\mathbf{R}^2} \frac{A(\mathbf{x})}{\lambda |\mathbf{C_x}|} e^{i\frac{2\pi}{\lambda}|\mathbf{C_x}|}\, d\mathbf{x} \right] \left[\int_{\mathbf{R}^2} \frac{A(\mathbf{z})}{\lambda |\mathbf{C_z}|} e^{i\frac{2\pi}{\lambda}|\mathbf{C_z}|}\, d\mathbf{z} \right]^*.$$

Substituting $\mathbf{x}$ in place of $\mathbf{z}$ in the second integral above, we get

$$I(\mathbf{u}, D) = \left| \int_{\mathbf{R}^2} \frac{A(\mathbf{x})}{\lambda |\mathbf{C_x}|} e^{i\frac{2\pi}{\lambda}|\mathbf{C_x}|} d\mathbf{x} \right|^2. \tag{1.20}$$

Formula (1.20) is a classic formula in diffraction theory. For most of the rest of this chapter we will examine (1.20) for various types of apertures.

REMARK 1.21   A rigorous treatment of our subject (see [Go, Chapter 3] or [Bo-W]) would include an *obliquity factor* in the integrand of (1.20). However, for the type of illumination illustrated in Figure 6.2, this obliquity factor is approximately 1. Moreover, this same approximation by 1 is a standard approximation in diffraction theory.   ∎

2   Fresnel diffraction

In this section we will discuss *Fresnel diffraction*, which is one of the fundamental types of diffraction in optics. Fresnel diffraction is based on an integral approximating the one in formula (1.20) for moderate distances D from the aperture.

Formula (1.20) says that the intensity $I(\mathbf{u}, D)$ is given by

$$I(\mathbf{u}, D) = \left| \int_{\mathbf{R}^2} \frac{A(\mathbf{x})}{\lambda |\mathbf{C_x}|} e^{i \frac{2\pi}{\lambda} |\mathbf{C_x}|} \, d\mathbf{x} \right|^2 . \tag{2.1}$$

Since $\mathbf{C_x}$ is the line segment from $(\mathbf{x}, 0)$ to $(\mathbf{u}, D)$ we have

$$|\mathbf{C_x}| = \left[|\mathbf{u} - \mathbf{x}|^2 + D^2 \right]^{\frac{1}{2}} . \tag{2.2}$$

We will now show that $\lambda |\mathbf{C_x}|$ in the denominator inside the integral in (2.1) can be replaced by λD. To show this we look at the ratio

$$\frac{\lambda |\mathbf{C_x}|}{\lambda D} = \left[1 + |\mathbf{u} - \mathbf{x}|^2 / D^2 \right]^{\frac{1}{2}} \tag{2.3}$$

and *we assume that* $|\mathbf{u} - \mathbf{x}|/D$ *is small enough that*

$$\left[1 + |\mathbf{u} - \mathbf{x}|^2 / D^2 \right]^{\frac{1}{2}} \approx 1. \tag{2.4}$$

In fact, the approximation in (2.4) will hold to within 95% accuracy provided

$$|\mathbf{u} - \mathbf{x}|/D \leq 0.32.$$

For a small aperture (relative to D), this inequality is not too restrictive on what **u**-values we allow in (2.1).

Because of (2.4) and (2.3) we can say that $\lambda |\mathbf{C_x}| \approx \lambda D$ and then (2.1) simplifies to

$$I(\mathbf{u}, D) \approx \left| \frac{1}{\lambda D} \int_{\mathbf{R}^2} A(\mathbf{x}) e^{i \frac{2\pi}{\lambda} |\mathbf{C_x}|} \, d\mathbf{x} \right|^2 . \tag{2.5}$$

We will now show how to approximate the phase factor $e^{i 2\pi |\mathbf{C_x}|/\lambda}$ in the integrand in (2.5). Here we have to approach the approximation differently than above, since for most light (e.g., visible light) the reciprocal of the wavelength, $1/\lambda$, is *enormous*. Consequently, $2\pi D/\lambda$ is a terrible approximation of $2\pi |\mathbf{C_x}|/\lambda$. To obtain a useful approximation we multiply (2.3) by $2\pi/\lambda$, obtaining

$$\frac{2\pi}{\lambda D} |\mathbf{C_x}| = \frac{2\pi}{\lambda} \left[1 + |\mathbf{u} - \mathbf{x}|^2 / D^2 \right]^{\frac{1}{2}} . \tag{2.6}$$

Using the series expansion from calculus

$$(1 + a)^{\frac{1}{2}} = 1 + \frac{1}{2} a - \frac{1}{8} a^2 + \cdots \tag{2.7}$$

to rewrite the right side of (2.6), we have

$$\frac{2\pi}{\lambda D} |\mathbf{C_x}| = \frac{2\pi}{\lambda} + \frac{\pi}{\lambda D^2} |\mathbf{u} - \mathbf{x}|^2 - \frac{\pi}{4\lambda D^4} |\mathbf{u} - \mathbf{x}|^4 + \cdots . \tag{2.8}$$

Now, let's assume that $|\mathbf{u} - \mathbf{x}|/D$ is small enough that $\pi|\mathbf{u} - \mathbf{x}|^4/4\lambda D^4$, and higher power terms, can be neglected in (2.8). Then (2.8) simplifies to

$$\frac{2\pi}{\lambda D}|C_\mathbf{x}| \approx \frac{2\pi}{\lambda} + \frac{\pi}{\lambda D^2}|\mathbf{u} - \mathbf{x}|^2. \tag{2.9}$$

Multiplying both sides of (2.9) by D yields

$$\frac{2\pi}{\lambda}|C_\mathbf{x}| \approx \frac{2\pi}{\lambda}D + \frac{\pi}{\lambda D}|\mathbf{u} - \mathbf{x}|^2. \tag{2.10}$$

From (2.10) we obtain

$$e^{i\frac{2\pi}{\lambda}|C_\mathbf{x}|} \approx e^{i\frac{2\pi}{\lambda}D}e^{\frac{i\pi}{\lambda D}|\mathbf{u}-\mathbf{x}|^2}. \tag{2.11}$$

Substituting the right side of (2.11) into (2.5), and factoring the constant outside the integral, we have

$$I(\mathbf{u}, D) \approx \left| e^{i\frac{2\pi}{\lambda}D}\frac{1}{\lambda D}\int_{\mathbf{R}^2} A(\mathbf{x})e^{\frac{i\pi}{\lambda D}|\mathbf{u}-\mathbf{x}|^2}\,d\mathbf{x} \right|^2. \tag{2.12}$$

We have now established two things. First, since $\left|e^{i\frac{2\pi}{\lambda}D}\right| = 1$ we have

$$I(\mathbf{u}, D) \approx \left| \frac{1}{\lambda D}\int_{\mathbf{R}^2} A(\mathbf{x})e^{\frac{i\pi}{\lambda D}|\mathbf{u}-\mathbf{x}|^2}\,d\mathbf{x} \right|^2. \tag{2.13}$$

Formula (2.13) is the *Fresnel diffraction integral*.

Second, all of our approximations remain valid for the integral describing $\psi(\mathbf{u}, D, t)$ in (1.17). Thus, from (1.17) we obtain

$$\psi(\mathbf{u}, D, t) \approx \frac{e^{i\frac{2\pi}{\lambda}D}}{\lambda D}\int_{\mathbf{R}^2} A(\mathbf{x})\psi(\mathbf{x}, 0, t)e^{\frac{i\pi}{\lambda D}|\mathbf{u}-\mathbf{x}|^2}\,d\mathbf{x}. \tag{2.14}$$

Formula (2.14) will prove useful later, when we discuss imaging by a lens.

Here is an example of how *FAS* can be used to analyze the Fresnel diffraction integral (2.13).

Example 2.15
Suppose that coherent light of wavelength $\lambda = 5 \times 10^{-4}$ mm (greenish-blue) light is shined onto a square aperture with aperture function $A(\mathbf{x}) = A(x, y)$ given by

$$A(x, y) = \begin{cases} 1 & \text{if } |x| < 1 \text{ mm and } |y| < 1 \text{ mm} \\ 0 & \text{if } |x| > 1 \text{ mm or } |y| > 1 \text{ mm.} \end{cases}$$

Graph $I(\mathbf{u}, D)$ for $D = 100$ mm, 120 mm, and 150 mm. □

SOLUTION It is not too difficult to see that

$$A(x, y) = \text{rec}\left(\frac{x}{2}\right)\text{rec}\left(\frac{y}{2}\right). \tag{2.16}$$

For $\mathbf{u} = (u, v)$ we see that

$$e^{\frac{i\pi}{\lambda D}|\mathbf{u}-\mathbf{x}|^2} = e^{\frac{i\pi}{\lambda D}\left[(u-x)^2+(v-y)^2\right]}$$

$$= e^{\frac{i\pi}{\lambda D}(u-x)^2} e^{\frac{i\pi}{\lambda D}(v-y)^2}. \tag{2.17}$$

Thus, using (2.16) and (2.17) in the integral in (2.13), we see that the integral over $\mathbf{R}^2$ separates into iterated integrals from $-\infty$ to ∞ in each variable x and y. Therefore, we obtain

$$I(\mathbf{u}, D) = \left| \frac{1}{(\lambda D)^{\frac{1}{2}}} \int_{-\infty}^{\infty} \mathrm{rec}\left(\frac{x}{2}\right) e^{\frac{i\pi}{\lambda D}(u-x)^2} \, dx \right|^2$$

$$\times \left| \frac{1}{(\lambda D)^{\frac{1}{2}}} \int_{-\infty}^{\infty} \mathrm{rec}\left(\frac{y}{2}\right) e^{\frac{i\pi}{\lambda D}(v-y)^2} \, dy \right|^2. \tag{2.18}$$

From (2.18) we see that it suffices to graph $I_1(u, D)$, defined by

$$I_1(u, D) = \left| \frac{1}{(\lambda D)^{\frac{1}{2}}} \int_{-\infty}^{\infty} \mathrm{rec}\left(\frac{x}{2}\right) e^{\frac{i\pi}{\lambda D}(u-x)^2} dx \right|^2 \tag{2.19}$$

since the second factor in (2.18) is just $I_1(v, D)$, which is obtained from $I_1(u, D)$ by changing variables from x to y and u to v. In particular, we have

$$I(\mathbf{u}, D) = I_1(u, D)I_1(v, D). \tag{2.20}$$

Now, the integral in (2.19)

$$\frac{1}{(\lambda D)^{\frac{1}{2}}} \int_{-\infty}^{\infty} \mathrm{rec}\left(\frac{x}{2}\right) e^{\frac{i\pi}{\lambda D}(u-x)^2} \, dx \tag{2.21}$$

is just a convolution integral of the same form as we described in Section 9 of Chapter 5. In fact, if we put

$$\lambda D = 4\pi t(\hbar/2m) \tag{2.22}$$

then (2.21) differs from $(f * {}_tF)(u)$ in formula (9.4) of Chapter 5, for $f(x) = \mathrm{rec}\,(x/2)$, by only a factor of $i^{1/2}$. Since $|i^{1/2}| = 1$, we will ignore this difference, *which does not occur anyway when we graph the expression in* (2.19).

Using *FresCos* and *FresSin* filters with wave constant λD and time constant $1/4\pi \approx .079477471$ [here we apply (2.22)], we obtain graphs of

$$I_1(u, D)^{\frac{1}{2}} = \left| \frac{1}{(\lambda D)^{\frac{1}{2}}} \int_{-\infty}^{\infty} \mathrm{rec}\left(\frac{x}{2}\right) e^{\frac{i\pi}{\lambda D}(u-x)^2} \, dx \right| \tag{2.23}$$

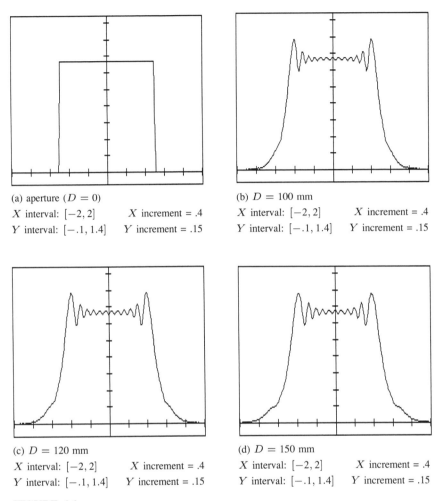

(a) aperture ($D = 0$)
X interval: $[-2, 2]$ X increment = .4
Y interval: $[-.1, 1.4]$ Y increment = .15

(b) $D = 100$ mm
X interval: $[-2, 2]$ X increment = .4
Y interval: $[-.1, 1.4]$ Y increment = .15

(c) $D = 120$ mm
X interval: $[-2, 2]$ X increment = .4
Y interval: $[-.1, 1.4]$ Y increment = .15

(d) $D = 150$ mm
X interval: $[-2, 2]$ X increment = .4
Y interval: $[-.1, 1.4]$ Y increment = .15

FIGURE 6.4
Graphs of the intensity function $I_1(u, D)^{1/2}$ from Example 2.15.

for $\lambda = 5 \times 10^{-4}$ and $D = 100$, 120, and 150. These graphs are shown in Figure 6.4. By choosing *Graphs* and *Sum of Squares* from the Display menu of *FAS*, and pressing n when asked whether to normalize, then the square of the expression in (2.23) will be graphed. This yields the graphs for $I_1(u, D)$ shown in Figure 6.5.

By changing variables in (2.19) from x to y and from u to v, we obtain $I_1(v, D)$, the other factor in formula (2.18) for $I(\mathbf{u}, D)$. For $\lambda = 5 \times 10^{-4}$ mm and $D = 150$ mm we will obtain an intensity function $I(\mathbf{u}, D)$ similar to the graph shown in Figure 6.6(a). This intensity function will yield a diffraction

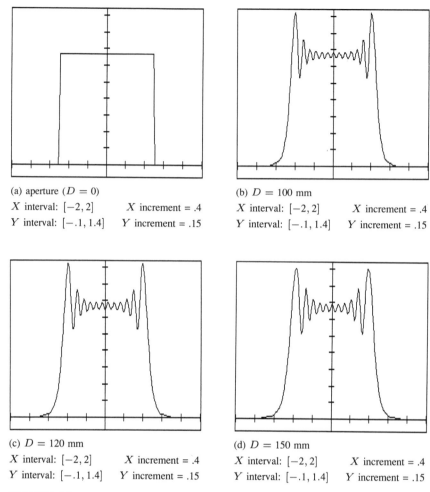

(a) aperture ($D = 0$)
X interval: $[-2, 2]$ X increment = .4
Y interval: $[-.1, 1.4]$ Y increment = .15

(b) $D = 100$ mm
X interval: $[-2, 2]$ X increment = .4
Y interval: $[-.1, 1.4]$ Y increment = .15

(c) $D = 120$ mm
X interval: $[-2, 2]$ X increment = .4
Y interval: $[-.1, 1.4]$ Y increment = .15

(d) $D = 150$ mm
X interval: $[-2, 2]$ X increment = .4
Y interval: $[-.1, 1.4]$ Y increment = .15

FIGURE 6.5
Graphs of the intensity function $I_1(u, D)$ from Example 2.15.

pattern similar to the one sketched in Figure 6.6(b). We see in Figure 6.6(b) the classic checkerboard (or plaid) pattern of a Fresnel diffraction pattern from a square aperture. ∎

REMARK 2.24 It is clear from Figure 6.6(b) that the Fresnel diffraction pattern represents a *distorted image* of the original aperture. Also, from Figure 6.5(d) one can gather some *quantitative* information about the location of the *bright border* around the edge of the Fresnel diffraction pattern, which is reminiscent of Gibbs' phenomenon. ∎

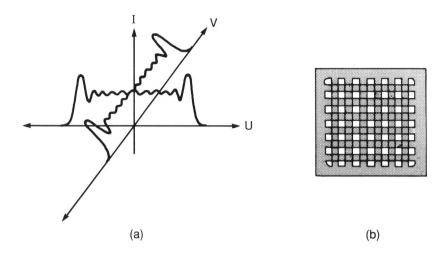

(a) (b)

FIGURE 6.6
Fresnel diffraction from a square aperture. (a) Graph of the intensity $I(u, v, D)$.
(b) Graph of the diffraction pattern (negative image).

As we can see from the example above, *FAS* can be used to graph the individual factors of the intensity function $I(\mathbf{u}, D)$ whenever the aperture function factors in the following way:

$$A(\mathbf{x}) = A_1(x)A_2(y). \qquad (2.25)$$

When this happens, we have

$$I(u, v, D) = I_1(u, D)I_2(v, D) \qquad (2.26)$$

where

$$I_1(u, D) = \left| \frac{1}{(\lambda D)^{\frac{1}{2}}} \int_{-\infty}^{\infty} A_1(x) e^{\frac{i\pi}{\lambda D}(u-x)^2} dx \right|^2 \qquad (2.27)$$

$$I_2(v, D) = \left| \frac{1}{(\lambda D)^{\frac{1}{2}}} \int_{-\infty}^{\infty} A_2(y) e^{\frac{i\pi}{\lambda D}(v-y)^2} dy \right|^2. \qquad (2.28)$$

In such a situation, we might call the aperture function A *separable*. There are a number of separable aperture functions that are important cases for study, for example, aperture functions for rectangular apertures or aperture functions that are Gaussian exponentials. These examples, and the case of *edge diffraction*, are treated in the exercises.

REMARK 2.29 Equation (2.22) represents more than just a mathematical convenience. Because quantum particles (like the electrons we dealt with earlier in the text) have wavelengths, these particles diffract through apertures in much the same way as light does. In fact, everything we discuss in this chapter on coherency, diffraction, and imaging applies to *electron optics*, as in *electron microscopy*, and *electron diffraction*. Also, *neutron diffraction spectroscopy* uses similar ideas. ∎

3 Fraunhofer diffraction

In this section we begin our discussion of *Fraunhofer diffraction*, which plays a very important role in applications of diffraction theory. This type of diffraction occurs as a limiting case of Fresnel diffraction, when the distance D tends to ∞ or the dimensions of the aperture tend to 0.

Fraunhofer diffraction theory is essential for understanding the diffraction of X-rays from crystals. X-ray diffraction photographs from crystals are used to determine the underlying crystal structure. For example, the structure of DNA as well as many protein structures have been determined from X-ray diffraction patterns. Another important application is Fraunhofer diffraction from diffraction gratings. This diffraction plays a role in physical chemistry, and it also provides a way of identifying the chemical constituents of stars.

We will now show how Fraunhofer diffraction arises as a limiting case of Fresnel diffraction. From formula (2.13) we have for the diffracted light intensity

$$I(\mathbf{u}, D) \approx \left| \frac{1}{\lambda D} \int_{\mathbf{R}^2} A(\mathbf{x}) e^{\frac{i\pi}{\lambda D}|\mathbf{u}-\mathbf{x}|^2} \, d\mathbf{x} \right|^2 . \tag{3.1}$$

Expanding the exponential in (3.1), we get

$$e^{\frac{i\pi}{\lambda D}|\mathbf{u}-\mathbf{x}|^2} = e^{\frac{i\pi}{\lambda D}|\mathbf{u}|^2} e^{\frac{-i2\pi}{\lambda D}\mathbf{u}\cdot\mathbf{x}} e^{\frac{i\pi}{\lambda D}|\mathbf{x}|^2} . \tag{3.2}$$

Recall the assumption in (1.18) that

$$A(\mathbf{x}) = 0, \qquad \text{for } |\mathbf{x}| > R. \tag{3.3}$$

We will show that

$$e^{\frac{i\pi}{\lambda D}|\mathbf{x}|^2} \approx 1, \qquad \text{for } |\mathbf{x}| \le R \tag{3.4}$$

in either of the following two cases:

$$D \to \infty \qquad \text{(far-field case)} \qquad (3.4\text{a})$$

$$R \to 0 \qquad \text{(small-aperture case)}. \qquad (3.4\text{b})$$

For the case of (3.4a), we observe that, for $|x| \le R$, as $D \to \infty$

$$0 \le \frac{\pi}{\lambda D}|\mathbf{x}|^2 \le \frac{\pi R^2}{\lambda D} \approx 0. \qquad (3.5)$$

Hence,

$$e^{\frac{i\pi}{\lambda D}|\mathbf{x}|^2} \approx e^{i0} = 1$$

so (3.4) holds. On the other hand, for the case of (3.4b) we still have (3.5) *provided R is sufficiently small* (and *D* stays *fixed*). Thus, (3.4) holds again.

Now, assuming that (3.4) holds, we obtain from (3.2) that

$$e^{\frac{i\pi}{\lambda D}|\mathbf{u}-\mathbf{x}|^2} \approx e^{\frac{i\pi}{\lambda D}|\mathbf{u}|^2} e^{-i\frac{2\pi}{\lambda D}\mathbf{u}\cdot\mathbf{x}}, \qquad \text{for } |x| \le R. \qquad (3.6)$$

Since we have (3.3), the integral in (3.1) is actually taken over only those **x** for which $|\mathbf{x}| \le R$. So we can use (3.6) to simplify the integral in (3.1). Substituting the right side of (3.6) into the integrand in (3.1), and factoring the exponential involving $|\mathbf{u}|^2$ outside the integral, we obtain

$$I(\mathbf{u}, D) \approx \left| \frac{e^{\frac{i\pi}{\lambda D}|\mathbf{u}|^2}}{\lambda D} \int_{\mathbf{R}^2} A(\mathbf{x}) e^{-i\frac{2\pi}{\lambda D}\mathbf{u}\cdot\mathbf{x}}\, d\mathbf{x} \right|^2. \qquad (3.7)$$

Since the complex exponential involving $|\mathbf{u}|^2$ has magnitude 1, formula (3.7) becomes

$$I(\mathbf{u}, D) \approx \left| \frac{1}{\lambda D} \int_{\mathbf{R}^2} A(\mathbf{x}) e^{-i\frac{2\pi}{\lambda D}\mathbf{u}\cdot\mathbf{x}}\, d\mathbf{x} \right|^2. \qquad (3.8)$$

Formula (3.8) is the *Fraunhofer diffraction formula*. The integral in (3.8) can be viewed as a two-dimensional Fourier transform. If we adopt the notation

$$\hat{A}(\mathbf{u}) = \int_{\mathbf{R}^2} A(\mathbf{x}) e^{-i2\pi\mathbf{u}\cdot\mathbf{x}}\, d\mathbf{x} \qquad (3.9)$$

then (3.8) can be written as

$$I(\mathbf{u}, D) \approx \left| \frac{1}{\lambda D} \hat{A}\left(\frac{1}{\lambda D}\mathbf{u} \right) \right|^2 \qquad (3.10)$$

which shows that *the intensity function from Fraunhofer diffraction is the modulus-squared of the (scaled) Fourier transform of the aperture function.*

Example 3.11
Discuss Fraunhofer diffraction from a rectangular aperture, having aperture function for $\mathbf{x} = (x, y)$

$$A(\mathbf{x}) = \begin{cases} 1 & \text{if } |x| < .1 \text{ mm and } |y| < .05 \text{ mm} \\ 0 & \text{if } |x| > .1 \text{ mm or } |y| > .05 \text{ mm.} \end{cases} \quad \square$$

SOLUTION   It is not too hard to see that $A(x, y) = \text{rec}\,(5x)\,\text{rec}\,(10y)$. Substituting this expression for $A(x, y)$ into (3.9) and separating into two integrals of x and y, we obtain [for $\mathbf{u} = (u, v)$ and $\mathbf{u} \cdot \mathbf{x} = ux + vy$]

$$\begin{aligned}
\hat{A}(u, v) &= \int_{\mathbf{R}^2} \text{rec}\,(5x)\,\text{rec}\,(10y) e^{-i2\pi(ux+vy)}\, dx\, dy \\
&= \int_{-\infty}^{\infty} \text{rec}\,(5x) e^{-i2\pi ux}\, dx \int_{-\infty}^{\infty} \text{rec}\,(10y) e^{-i2\pi vy}\, dy \\
&= \frac{1}{5}\,\text{sinc}\left(\frac{u}{5}\right) \frac{1}{10}\,\text{sinc}\left(\frac{v}{10}\right).
\end{aligned}$$

Thus, from (3.10) we obtain

$$I(u, v, D) \approx \frac{1}{(50\lambda D)^2}\text{sinc}^2\left(\frac{u}{5\lambda D}\right) \text{sinc}^2\left(\frac{v}{10\lambda D}\right) \tag{3.12}$$

In Figure 6.7(a) we have graphed the two separate factors of $I(u, v, D)$. Notice that the positions of *zero intensity*

$$u = \pm 5\lambda D,\ \pm 10\lambda D, \ldots \qquad \text{and} \qquad v = \pm 10\lambda D,\ \pm 20\lambda D, \ldots$$

are proportional to λD. The graph of the diffraction pattern determined by the intensity function $I(u, v, D)$ is shown in Figure 6.7(b).

The pattern in Figure 6.7(b) is the classic Fraunhofer diffraction pattern from a rectangular aperture (compare it to the photograph in Figure 6.8).   ∎

4   Circular apertures

In this section we will describe Fraunhofer diffraction from circular apertures. The diffraction integral for a circular aperture is usually described in terms of Bessel functions (see [Wa, Chapter 7.6]). The advantage of the method described below is that no special knowledge of Bessel functions is needed.

For a circular aperture, the aperture function is

$$A(\mathbf{x}) = \begin{cases} 1 & \text{if } |\mathbf{x}| < R \\ 0 & \text{if } |\mathbf{x}| > R \end{cases} \tag{4.1}$$

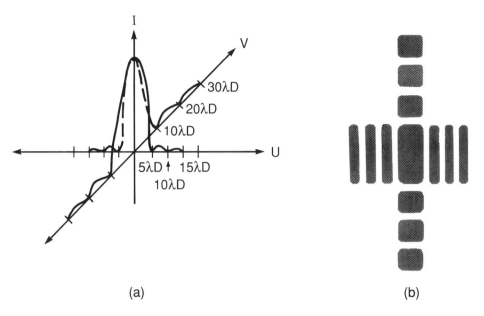

(a)   (b)

FIGURE 6.7
Fraunhofer diffraction by a rectangular aperture. (a) Graph of the intensity $I(u, v, D)$. (b) Graph of the diffraction pattern (negative image).

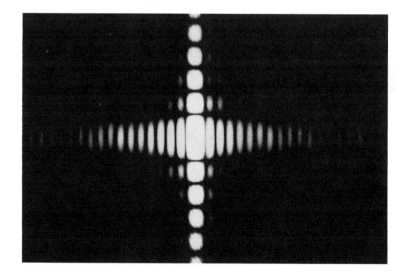

FIGURE 6.8
Photograph of a Fraunhofer diffraction pattern from a rectangular aperture (reproduced with permission from [Wa, p. 291]).

where R is the radius of the aperture. If we use (4.1) in formula (3.9) we have

$$\hat{A}(\mathbf{u}) = \int_{|x|<R} e^{-i2\pi\mathbf{u}\cdot\mathbf{x}} \, d\mathbf{x}. \qquad (4.2)$$

Now, if we change to *polar coordinates*

$$\mathbf{x} = (r\cos\theta, r\sin\theta), \qquad \mathbf{u} = (\rho\cos\phi, \rho\sin\phi) \qquad (4.3)$$

then (4.2) becomes

$$\hat{A}(\rho\cos\phi, \rho\sin\phi) = \int_0^{2\pi} \int_0^R e^{-i2\pi r\rho(\cos\theta\cos\phi+\sin\theta\sin\phi)} \, r \, dr \, d\theta$$

$$= \int_0^R r \int_0^{2\pi} \left[e^{-i2\pi r\rho\cos(\theta-\phi)} \, d\theta \right] dr. \qquad (4.4)$$

Substituting θ for $\theta - \phi$, we obtain

$$\hat{A}(\rho\cos\phi, \rho\sin\phi) = \int_0^R r \left[\int_{-\phi}^{2\pi-\phi} e^{-i2\pi r\rho\cos\theta} \, d\theta \right] dr. \qquad (4.5)$$

Because of the periodicity of $\cos\theta$, the inner integrand in (4.5) has period 2π. Therefore, it can be integrated over the interval $[-\pi, \pi]$ instead of over the interval $[-\phi, 2\pi - \phi]$. Thus,

$$\hat{A}(\rho\cos\phi, \rho\sin\phi) = \int_0^R r \left[\int_{-\pi}^{\pi} e^{-i2\pi r\rho\cos\theta} \, d\theta \right] dr. \qquad (4.6)$$

It is important to note that the right side of (4.6) has no dependence on ϕ. It is just a function of ρ (the variables r and θ being integrated out). For simplicity, then, we shall just write $\hat{A}(\rho)$ instead of $\hat{A}(\rho\cos\phi, \rho\sin\phi)$. And so, formula (4.6) can be rewritten as

$$\hat{A}(\rho) = \int_0^R r \left[\int_{-\pi}^{\pi} e^{-i2\pi r\rho\cos\theta} \, d\theta \right] dr. \qquad (4.7)$$

Substituting $s = 2\pi\rho r$ into the outer integral in (4.7) we get

$$\hat{A}(\rho) = \frac{1}{(2\pi\rho)^2} \int_0^{2\pi\rho R} s \left[\int_{-\pi}^{\pi} e^{-is\cos\theta} \, d\theta \right] ds. \qquad (4.8)$$

In order to simplify (4.8) we define the function $H(x)$ by

$$H(x) = \int_0^x s \left[\int_{-\pi}^{\pi} e^{-is\cos\theta} \, d\theta \right] ds. \qquad (4.9)$$

We will show at the end of this section why $H(x)$ satisfies the following identity

$$H(x) = 2x^2 \int_{-1}^1 \left[1 - w^2 \right]^{\frac{1}{2}} e^{-ixw} \, dw. \qquad (4.10)$$

Making use of (4.10) for now, we obtain from (4.8)

$$\hat{A}(\rho) = \frac{1}{(2\pi\rho)^2} H(2\pi\rho R)$$

$$= 2R^2 \int_{-1}^{1} \left[1 - w^2\right]^{\frac{1}{2}} e^{-i2\pi\rho R w} \, dw. \tag{4.11}$$

Making a final change of variables, $x = Rw$, we can rewrite (4.11) as

$$\hat{A}(\rho) = \int_{-R}^{R} 2R \left[1 - \left(\frac{x}{R}\right)^2\right]^{\frac{1}{2}} e^{-i2\pi\rho x} \, dx. \tag{4.12}$$

The integral on the right side of (4.12) is a one-dimensional Fourier transform. If we define the function $a(x)$ by

$$a(x) = \begin{cases} 2R\left[1 - \left(\frac{x}{R}\right)^2\right]^{\frac{1}{2}} & \text{for } |x| \leq R \\ 0 & \text{for } |x| > R \end{cases} \tag{4.13}$$

then we can express (4.12) as

$$\hat{A}(\rho) = \hat{a}(\rho). \tag{4.14}$$

Using (4.14) we can express the intensity function $I(\mathbf{u}, D)$ as follows:

$$I(\mathbf{u}, D) = \left| \frac{1}{\lambda D} \hat{a} \left(\frac{\rho}{\lambda D} \right) \right|^2, \qquad (\rho = |\mathbf{u}|). \tag{4.15}$$

REMARK 4.16 Formula (4.14) shows that the *radial dependence* of the transform $\hat{A}$ can be determined by Fourier transforming the function $a(x)$ in formula (4.13). *When using FAS to do this, it is best to use the following formula for* $a(x)$:

$$a(x) = 2R \operatorname{rec}(x/2R) \operatorname{sqr}(\operatorname{abs}(1 - (x/R) \wedge 2)). \tag{4.17}$$

This formula for $a(x)$ avoids possible run-time errors that can occur when using the sqr function. ∎

Example 4.18
Suppose that light of wavelength λ illuminates a circular aperture of radius 1 mm. Describe the Fraunhofer diffraction pattern that results. ⬚

SOLUTION The advantage of (4.15) is that *FAS* can be used to approximate it. Assuming units of millimeters, we use *FAS* to compute the *power spectrum*

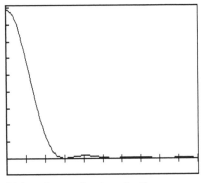

(a) intensity as a function of radius

X interval: $[0, 2\lambda D]$ X increment = $.2\lambda D$

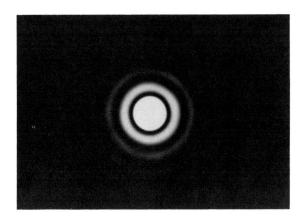

(b) Photograph of diffraction pattern

FIGURE 6.9
Diffraction by a circular aperture of radius 1. (Photograph reproduced with permission from [Wa, p. 294].)

$|\hat{a}(\rho)|^2$ where [see formula (4.17)]

$$a(x) = 2 \operatorname{rec}(x/2) \operatorname{sqr}(\operatorname{abs}(1 - x \wedge 2)). \tag{4.19}$$

By labeling coordinates by multiples of λD, we obtain the graph of $I(\mathbf{u}, D)$. This graph is shown in Figure 6.9(a) as a function of the radial coordinate ρ. To obtain Figure 6.9(a) we used 1024 points and an initial interval of $[-8, 8]$. Notice that it corresponds nicely to the diffraction pattern shown in Figure 6.9(b). This pattern is called an *Airy pattern*. The alternating dark and bright rings are called *Airy rings*. ∎

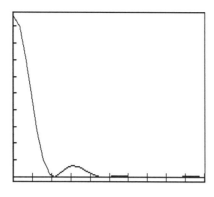

X interval: $[0, .5\lambda D]$ X increment = $.05\lambda D$

FIGURE 6.10
Diffraction by an annulus (see Example 4.20). Intensity as a function of radius.

Example 4.20
Suppose that an aperture has an aperture function

$$A(\mathbf{x}) = \begin{cases} 1 & \text{if } 2 \text{ mm} < |\mathbf{x}| < 5 \text{ mm} \\ 0 & \text{if } |\mathbf{x}| < 2 \text{ mm or } |x| > 5 \text{ mm.} \end{cases} \qquad (4.21)$$

Graph the intensity function $I(\mathbf{u}, D)$. □

SOLUTION This aperture can be thought of as a circular aperture of radius 5 mm that has an opaque, circular region (radius 2 mm) at its center. The aperture function is the difference of two circular aperture functions (one using 5 mm and the other using 2 mm). Thus, using formula (4.17), we write $a(x)$ as

$$a(x) = 10 \operatorname{rec}(x/10) \operatorname{sqr}(\operatorname{abs}(1-(x/5)\wedge 2)) - 4 \operatorname{rec}(x/4) \operatorname{sqr}(\operatorname{abs}(1-(x/2)\wedge 2)).$$

Using this formula, we compute the *power spectrum* $|\hat{a}(\rho)|^2 = I$ and label coordinates by multiples of λD. See Figure 6.10. ∎

We conclude this section by establishing (4.10). We begin by showing that the *derivatives* of both sides of (4.10) are equal. That is, that

$$H'(x) = \frac{d}{dx}\left[2x^2 \int_{-1}^{1} [1-w^2]^{\frac{1}{2}} e^{-ixw}\, dw\right]. \qquad (4.22)$$

Computing $H'(x)$ is done by applying the Fundamental Theorem of Calculus to (4.9); in this way we obtain

$$H'(x) = x \int_{-\pi}^{\pi} e^{-ix\cos\theta}\, d\theta.$$

Because $\cos\theta$ is even, we then have

$$H'(x) = 2x \int_0^\pi e^{-ix\cos\theta}\, d\theta.$$

Substituting $w = \cos\theta$, we get

$$H'(x) = 2x \int_{-1}^1 \left[1 - w^2\right]^{-\frac{1}{2}} e^{-ixw}\, dw. \tag{4.23}$$

We now turn to the right side of (4.22). Using the product rule and also differentiating inside the integral, we have

$$\frac{d}{dx}\left[2x^2 \int_{-1}^1 \left[1 - w^2\right]^{\frac{1}{2}} e^{-ixw}\, dw\right]$$

$$= 4x \int_{-1}^1 \left[1 - w^2\right]^{\frac{1}{2}} e^{-ixw}\, dw$$

$$- 2ix^2 \int_{-1}^1 w \left[1 - w^2\right]^{\frac{1}{2}} e^{-ixw}\, dw. \tag{4.24}$$

Bringing $-ix$ inside the last integral above, as a factor on e^{-ixw}, we obtain

$$\frac{d}{dx}\left[2x^2 \int_{-1}^1 \left[1 - w^2\right]^{\frac{1}{2}} e^{-ixw}\, dw\right]$$

$$= 4x \int_{-1}^1 \left[1 - w^2\right]^{\frac{1}{2}} e^{-ixw}\, dw + 2x \int_{-1}^1 w \left[1 - w^2\right]^{\frac{1}{2}} \frac{d}{dw}\left(e^{-ixw}\right)\, dw.$$

Integrating by parts in the last integral above, and then combining integrals, we have

$$\frac{d}{dx}\left[2x^2 \int_{-1}^1 \left[1 - w^2\right]^{\frac{1}{2}} e^{-ixw}\, dw\right]$$

$$= 4x \int_{-1}^1 \left\{\left[1 - w^2\right]^{\frac{1}{2}} - \frac{1}{2}\left[1 - w^2\right]^{\frac{1}{2}} + \frac{1}{2}w^2 \left[1 - w^2\right]^{-\frac{1}{2}}\right\} e^{-ixw}\, dw.$$

Performing simple algebra then yields

$$\frac{d}{dx}\left[2x^2 \int_{-1}^1 \left[1 - w^2\right]^{\frac{1}{2}} e^{-ixw}\, dw\right] = 2x \int_{-1}^1 \left[1 - w^2\right]^{-\frac{1}{2}} e^{-ixw}\, dw. \tag{4.25}$$

Comparing (4.25) with (4.23) we see that (4.22) holds. Consequently,

$$H(x) = C + 2x^2 \int_{-1}^1 \left[1 - w^2\right]^{\frac{1}{2}} e^{-ixw}\, dw \tag{4.26}$$

where C is an unknown constant. Putting $x = 0$ in both sides of (4.26) we get $0 = C + 0$. Hence $C = 0$, which proves (4.10).

5 Interference

In this section we will examine Fraunhofer diffraction from two identical apertures, separated by some finite distance. This situation results in an *interference* phenomenon that is easily described using Fourier transforms.

Suppose two identical apertures, each having aperture function $A_0(\mathbf{x})$, are separated from each other by a finite distance (see Figure 6.11). Assuming that the apertures are separated along the horizontal x-axis, we can write the aperture function $A(\mathbf{x})$ as

$$A(x,y) = A_0\left(x - \frac{1}{2}\delta, y\right) + A_0\left(x + \frac{1}{2}\delta, y\right) \tag{5.1}$$

where δ is a positive constant.

Fourier transforming both sides of (5.1), and applying shifting and linearity properties, we obtain

$$\hat{A}(u,v) = \hat{A}_0(u,v)e^{-i\pi\delta u} + \hat{A}_0(u,v)e^{+i\pi\delta u}$$
$$= \hat{A}_0(u,v)\,2\cos(\pi\delta u).$$

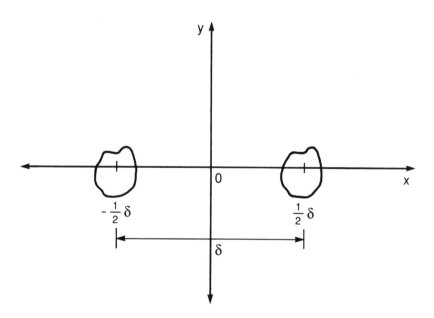

FIGURE 6.11
Two identical apertures.

Thus, the intensity function $I(u, v, D)$ is given by

$$I(u, v, D) = \left| \frac{1}{\lambda D} \hat{A}_0 \left(\frac{u}{\lambda D}, \frac{v}{\lambda D} \right) \right|^2 \left[4 \cos^2 \frac{\pi \delta u}{\lambda D} \right]. \tag{5.2}$$

Or, using $I_0(u, v, D)$ to denote the intensity from a single aperture,

$$I(u, v, D) = I_0(u, v, D) \left[4 \cos^2 \frac{\pi \delta u}{\lambda D} \right]. \tag{5.3}$$

The factor $4 \cos^2(\pi \delta u / \lambda D)$ is the *interference factor*. It will be *zero* (destructive interference) when

$$u = \pm \frac{\lambda D}{2\delta}, \ \pm \frac{3 \lambda D}{2\delta}, \ldots, \ \pm \frac{(2k+1)\lambda D}{2\delta}, \ldots, \tag{5.4}$$

and it will acheive a maximum value of 4 (constructive interference) when

$$u = 0, \ \pm \frac{\lambda D}{\delta}, \ \pm \frac{2 \lambda D}{\delta}, \ldots, \ \pm \frac{n \lambda D}{\delta}, \ldots. \tag{5.5}$$

Since the equations in (5.4) represent *vertical lines* in the u-v plane, it follows from (5.3) that the diffraction pattern from the two separated apertures will look like the pattern from a single aperture *overlayed with alternating dark vertical strips of low intensity* (interference fringes). See, for example, Figure 6.12, where we have shown the diffraction pattern from two horizontally separated circular apertures. An important fact to notice is that the distance between the interference fringes is *inversely* proportional to the distance between the apertures.

In the next section we will examine a more complicated interference phenomenon, the interference from diffraction gratings. The basic principle, however, is contained in the discussion we have given above of interference from two apertures.

6   Diffraction gratings

In this section we begin a discussion of diffraction gratings. Diffraction gratings are an important tool in the *spectral analysis* of light, the separation of light into its component frequencies.

Diffraction gratings are an example of a *linear array* of apertures. Suppose that M identical apertures are evenly spaced by a distance δ along the x-axis, as shown in Figure 6.13. If we choose the origin of coordinates $(0,0)$ to be within the left-most aperture and denote the aperture function for this aperture

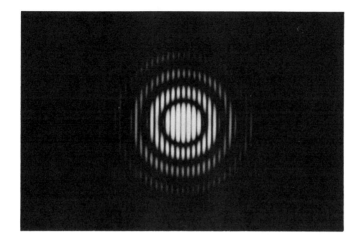

FIGURE 6.12
Diffraction pattern from two horizontally separated, identical circular apertures (reproduced with permission from [Wa, p. 296]).

FIGURE 6.13
Linear array of identical apertures.

by $A_0(\mathbf{x})$, then the aperture function $A(\mathbf{x})$ for the entire array is given by

$$A(x, y) = \sum_{m=0}^{M-1} A_0(x - m\delta, y). \tag{6.1}$$

Consequently, by linearity and shifting properties, the Fourier transform $\hat{A}$ satisfies

$$\hat{A}(u, v) = \sum_{m=0}^{M-1} \hat{A}_0(u, v) e^{-i2\pi m\delta u}. \tag{6.2}$$

Factoring out $\hat{A}_0(u, v)$ we get

$$\hat{A}(u, v) = \hat{A}_0(u, v) \sum_{m=0}^{M-1} e^{-i2\pi m\delta u}. \tag{6.3}$$

The sum on the right side of (6.3) can be treated as a finite geometric series, and its exact sum can be found by using (1.10) from Chapter 2 with $r = e^{-i2\pi\delta u}$. Doing this, we have

$$\sum_{m=0}^{M-1} e^{-i2\pi m\delta u} = \sum_{m=0}^{M-1} \left[e^{-i2\pi\delta u} \right]^m$$

$$= \frac{1 - e^{-i2\pi M\delta u}}{1 - e^{-i2\pi\delta u}}.$$

By factoring out $e^{-i\pi M\delta u}$ from the numerator of the fraction above and $e^{-i\pi\delta u}$ from the denominator, we obtain

$$\sum_{m=0}^{M-1} e^{-i2\pi m\delta u} = \frac{e^{-i\pi M\delta u}}{e^{-i\pi\delta u}} \frac{e^{i\pi M\delta u} - e^{-i\pi M\delta u}}{e^{i\pi\delta u} - e^{-i\pi\delta u}}$$

$$= e^{-i\pi(M-1)\delta u} \frac{\sin(\pi M\delta u)}{\sin(\pi\delta u)}.$$

Using this last result in (6.3), we have

$$\hat{A}(u, v) = \hat{A}_0(u, v) \frac{\sin(\pi M\delta u)}{\sin(\pi\delta u)} e^{-i\pi(M-1)\delta u}. \tag{6.4}$$

The phase factor $e^{-i\pi(M-1)\delta u}$ in (6.4) is a result of our choice of origin of coordinates. In fact, by the shifting property, if we shift this origin by $(M-1)\delta/2$ in the x-direction, then *the phase factor will be canceled* and we will have

$$\hat{A}(u, v) = \hat{A}_0(u, v) \frac{\sin(\pi M\delta u)}{\sin(\pi\delta u)}. \tag{6.5}$$

In any case, since $|e^{-i\pi(M-1)\delta u}| = 1$, we will have for the intensity function $I(u, v, D)$ of the array

$$I(u, v, D) = I_0(u, v, D) \frac{\sin^2(\pi M\delta u/\lambda D)}{\sin^2(\pi\delta u/\lambda D)} \tag{6.6}$$

where

$$I_0(u, v, D) = \left| \frac{1}{\lambda D} \hat{A}_0 \left(\frac{u}{\lambda D}, \frac{v}{\lambda D} \right) \right|^2 \tag{6.7}$$

is the intensity function from one aperture of the array.

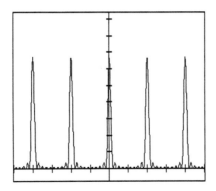

FIGURE 6.14
Graph of the structure factor from a linear array of 10 identical apertures.

Formula (6.6) says that the diffraction pattern intensity from a linear array of M apertures is equal to the diffraction pattern intensity from a single aperture (called the *form factor* or *shape function*) multiplied by a *structure factor* $\mathcal{S}_M(u)$ defined by

$$\mathcal{S}_M(u) = \frac{\sin^2(\pi M \delta u/\lambda D)}{\sin^2(\pi \delta u/\lambda D)} = M^2 \frac{\text{sinc}^2(M\delta u/\lambda D)}{\text{sinc}^2(\delta u/\lambda D)} \ . \qquad (6.8)$$

[The second formula for $\mathcal{S}_M(u)$ is the one to use for graphing by *FAS*.]

In Figure 6.14 we have graphed the structure factor $\mathcal{S}_M(u)$. The pattern created *in the u-v plane* would be a series of vertical strips of high intensity, M^2, with some side fringes from the lower intensity peaks.

A common type of linear array is a *slit diffraction grating*. The slit diffraction grating is a linear array of M vertical slits. In this case, the shape function $I_0(u, v, D)$ has a pattern like the one shown in Figure 6.37(a). When M is very large, the high intensity vertical strips from the structure factor cross the horizontal band from the shape factor, creating a sequence of thin vertical line segments (see Figure 6.15). In this case, *the graph of $I(u, 0, D) = I_0(u, 0, D)\mathcal{S}_M(u)$ along the u-axis is a good indicator of the form of $I(u, v, D)$.* We will refer to $I(u, 0, D)\mathcal{S}_M(u)$ as the *instrument function* for the diffraction grating. And we label the instrument function by $\mathcal{I}(u)$, thus

$$\mathcal{I}(u) = I(u, 0, D)\mathcal{S}_M(u), \qquad \textit{(instrument function)}. \qquad (6.9)$$

Notice, in particular, that the thin lines corresponding to the main peaks in

$$
\cdots \quad \Big| \quad\quad \Big| \quad\quad \Big| \quad\quad \Big| \quad\quad \Big| \quad\quad \Big| \quad\quad \Big| \quad \cdots
$$

$$
-\frac{3\lambda D}{\delta} \qquad -\frac{2\lambda D}{\delta} \qquad -\frac{\lambda D}{\delta} \qquad 0 \qquad \frac{\lambda D}{\delta} \qquad \frac{2\lambda D}{\delta} \qquad \frac{3\lambda D}{\delta}
$$

FIGURE 6.15
Diffraction pattern from a linear array of vertical slits.

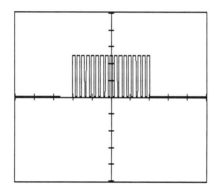

FIGURE 6.16
One-dimensional 17-slit diffraction grating.

the instrument function are located at the points

$$
u = 0,\ \pm\frac{\lambda D}{\delta},\ \pm\frac{2\lambda D}{\delta}, \ldots,\ \pm\frac{m\lambda D}{\delta}, \ldots.
$$

We now show how to use *FAS* to model the slit diffraction grating, and generalizations of it.

Example 6.10   SLIT DIFFRACTION GRATING
Use *FAS* to graph the instrument function for a slit diffraction grating.   ☐

SOLUTION   First, we show how to generate a graph like Figure 6.16, which is a good model for a 17-slit diffraction grating. Here is the procedure for producing such a graph.

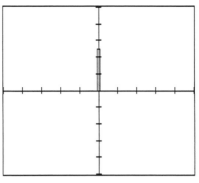

(a) unit function

X interval: $[-32, 32]$ X increment = 6.4
Y interval: $[-2, 2]$ Y increment = .4

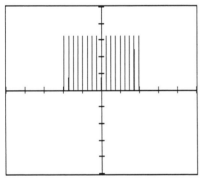

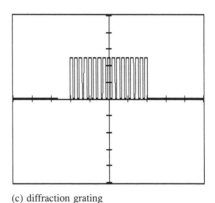

(b) comb function (17 spikes)

X interval: $[-32, 32]$ X increment = 6.4
Y interval: $[-6402, 6402]$ Y increment = 1280.2

(c) diffraction grating

X interval: $[-32, 32]$ X increment = 6.4
Y interval: $[-2, 2]$ Y increment = .4

FIGURE 6.17
Creation of a one-dimensional diffraction grating.

Step 1: Choose *Conv* from the initial menu. Then choose *convolution*, an interval of $[-32, 32]$, and 4096 points.

Step 2: For the first function, enter the function $\text{rec}(x)$ and the graph shown in Figure 6.17(a) results. This function is called the *unit function*.

Step 3: For the second function, choose *Point* from the function menu. Enter 17 for the number of integers to assign values to, and then enter the

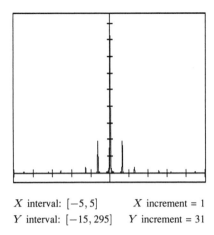

X interval: $[-5, 5]$ X increment = 1
Y interval: $[-15, 295]$ Y increment = 31

FIGURE 6.18
Instrument function from 17-slit grating.

following data:

Integer Number? 0	Function Value? 4096
Integer Number? 100	Function Value? 4096
Integer Number? −100	Function Value? 4096
Integer Number? 200	Function Value? 4096
Integer Number? −200	Function Value? 4096

| Integer Number? 800 | Function Value? 4096 |
| Integer Number? −800 | Function Value? 4096 |

The graph shown in Figure 6.17(b) results. This is called a *comb function*. You might save this comb function, since it will be used for other examples.

Step 4: Press y when asked if you want to divide by the length of the interval. A graph like the one shown in Figure 6.17(c) results. Save this graph for use in the *Four T* procedure.

If you compute the *power spectrum* of the function obtained at the end of the procedure just described, then you get a graph like the one shown in Figure 6.18. This is the graph of the instrument function $\mathcal{I}(u)$.

To see the relation of the form factor to the instrument function, you can proceed as follows. Save the instrument function, then compute the power spectrum of $17\text{rec}(x)$ where 17 is the number of slits. Save this graph, and then using

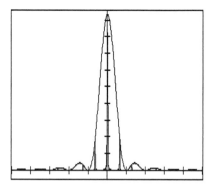

FIGURE 6.19
Form factor and instrument function from a 17-slit grating.

the *Graph* procedure, plot these last two graphs together. This results in the graphs shown in Figure 6.19. ∎

7 Spectral analysis with diffraction gratings

The importance of diffraction gratings consists in their ability to *spatially separate* light of distinct wavelengths. This process is called *spectral analysis*.

Suppose that light composed of two distinct wavelengths, say $\lambda_1 < \lambda_2$, illuminates in a coherent way a diffraction grating consisting of M vertical slits. We shall assume that, although the spatial coherency condition (1.15) is satisfied, the light is *time incoherent*. By time incoherency we mean that the *intensities from the separate wavelengths simply add together* with no interference. (A mathematical formulation of this is similar to the one we used to define spatial incoherency in Section 1, but the details would take us too far afield; see [Bo-W].) As can be seen from Figure 6.15, the location of the lines (intensity peaks) in the diffraction pattern are *wavelength dependent*. Thus, as shown in Figure 6.20, these wavelengths correspond to distinct lines, *spectral lines*, in the diffraction pattern.

The ability to separate distinct wavelengths depends on how close together the wavelengths are. Since each spectral line has a certain width, it is not possible to separate wavelengths that are much closer together than this width. One criterion that is used to determine when separation of wavelengths is possible is *Rayleigh's criterion*. Rayleigh's criterion says that to insure separation one should have the intensity peak for the larger wavelength λ_2 lying to the right of

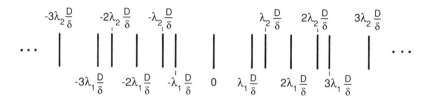

FIGURE 6.20
Spectral lines from two wavelengths $(\lambda_1 < \lambda_2)$.

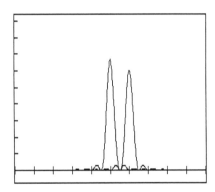

FIGURE 6.21
Intensity functions for two spectral lines.

the first zero value of the intensity function for the smaller wavelength λ_1 (see Figure 6.21).

For a vertical slit grating there are several wavelength-dependent spectral lines. Notice that in Figure 6.15 there are spectral lines at $\lambda D/\delta$, $2\lambda D/\delta$, ... Each of these lines will give rise to wavelength separation (the one at 0 will not, because it does not depend on wavelength). Suppose we concentrate on the line at $2\lambda D/\delta$, which generates the *second-order spectrum*. By Rayleigh's criterion, there will be a separation of λ_1 and λ_2 if

$$\frac{2\lambda_2 D}{\delta} \geq \frac{2\lambda_1 D}{\delta} + \frac{\lambda_1 D}{M\delta}$$

since $2\lambda_1 D/\delta + \lambda_1 D/M\delta$ is the position of the first zero to the right of the intensity peak for λ_1. Thus, we will have separation if

$$\frac{\lambda_2 - \lambda_1}{\lambda_1} \geq \frac{1}{2M} \, . \tag{7.1}$$

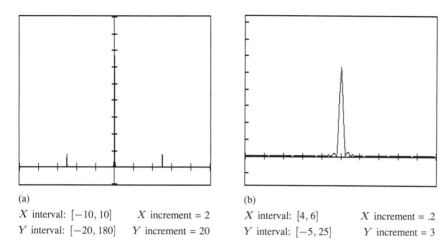

(a)

| X interval: $[-10, 10]$ | X increment = 2 |
| Y interval: $[-20, 180]$ | Y increment = 20 |

(b)

| X interval: $[4, 6]$ | X increment = .2 |
| Y interval: $[-5, 25]$ | Y increment = 3 |

FIGURE 6.22
Diffraction from a sinusoidal grating. (a) Graph of instrument function.
(b) Closeup of spectral line.

Formula (7.1) describes the minimum *relative separation* needed in the second-order spectrum between λ_1 and λ_2 according to Rayleigh's criterion. Notice that the magnitude of this minumum relative separation is decreased if M is increased. When the slit distance δ is kept constant, increasing M corresponds to increasing the length ($\approx M\delta$) of the grating. This gives the general rule that *the resolution* (ability to separate wavelengths) *of a vertical slit grating is proportional to the length of the grating.*

There are several other criterion for measuring resolution of diffraction gratings, for example, *Sparrow's criterion* (see [Mi-T]). We will not discuss these other criterion here.

Vertical slit gratings are not the only type of diffraction grating. Another popular type is the *sinusoidal* (or *holographic*) grating. This grating has an aperture function of the form (here we describe just the x-direction component, as discussed in Section 6)

$$A(x) = \text{rec}\left(\frac{x}{\beta}\right)\left[(1 - \alpha) + \alpha\cos(2\pi Nx)\right]. \tag{7.2}$$

Here β is a positive constant, α is a constant satisfying $0 < \alpha < 0.5$, and N is a positive integer called the *frequency* of the grating.

For example, in Figure 6.22(a) and (b) we graph the instrument function resulting from choosing $\beta = 20$, $\alpha = 0.4$, and $N = 5$. The main feature of the instrument function for the sinusoidal grating is that essentially all the light away from the central peak at the origin is chanelled into *one* intense peak on each side (instead of several peaks on each side for the vertical slit grating). This feature, which removes the difficulty of distinguishing several spectral orders and also can yield more intense spectral lines, has made sinusoidal gratings very popular.

We will now work out the exact transform of the function $A(x)$ in (7.2). Rewriting $\cos(2\pi Nx)$ as a sum of exponentials, we have

$$A(x) = (1 - \alpha)\mathrm{rec}\left(\frac{x}{\beta}\right) + \frac{\alpha}{2}\mathrm{rec}\left(\frac{x}{\beta}\right)e^{i2\pi Nx} + \frac{\alpha}{2}\mathrm{rec}\left(\frac{x}{\beta}\right)e^{-i2\pi Nx}.$$

Applying the scaling property to transform $\mathrm{rec}\,(x/\beta)$ and using the modulation property, we have

$$\hat{A}(u) = (1 - \alpha)\beta\mathrm{sinc}\,(\beta u) + \frac{\alpha\beta}{2}\mathrm{sinc}\,(\beta(u - N))$$

$$+ \frac{\alpha\beta}{2}\mathrm{sinc}\,(\beta(u + N)). \qquad (7.3)$$

From (7.3) we can approximate $|\hat{A}(u)|^2$ by *ignoring* all the nonsquare terms. The approximation

$$|\hat{A}(u)|^2 \approx (1 - \alpha)^2\beta^2\mathrm{sinc}^2\,(\beta u) + \frac{\alpha^2\beta^2}{4}\mathrm{sinc}^2\,(\beta(u - N))$$

$$+ \frac{\alpha^2\beta^2}{4}\mathrm{sinc}^2\,(\beta(u + N)) \qquad (7.4)$$

will be valid when N is large enough that the cross terms are approximately zero.

From (7.4) we obtain for the instrument function $\mathcal{I}(u)$

$$\mathcal{I}(u) \approx (1 - \alpha)^2\beta^2\mathrm{sinc}^2\left(\frac{\beta u}{\lambda D}\right) + \frac{\alpha^2\beta^2}{4}\mathrm{sinc}^2\left(\beta\left(\frac{u}{\lambda D} - N\right)\right)$$

$$+ \frac{\alpha^2\beta^2}{4}\mathrm{sinc}^2\left(\beta\left(\frac{u}{\lambda D} + N\right)\right). \qquad (7.5)$$

In Figure 6.23 we show this instrument function near $u = N\lambda D$. For u near $N\lambda D$ there is also the approximation

$$\mathcal{I}(u) \approx \frac{\alpha^2\beta^2}{4}\mathrm{sinc}^2\left(\beta\left(\frac{u}{\lambda D} - N\right)\right), \qquad (u \approx N\lambda D). \qquad (7.6)$$

Using the spectral lines to the right of 0 created by this instrument function, we have by Rayleigh's criterion that

$$\frac{\lambda_2 - \lambda_1}{\lambda_1} \geq \frac{1}{N\beta} \qquad (7.7)$$

is needed to insure separation of distinct wavelengths. Thus, *the resolution of a sinusoidal grating is increased by either increasing the length β of the grating or by increasing the frequency N of the grating.*

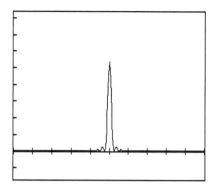

FIGURE 6.23
Graph of instrument function for a sinusoidal grating, centered on $N\lambda D$.

REMARK 7.8   Besides having advantages in spectral analysis, sinusoidal gratings are easily manufactured. Here is a brief outline of the process by which they are made. By exposing a photoreactive substance to two interfering laser beams, a sinusoidal interference pattern (as described in Section 5) of light intensity is converted into a sinusoidal pattern of chemically transformed material. This transformed material is then sloughed off with a solvent, creating a wafer whose surface profile is also sinusoidal. By using a material that partially absorbs light, a transmittance intensity similar to the sinusoid $(1 - \alpha) + \alpha \cos(2\pi N x)$ is created. The value of α depends on the light absorbency of the material. Finally, by placing the wafer in an opaque holder, an aperture function like (7.2) is produced.   ∎

8   The phase transformation induced by a thin lens

In this section we begin our discussion of the effect of a thin lens on monochromatic light. Understanding the effect of lenses is obviously of fundamental importance in optics.

To understand the effect of a lens, we must modify our discussion in Sections 1 and 2. Suppose that light travels along a path **C** from the point $(\mathbf{x}, 0)$ to the point $(\mathbf{u}, D + \epsilon)$, where $(\mathbf{u}, D + \epsilon)$ is located at the *exit plane* of a thin lens (see Figure 6.24). As shown in Figure 6.24, we will assume that the lens is thin enough that the light path from the entrance plane to the exit plane is horizontal (parallel to the optic axis). In particular, the path **C** intersects the entrance plane at the point $(\mathbf{u}, D)$.

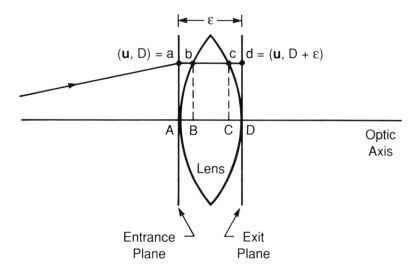

FIGURE 6.24
Light ray through a thin lens.

We now need to reexamine the following quantity from (1.1):

$$\frac{1}{\lambda|\mathbf{C}|} e^{i\frac{2\pi}{\lambda}\int_{\mathbf{C}} \eta(s)\, ds} \tag{8.1}$$

If we let $\mathbf{C}_1$ be the path from $(\mathbf{x}, 0)$ to $(\mathbf{u}, D)$ and let $\mathbf{C}_2$ be the straight line from $(\mathbf{u}, D)$ to $(\mathbf{u}, D + \epsilon)$, then

$$\frac{1}{\lambda|\mathbf{C}|} e^{i\frac{2\pi}{\lambda}\int_{\mathbf{C}} \eta(s)\, ds} = \frac{1}{\lambda|\mathbf{C}|} e^{i\frac{2\pi}{\lambda}\left[\int_{\mathbf{C}_1} \eta(s)\, ds + \int_{\mathbf{C}_2} \eta(s)\, ds\right]}$$

$$= \frac{1}{\lambda|\mathbf{C}|} e^{i\frac{2\pi}{\lambda}\int_{\mathbf{C}_1} \eta(s)\, ds}\, e^{i\frac{2\pi}{\lambda}\int_{\mathbf{C}_2} \eta(s)\, ds}. \tag{8.2}$$

We now assume that the index of refraction $\eta(s)$ is the constant 1 outside the lens, and that within the lens $\eta(s)$ is a constant, which we denote by η. For example, η might be the index of refraction of glass (relative to air). In any case, we require that $\eta > 1$.

Since $\eta(s) = 1$ over $\mathbf{C}_1$ we have

$$\frac{1}{\lambda|\mathbf{C}|} e^{i\frac{2\pi}{\lambda}\int_{\mathbf{C}_1} \eta(s)\, ds} = \frac{1}{\lambda|\mathbf{C}|} e^{i\frac{2\pi}{\lambda}|\mathbf{C}_1|}.$$

Making the same approximation as we described in discussing Fresnel diffraction in Section 2, we get

$$\frac{1}{\lambda|\mathbf{C}|}e^{i\frac{2\pi}{\lambda}|\mathbf{C}_1|} \approx \frac{e^{i\frac{2\pi}{\lambda}D}}{\lambda|\mathbf{C}|}e^{\frac{i\pi}{\lambda D}|\mathbf{u}-\mathbf{x}|^2}.$$

Thus, (8.2) becomes

$$\frac{1}{\lambda|\mathbf{C}|}e^{i\frac{2\pi}{\lambda}\int_{\mathbf{C}}\eta(s)\,ds} \approx \frac{e^{i\frac{2\pi}{\lambda}D}}{\lambda|\mathbf{C}|}e^{\frac{i\pi}{\lambda D}|\mathbf{u}-\mathbf{x}|^2}e^{i\frac{2\pi}{\lambda}\int_{\mathbf{C}_2}\eta(s)\,ds}. \tag{8.3}$$

Since $|\mathbf{C}| = |\mathbf{C}_1| + |\mathbf{C}_2| = |\mathbf{C}_1| + \epsilon$, we obtain (by similar reasoning as we used in Section 2, since ϵ is very small)

$$\frac{1}{\lambda|\mathbf{C}|} = \frac{1}{\lambda|\mathbf{C}_1| + \lambda\epsilon} \approx \frac{1}{\lambda D + \lambda\epsilon} \approx \frac{1}{\lambda D}.$$

So (8.3) becomes

$$\frac{1}{\lambda|\mathbf{C}|}e^{i\frac{2\pi}{\lambda}\int_{\mathbf{C}}\eta(s)\,ds} \approx \frac{e^{i\frac{2\pi}{\lambda}D}}{\lambda D}e^{\frac{i\pi}{\lambda D}|\mathbf{u}-\mathbf{x}|^2}e^{i\frac{2\pi}{\lambda}\int_{\mathbf{C}_2}\eta(s)\,ds}. \tag{8.4}$$

Now we examine the last factor in (8.4),

$$e^{i\frac{2\pi}{\lambda}\int_{\mathbf{C}_2}\eta(s)\,ds}. \tag{8.5}$$

We break up the line segment into three parts (see Figure 6.24). The segment from a to b (denoted by ab), the segment bc through the lens itself, and the segment cd. Therefore,

$$e^{i\frac{2\pi}{\lambda}\int_{\mathbf{C}_2}\eta(s)\,ds} = e^{i\frac{2\pi}{\lambda}\left(\int_{ab}1\,ds + \int_{bc}\eta\,ds + \int_{cd}1\,ds\right)}$$

$$= e^{i\frac{2\pi}{\lambda}(|ab| + \eta|bc| + |cd|)}. \tag{8.6}$$

By projecting down vertically to the optic axis (see Figure 6.24), we have $|ab| = |AB|$, $|bc| = |BC|$, and $|cd| = |CD|$. Thus, from (8.6) we get

$$e^{i\frac{2\pi}{\lambda}\int_{\mathbf{C}_2}\eta(s)\,ds} = e^{i\frac{2\pi}{\lambda}(|AB| + \eta|BC| + |CD|)}. \tag{8.7}$$

We now determine each of the lengths on the right side of (8.7). We assume, as shown in Figure 6.24, that the lens has the shape of a *double convex lens*. This is a very common type of lens, although our analysis can be adapted to cover any of the basic lens shapes. Let's suppose that the front and back surfaces of the lens are both spherical with radii of curvature R_1 and R_2, respectively (see Figure 6.25).

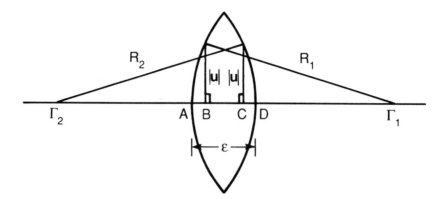

FIGURE 6.25
Geometry of a double convex lens with spherical surfaces.

Denoting the center of curvature for the front surface by Γ_1 and the back surface by Γ_2, we see from Figure 6.25 that

$$|B\Gamma_1| = \left[R_1^2 - |\mathbf{u}|^2\right]^{\frac{1}{2}}, \qquad |\Gamma_2 C| = \left[R_2^2 - |\mathbf{u}|^2\right]^{\frac{1}{2}}$$

$$|A\Gamma_1| = R_1, \qquad |\Gamma_2 D| = R_2, \qquad |AD| = \epsilon.$$

From all these equalities it is not hard to deduce that

$$|AB| = R_1 - \left[R_1^2 - |\mathbf{u}|^2\right]^{\frac{1}{2}}$$
$$|CD| = R_2 - \left[R_2^2 - |\mathbf{u}|^2\right]^{\frac{1}{2}}$$
$$|BC| = \epsilon - |AB| - |CD|. \tag{8.8}$$

We now make the approximations

$$\left[R_1^2 - |\mathbf{u}|^2\right]^{\frac{1}{2}} \approx R_1 - \frac{|\mathbf{u}|^2}{2R_1}, \qquad \left[R_2^2 - |\mathbf{u}|^2\right]^{\frac{1}{2}} \approx R_2 - \frac{|\mathbf{u}|^2}{2R_2} \tag{8.9}$$

which are similar to the approximations we made in discussing Fresnel diffraction in Section 2. These approximations will certainly be valid when R_1 and R_2 are relatively large in comparison to the radius of the lens.

Using (8.9) in (8.8), we obtain

$$|AB| \approx \frac{|\mathbf{u}|^2}{2R_1}, \qquad |CD| \approx \frac{|\mathbf{u}|^2}{2R_2}, \qquad |BC| = \epsilon - |AB| - |CD|. \tag{8.10}$$

Employing (8.10) we have

$$|AB| + \eta|BC| + |CD| = \epsilon\eta + (1 - \eta)\left(|AB| + |CD|\right)$$

$$\approx \epsilon\eta + (1 - \eta)\left[\frac{1}{R_1} + \frac{1}{R_2}\right]\frac{|\mathbf{u}|^2}{2}. \qquad (8.11)$$

Before we make use of this result, we make the following definition. For a *double convex lens* with spherical surfaces having radii of curvature R_1 and R_2, we define the *focal length* f via the formula

$$\frac{1}{f} = (\eta - 1)\left[\frac{1}{R_1} + \frac{1}{R_2}\right]. \qquad (8.12)$$

From (8.12), we can express (8.11) as

$$|AB| + \eta|BC| + |CD| \approx \epsilon\eta - \frac{|\mathbf{u}|^2}{2f}. \qquad (8.13)$$

Combining (8.13) with (8.7) we have

$$e^{i\frac{2\pi}{\lambda}\int_{C_2}\eta(s)\,ds} \approx e^{i\frac{2\pi}{\lambda}\epsilon\eta}e^{\frac{-i\pi}{\lambda f}|\mathbf{u}|^2}. \qquad (8.14)$$

Substituting the right side of (8.14) in place of the last factor in (8.4) yields

$$\frac{1}{\lambda|\mathbf{C}|}e^{i\frac{2\pi}{\lambda}\int_C\eta(s)\,ds} \approx \frac{e^{i\frac{2\pi}{\lambda}(D+\epsilon\eta)}}{\lambda D}e^{\frac{i\pi}{\lambda D}|\mathbf{u}-\mathbf{x}|^2}e^{\frac{-i\pi}{\lambda f}|\mathbf{u}|^2}. \qquad (8.15)$$

As a final simplification, we define δ to be the constant $(2\pi/\lambda)(D + \epsilon\eta)$, and then (8.15) becomes

$$\frac{1}{\lambda|\mathbf{C}|}e^{i\frac{2\pi}{\lambda}\int_C\eta(s)\,ds} \approx \frac{e^{i\delta}}{\lambda D}e^{\frac{i\pi}{\lambda D}|\mathbf{u}-\mathbf{x}|^2}e^{\frac{-i\pi}{\lambda f}|\mathbf{u}|^2}. \qquad (8.16)$$

Using (8.16) we can express the light $\psi(\mathbf{u}, D + \epsilon, t)$ at the exit plane of the lens as

$$\psi(\mathbf{u}, D + \epsilon, t) \approx \frac{e^{i\delta}}{\lambda D}\int_{\mathbf{R}^2}A(\mathbf{x})\psi(\mathbf{x}, 0, t)e^{\frac{i\pi}{\lambda D}|\mathbf{u}-\mathbf{x}|^2}\,d\mathbf{x}\,e^{\frac{-i\pi}{\lambda f}|\mathbf{u}|^2}P(\mathbf{u}). \qquad (8.17)$$

The aperture function $A(\mathbf{x})$ has been introduced in Section 1. The factor $P(\mathbf{u})$ is the *pupil function of the lens*, whose purpose is mainly to express the fact that the lens has a *finite extent*. Most often, we will assume that

$$P(\mathbf{u}) = \begin{cases} 1 & \text{if } \mathbf{u} \text{ lies inside the lens aperture} \\ 0 & \text{if } \mathbf{u} \text{ lies outside the lens aperture.} \end{cases} \qquad (8.18)$$

Formula (8.18) describes a lens with *no aberrations* (this is also called the

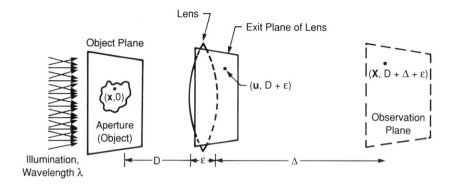

FIGURE 6.26
Geometry of single lens imaging.

diffraction limited case). The pupil function plays the role of an aperture function when light diffracts through the lens aperture. When there are no aberrations, then the only limitation to perfect imaging is the finite extent of the lens, through which light from the original aperture must diffract. (Since we are dealing with monochromatic light we are ignoring *chromatic aberration*, which affects all simple lenses.)

In subsequent sections, we will examine the profound implications of formula (8.17).

9 Imaging with a single lens

In this section, we will derive the general formula for imaging with a single lens. This derivation shows that there is a *wave optical* reason for the classic lens equation. In subsequent sections we will discuss two specific cases of the imaging formula, the cases of *coherent* and *incoherent* illumination, which are the two fundamental types of illumination.

Suppose that we have the setup shown in Figure 6.26, a lens situated D units in front of the aperture and an observation plane located Δ units behind the lens. From formula (8.17) we have that the light emergent from the exit plane of the lens is described by

$$\psi(\mathbf{u}, D + \epsilon, t) \approx \frac{e^{i\delta}}{\lambda D} \int_{\mathbf{R}^2} A(\mathbf{x})\psi(\mathbf{x}, 0, t)e^{\frac{i\pi}{\lambda D}|\mathbf{u}-\mathbf{x}|^2}\, d\mathbf{x}\, e^{\frac{-i\pi}{\lambda f}|\mathbf{u}|^2} P(\mathbf{u}). \qquad (9.1)$$

NOTATION 9.2
Our formulas in this section will be very unwieldy unless we do something to abbreviate them. Accordingly, we adopt the following conventions. We will write

$\int$ instead of $\int_{\mathbf{R}^2}$; all integrals will be over $\mathbf{R}^2$. Also, instead of $\psi(\mathbf{x}, 0, t)$ we will just write $\psi(\mathbf{x}, t)$, the vector $\mathbf{x}$ being used to indicate $(\mathbf{x}, 0)$. Similarly, we shall write $\psi(\mathbf{u}, t)$ in place of $\psi(\mathbf{u}, D+\epsilon, t)$ and $\psi(\mathbf{X}, t)$ in place of $\psi(\mathbf{X}, D+\Delta+\epsilon, t)$. The function $\psi(\mathbf{X}, t)$ represents the light at the observation plane.

With the notation described in (9.2), formula (9.1) simplifies to

$$\psi(\mathbf{u}, t) \approx \frac{e^{i\delta}}{\lambda D} \int A(\mathbf{x})\psi(\mathbf{x}, t)e^{\frac{i\pi}{\lambda D}|\mathbf{u}-\mathbf{x}|^2}\, d\mathbf{x}\, e^{\frac{-i\pi}{\lambda f}|\mathbf{u}|^2}P(\mathbf{u}). \tag{9.3}$$

Applying the Fresnel diffraction formula (2.14) to $\psi(\mathbf{u}, t)$ instead of $A(\mathbf{x})\psi(\mathbf{x}, 0, t)$, we can express the light $\psi(\mathbf{X}, t)$ at the observation plane by

$$\psi(\mathbf{X}, t) \approx \frac{e^{i\frac{2\pi}{\lambda}\Delta}}{\lambda\Delta} \int \psi(\mathbf{u}, t)e^{\frac{i\pi}{\lambda\Delta}|\mathbf{X}-\mathbf{u}|^2}\, d\mathbf{u}. \tag{9.4}$$

Our next step will be to substitute the right side of (9.3) into the integrand in (9.4). To simplify notation we define $E(\mathbf{x}, \mathbf{u}, \mathbf{X})$ by

$$E(\mathbf{x}, \mathbf{u}, \mathbf{X}) = e^{\frac{i\pi}{\lambda D}|\mathbf{u}-\mathbf{x}|^2} e^{\frac{-i\pi}{\lambda f}|\mathbf{u}|^2} e^{\frac{i\pi}{\lambda\Delta}|\mathbf{X}-\mathbf{u}|^2}. \tag{9.5}$$

This quantity is what we get if we combine all the complex exponentials involved in (9.3) and (9.4). Now, substituting the right side of (9.3) into the integrand in (9.4) and combining all the complex exponentials so that we get $E(\mathbf{x}, \mathbf{u}, \mathbf{X})$, we obtain

$$\psi(\mathbf{X}, t) \approx \frac{e^{i\frac{2\pi}{\lambda}\Delta}e^{i\delta}}{\lambda^2 D\Delta} \int\int A(\mathbf{x})\psi(\mathbf{x}, t)E(\mathbf{x}, \mathbf{u}, \mathbf{X})P(\mathbf{u})\, d\mathbf{x}\, d\mathbf{u}. \tag{9.6}$$

To save further space, we define c by $c = (2\pi/\lambda)\Delta + \delta$, so that

$$\psi(\mathbf{X}, t) \approx \frac{e^{ic}}{\lambda^2 D\Delta} \int\int A(\mathbf{x})\psi(\mathbf{x}, t)E(\mathbf{x}, \mathbf{u}, \mathbf{X})P(\mathbf{u})\, d\mathbf{x}\, d\mathbf{u}. \tag{9.7}$$

Our next step is to simplify the expression for $E(\mathbf{x}, \mathbf{u}, \mathbf{X})$ in (9.5), which will help to make (9.7) more meaningful. If we expand the quantities $|\mathbf{u} - \mathbf{x}|^2$ and $|\mathbf{X} - \mathbf{u}|^2$ in (9.5), and recombine exponentials we obtain

$$E(\mathbf{x}, \mathbf{u}, \mathbf{X}) = e^{\frac{i\pi}{\lambda}\left[\frac{1}{D}+\frac{1}{\Delta}-\frac{1}{f}\right]|\mathbf{u}|^2} e^{-i\frac{2\pi}{\lambda}\mathbf{u}\cdot\left(\frac{\mathbf{x}}{D}+\frac{\mathbf{X}}{\Delta}\right)} e^{\frac{i\pi}{\lambda D}|\mathbf{x}|^2} e^{\frac{i\pi}{\lambda\Delta}|\mathbf{X}|^2}. \tag{9.8}$$

The first exponential factor in (9.8) will be completely eliminated if we assume that

$$\frac{1}{D} + \frac{1}{\Delta} - \frac{1}{f} = 0. \tag{9.9}$$

Formula (9.9) is the classic *imaging equation* for a thin lens.

Let's assume that (9.9) holds. Then the first factor in (9.8) is just $e^0 = 1$. Hence (9.8) becomes

$$E(\mathbf{x}, \mathbf{u}, \mathbf{X}) = e^{-i\frac{2\pi}{\lambda}\mathbf{u}\cdot(\frac{\mathbf{x}}{D}+\frac{\mathbf{X}}{\Delta})}e^{\frac{i\pi}{\lambda D}|\mathbf{x}|^2}e^{\frac{i\pi}{\lambda\Delta}|\mathbf{X}|^2}. \tag{9.10}$$

Substituting the right side of (9.10) in place of $E(\mathbf{x}, \mathbf{u}, \mathbf{X})$ in (9.7) and factoring the complex exponential involving $|\mathbf{X}|^2$ outside the integrals, we get

$$\psi(\mathbf{X}, t) \approx \frac{e^{ic}}{\lambda^2 D\Delta}e^{\frac{i\pi}{\lambda\Delta}|\mathbf{X}|^2}$$
$$\int\int A(\mathbf{x})\psi(\mathbf{x}, t)P(\mathbf{u})e^{-i\frac{2\pi}{\lambda}\mathbf{u}\cdot(\frac{\mathbf{x}}{D}+\frac{\mathbf{X}}{\Delta})}e^{\frac{i\pi}{\lambda D}|\mathbf{x}|^2}\,d\mathbf{u}\,d\mathbf{x}. \tag{9.11}$$

We also interchanged the integrals to obtain (9.11).
Since

$$\int P(\mathbf{u})e^{-i\frac{2\pi}{\lambda}\mathbf{u}\cdot(\frac{\mathbf{x}}{D}+\frac{\mathbf{X}}{\Delta})}\,d\mathbf{u} = \hat{P}\left[\frac{\mathbf{x}}{\lambda D} + \frac{\mathbf{X}}{\lambda\Delta}\right]$$

we rewrite (9.11) as

$$\psi(\mathbf{X}, t) \approx \frac{e^{ic}}{\lambda^2 D\Delta}e^{\frac{i\pi}{\lambda\Delta}|\mathbf{X}|^2}\int A(\mathbf{x})\psi(\mathbf{x}, t)\hat{P}\left[\frac{\mathbf{x}}{\lambda D} + \frac{\mathbf{X}}{\lambda\Delta}\right]e^{\frac{i\pi}{\lambda D}|\mathbf{x}|^2}\,d\mathbf{x}. \tag{9.12}$$

Finally, suppose an observation of the intensity $I(\mathbf{X})$ defined by

$$I(\mathbf{X}) = \frac{1}{T}\int_0^T |\psi(\mathbf{X}, t)|^2\,dt$$

is made (for some large value of T). Then we obtain from (9.12)

$$I(\mathbf{X}) \approx \frac{1}{(\lambda^2 D\Delta)^2}\int\int\left\{A(\mathbf{x})e^{\frac{i\pi}{\lambda D}|\mathbf{x}|^2}\hat{P}\left[\frac{\mathbf{x}}{\lambda D} + \frac{\mathbf{X}}{\lambda\Delta}\right]\right\}$$
$$\times\left\{A(\mathbf{z})e^{\frac{i\pi}{\lambda D}|\mathbf{z}|^2}\hat{P}\left[\frac{\mathbf{z}}{\lambda D} + \frac{\mathbf{X}}{\lambda\Delta}\right]\right\}^*\Gamma(\mathbf{x}, \mathbf{z})\,d\mathbf{x}\,d\mathbf{z} \tag{9.13}$$

where $\Gamma(\mathbf{x}, \mathbf{z})$ is the *cross-correlation function* defined in (1.11).

Formula (9.13) is the *general equation for imaging with monochromatic light*. It is extremely complicated, and that's because we have made no assumptions about the initial illuminating light other than its being unpolarized and having a single wavelength λ.

In the next two sections we will discuss the two main methods for simplifying (9.13). These involve the assumption of either coherent or incoherent illumination.

10 Coherent imaging

In the previous section we derived the general formula for imaging. In this section we will discuss one of the two fundamental assumptions, *coherency of illumination*, that can be used to simplify that formula.

We assume that the light illuminating the aperture in Figure 6.26 is coherent. That is, the cross-correlation function $\Gamma(\mathbf{x}, \mathbf{z})$ satisfies the following condition:

$$\Gamma(\mathbf{x}, \mathbf{z}) = 1 \qquad [Coherency] \tag{10.1}$$

for all points $\mathbf{x}$ and $\mathbf{z}$ *within the aperture*. For $\mathbf{x}$ or $\mathbf{z}$ lying outside the aperture, we would have either $A(\mathbf{x}) = 0$ or $A(\mathbf{z}) = 0$, so we are free to use (10.1) for *all* $\mathbf{x}$ and $\mathbf{z}$ in (9.13). Doing so, we find that the two integrals in (9.13) *separate into integrals over* $\mathbf{x}$ *and* $\mathbf{z}$, as follows:

$$I(\mathbf{X}) \approx \frac{1}{(\lambda^2 D \Delta)^2} \left\{ \int A(\mathbf{x}) e^{\frac{i\pi}{\lambda D} |\mathbf{x}|^2} \hat{P} \left[\frac{\mathbf{x}}{\lambda D} + \frac{\mathbf{X}}{\lambda \Delta} \right] d\mathbf{x} \right\}$$
$$\times \left\{ \int A(\mathbf{z}) e^{\frac{i\pi}{\lambda D} |\mathbf{z}|^2} \hat{P} \left[\frac{\mathbf{z}}{\lambda D} + \frac{\mathbf{X}}{\lambda \Delta} \right] d\mathbf{z} \right\}^* .$$

Substituting $\mathbf{x}$ in place of $\mathbf{z}$ in the second integral above, we get

$$I(\mathbf{X}) \approx \left| \frac{1}{\lambda^2 D \Delta} \int A(\mathbf{x}) e^{\frac{i\pi}{\lambda D} |\mathbf{x}|^2} \hat{P} \left[\frac{\mathbf{x}}{\lambda D} + \frac{\mathbf{X}}{\lambda \Delta} \right] d\mathbf{x} \right|^2 . \tag{10.2}$$

To simplify (10.2) further, we make the *small object assumption* that, for all $\mathbf{x}$ lying in the aperture, $|\mathbf{x}|$ is small enough that

$$e^{\frac{i\pi}{\lambda D} |\mathbf{x}|^2} \approx 1, \qquad \text{for } \mathbf{x} \text{ within the aperture.} \tag{10.3}$$

Formula (10.3) is a natural approximation to make in the field of microscopy, for example. Since $A(\mathbf{x}) = 0$ for $\mathbf{x}$ outside the aperture, we can use (10.3) for *all* $\mathbf{x}$ in (10.2). Then, (10.2) simplifies to

$$I(\mathbf{x}) \approx \left| \frac{1}{\lambda^2 D \Delta} \int A(\mathbf{x}) \hat{P} \left[\frac{\mathbf{x}}{\lambda D} + \frac{\mathbf{X}}{\lambda \Delta} \right] d\mathbf{x} \right|^2 . \tag{10.4}$$

As a final simplification, we define the *degree of magnification* M by

$$M = \frac{\Delta}{D} . \tag{10.5}$$

We then make the substitution $\mathbf{x} = -\mathbf{s}/M$ (and $d\mathbf{x} = d\mathbf{s}/M^2$), so that (10.4) becomes

$$I(\mathbf{X}) \approx \left| \int \left\{ \frac{1}{M} A \left(\frac{-\mathbf{s}}{M} \right) \right\} \left\{ \frac{1}{(\lambda \Delta)^2} \hat{P} \left[\frac{1}{\lambda \Delta} (\mathbf{X} - \mathbf{s}) \right] \right\} d\mathbf{s} \right|^2 . \tag{10.6}$$

The integral in (10.6) is a two-dimensional convolution. The function

$$\frac{1}{M} A\left(\frac{-\mathbf{X}}{M}\right), \qquad (magnified,\ inverted\ object). \qquad (10.7)$$

is convoluted with

$$\frac{1}{(\lambda\Delta)^2} \hat{P}\left[\frac{1}{\lambda\Delta}\mathbf{X}\right], \qquad (point\ spread\ function). \qquad (10.8)$$

The interpretation of the function in (10.7) as a magnified, inverted object is fairly clear (the factor $1/M$ on the outside is to preserve the 2-Norm squared, which measures intensity). In the following four examples, we will show how *FAS* can be used to understand the role of the function in (10.8), which is called the *point spread function* (or *PSF*).

Example 10.9

Suppose that coherent light having wavelength $\lambda = 5 \times 10^{-4}$ mm illuminates an edge having aperture function

$$A(x, y) = \begin{cases} 1 & \text{if } x > 0 \\ 0 & \text{if } x < 0 \end{cases}$$

and this edge is imaged by a lens with a square aperture of side length 40 mm and focal length 1 m. Assume that $D = 2$ m and $\Delta = 2$ m. Describe the image that results. ⬚

SOLUTION We will use (10.6) to describe the image (even though, strictly speaking, the small object condition is not satisfied). By the change of scale property,

$$P(\lambda\Delta\mathbf{x}) \xrightarrow{\mathcal{F}} \frac{1}{(\lambda\Delta)^2} \hat{P}\left[\frac{1}{\lambda\Delta}\mathbf{X}\right] \qquad (10.10)$$

where, for this example,

$$P(\lambda\Delta\mathbf{x}) = P(\lambda\Delta x, \lambda\Delta y) = \begin{cases} 1 & \text{if } |x| < \frac{20}{\lambda\Delta} = 20 \text{ and } |y| < 20 \\ 0 & \text{if } |x| > 20 \text{ or } |y| > 20 \end{cases}$$

$$= \operatorname{rec}\left(\frac{x}{40}\right) \operatorname{rec}\left(\frac{y}{40}\right). \qquad (10.11)$$

It follows that

$$\frac{1}{(\lambda\Delta)^2} \hat{P}\left[\frac{1}{\lambda\Delta}\mathbf{X}\right] = 40\operatorname{sinc}(40X)\,40\operatorname{sinc}(40Y). \qquad (10.12)$$

When this last function is convoluted with $A(-X, -Y)$, the integral separates into integrals over X and Y separately. Since the integral along the Y-direction will always be

$$\int_{-\infty}^{\infty} 40 \operatorname{sinc}(40Y)\, dY = \operatorname{rec}\left(\frac{y}{40}\right)\Big|_{y=0}, \qquad \textit{(Fourier inversion)}$$

$$= 1$$

we only need to convolute $40 \operatorname{sinc}(40X)$ with

$$f(X) = \begin{cases} 1 & \text{for } X > 0 \\ 0 & \text{for } X < 0 \end{cases}$$

to obtain a representation of the image. (*Note*: By convoluting with the original edge function, rather than its inversion through the origin, we follow the standard procedure in optics.) Using *FAS* to perform this convolution, we get the results shown in Figure 6.27. In Figure 6.27(a) we have the graph of $40 \operatorname{sinc}(40X)$, and in Figure 6.27(b) we have the graph of the edge. The resulting image, obtained by convoluting these two functions (and then summing squares) is shown in Figure 6.27(c).

The graph shown in Figure 6.27(c) is a good prediction of the effects of coherent imaging. See [Go], p. 133] for a nice photograph illustrating this. In particular, the oscillation and Gibbs' phenomenon at the edge boundary which appear in Figure 6.27(c) are plainly visible in that photograph. Moreover, if the *image edge boundary* is located at the point of half-intensity, then we see from Figure 6.27(c) that this occurs *slightly to the right of the true object edge boundary*. This is a well-known phenomenon in coherent imaging. The reader might consult [Mi-T] for more information on these points. ∎

Example 10.13

Suppose that coherent light having wavelength $\lambda = 5 \times 10^{-4}$ mm illuminates a small square with aperture function

$$A(x,y) = \begin{cases} 1 & \text{if } |x| < .001 \text{ mm and } |y| < .001 \text{ mm} \\ 0 & \text{if } |x| > .001 \text{ mm or } |y| > .001 \text{ mm} \end{cases} \tag{10.14}$$

and this square is imaged by a lens having focal length of $1/11$ m with a square aperture of side length 20 mm. Assume that $D = .1$ m and $\Delta = 1$ m. Describe the image. ∎

SOLUTION In this example, the magnified inverted object is defined by

$$\frac{1}{10} A\left(\frac{-x}{10}, \frac{-y}{10}\right) = \begin{cases} 0.1 & \text{if } |x| < .01 \text{ mm and } |y| < .01 \text{ mm} \\ 0 & \text{if } |x| > .01 \text{ mm or } |y| > .01 \text{ mm.} \end{cases} \tag{10.15}$$

In this case, the convolution in (10.6) splits separately into two identical convolutions in x and y. In Figure 6.28 we show the results of the convolution

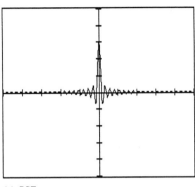

(a) *PSF*
X interval: [−1, 1]   X increment = .2
Y interval: [−64, 64]   Y increment = 12.8

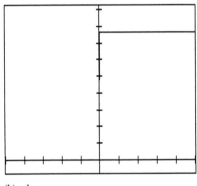

(b) edge
X interval: [−10, 10]   X increment = 2
Y interval: [−.1, 1.2]   Y increment = .13

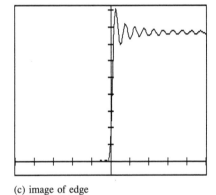

(c) image of edge
X interval: [−.5, .5]   X increment = .1
Y interval: [−.1, 1.2]   Y increment = .13

FIGURE 6.27
Coherent imaging of an edge.

with respect to x. The graph in Figure 6.28(a) is a sinc-type function that is the
Fourier transform of

$$P(.5x) = \begin{cases} 1 & \text{if } |x| < 20 \\ 0 & \text{if } |x| > 20 \end{cases}$$

$$= \text{rec}\left(\frac{x}{40}\right).$$

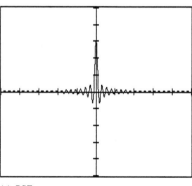

(a) *PSF*
X interval: $[-1, 1]$ X increment = .2
Y interval: $[-64, 64]$ Y increment = 12.8

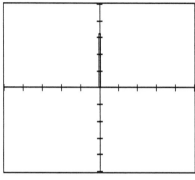

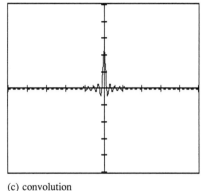

(b) point-like object
X interval: $[-1, 1]$ X increment = .2
Y interval: $[-.15, .15]$ Y increment = .03

(c) convolution
X interval: $[-1, 1]$ X increment = .2
Y interval: $[-64, 64]$ Y increment = 12.8

FIGURE 6.28
Coherent image of point-like object.

while Figure 6.28(b) shows the graph of

$$g(x) = \begin{cases} 0.1 & \text{if } |x| < .01 \\ 0 & \text{if } |x| > .01 \end{cases}$$
$$= 0.1 \operatorname{rec}(50x).$$

The convolution in Figure 6.28(c) is essentially just a reproduction of the function in Figure 6.28(a). For this reason, the function

$$\frac{1}{\lambda \Delta} \hat{P}\left[\frac{\mathbf{X}}{\lambda \Delta}\right]$$

is often called the point spread function. The effect of the *PSF* is to spread out a point, in this case a very small square or *pixel*. The convolution in (10.6) can be interpreted as a summation of the spreads of all the pixels composing the original image. (*Note*: The actual image will be the sum of squares of the convolution shown in Figure 6.28(c). We wanted to emphasize here, however, the reproduction of the *PSF* by a small point-like object.)   ∎

REMARK 10.16   These examples show some of the sensitivities of coherent imaging. Coherent imaging creates spread out images of very tiny features of an object; it also tends to brighten the edges of objects. For instance, dust particles will appear as *ringed objects* due to their tracing out of the *PSF* of the imaging lens (see Exercise 6.56). Other *false details* include enhancement and fringing of edges (as in our first example), and interference patterns due to the interference of nearby point spreads (see Exercise 6.57).   ∎

Our next two examples illustrate methods of image enhancement that are based on *modifying the transforms of objects*. These are just two examples taken from a wide variety of examples that belong to the category known as *spatial filtering* of images (i.e., modifying *transforms* of objects in order to enhance their images).

Example 10.17

In the field of microscopy there is a method known as *dark field imaging*. This method is used when intense background light obscures the details in an image. An opaque object (aperture stop) is placed over the center of the lens pupil. This aperture stop blocks out much of the background light and allows the previously obscured details to emerge.

In this example we describe a one-dimensional model for dark field imaging. In Figure 6.29(a) we show the graph of the following function (one-dimensional object):

$$A(x) = \exp(-(x/3) \wedge 10)(1 + .02\cos(6\pi x)). \qquad (10.18)$$

Its transform is shown in Figure 6.29(b). The central spike is due mostly to the transform of $\exp(-(x/3) \wedge 10)$ which forms a kind of *bright background* on which is superimposed the small oscillations of $.02\cos(6\pi x)$. The dark field method employs a pupil function of the form [see Figure 6.29(c)]

$$P_{\mathrm{DF}}(\lambda\Delta x) = \begin{cases} 0 & \text{if } 40 < |x| \\ 10 & \text{if } 3 < |x| < 40 \\ 0 & \text{if } |x| < 3. \end{cases} \qquad (10.19)$$

The value of 10 is assigned for $3 < |x| < 40$ in order to model an *increase in intensity* of illumination (so as to make the image light intense enough to

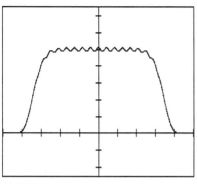

(a) object

X interval: $[-4, 4]$ $\quad\quad$ X increment = .8

Y interval: $[-.5, 1.5]$ $\quad$ Y increment = .2

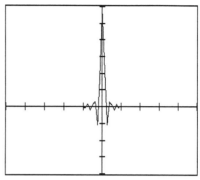

(b) transform of object function

X interval: $[-5, 5]$ $\quad\quad$ X increment = 1

Y interval: $[-4, 6]$ $\quad\quad$ Y increment = 1

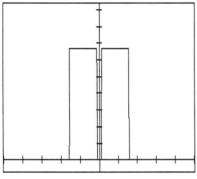

(c) dark field pupil function

X interval: $[-128, 128]$ X increment = 25.6

Y interval: $[-1, 14]$ $\quad\quad$ Y increment = 1.5

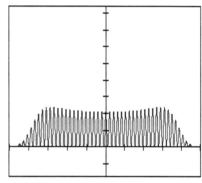

(d) dark field image

X interval: $[-4, 4]$ $\quad\quad$ X increment = .8

Y interval: $[-.5, 1.5]$ $\quad$ Y increment = .2

FIGURE 6.29
Dark field imaging.

be recorded, say by photographic film). After transforming this pupil function (using an interval of $[-128, 128]$ and 2048 points), convoluting with the object function in (10.18), and then summing squares, the dark field image shown in Figure 6.29(d) results.

A comparison of the *bright field image* using the unblocked pupil

$$P_{\text{BF}}(\lambda\Delta x) = \begin{cases} 0 & \text{if } |x| > 40 \\ 1 & \text{if } |x| < 40 \end{cases} \tag{10.20}$$

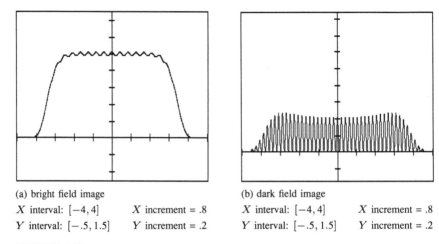

(a) bright field image
X interval: $[-4, 4]$ X increment = .8
Y interval: $[-.5, 1.5]$ Y increment = .2

(b) dark field image
X interval: $[-4, 4]$ X increment = .8
Y interval: $[-.5, 1.5]$ Y increment = .2

FIGURE 6.30
Comparison of bright field and dark field imaging of object shown in Figure 6.29(a).

is shown in Figure 6.30. In order to measure the visibility of the oscillations in the two images, one calculates the *relative contrast* between the peaks and valleys by

$$\text{contrast} = \frac{I_{max} - I_{min}}{I_{max} + I_{min}} . \tag{10.21}$$

For the bright field image in Figure 6.30(a) we get a contrast of .02, while for the dark field image the contrast is about 1. Thus, the dark field method has improved the image contrast by a factor of about 50.   ▢

There are problems associated with dark field imaging. As the reader might have observed, in the previous example the contrast was dramatically improved with dark field imaging, but the frequency of the oscillations *was increased by a factor of* 2. This is a *false detail* in the dark field image. It is this problem of false details that makes the interpretation of dark field images troublesome.

A method that alleviates some of the problems with false details is *gray field imaging*. In gray field imaging, the aperture stop *partially transmits* a small fraction of the light at the center of the transform, thus providing a gray background in the image (against which more details of the object are visible).

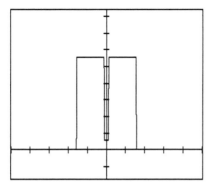

FIGURE 6.31
Gray field pupil function.

Example 10.22
Suppose that the object described by (10.18) is imaged but now a *gray field* pupil function of the form

$$P_{\mathrm{GF}}(\lambda \Delta x) = \begin{cases} 0 & \text{if } 40 < |x| \\ 10 & \text{if } 3 < |x| < 40 \\ 1 & \text{if } |x| < 3 \end{cases} \tag{10.23}$$

is used (see Figure 6.31).

In this case, the gray field image shown in Figure 6.32(b) results. The false detail no longer appears. Also, the contrast is about .2 which is a 10-fold improvement over the contrast in the bright field image. □

11 Fourier transforming properties of a lens

As we saw in section 10, with the dark field and gray field imaging examples, sometimes images can be enhanced by manipulating the transform of the object. This requires knowledge of the transform, however. While computer approximations can sometimes be used, it is an amazing fact that *nature produces the transform of the function automatically in the back focal plane of the imaging lens*! In this section we will derive this important property mathematically.

If we put $\Delta = f$ in formula (9.8), then we have

$$E(\mathbf{x}, \mathbf{u}, \mathbf{X}) = e^{\frac{i\pi}{\lambda D}|\mathbf{u}|^2} e^{-i\frac{2\pi}{\lambda}\mathbf{u}\cdot\left(\frac{\mathbf{x}}{D}+\frac{\mathbf{X}}{f}\right)} e^{\frac{i\pi}{\lambda D}|\mathbf{x}|^2} e^{\frac{i\pi}{\lambda f}|\mathbf{X}|^2}. \tag{11.1}$$

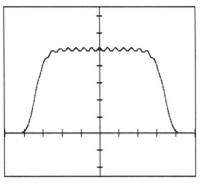 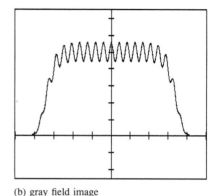

(a) bright field image

X interval: $[-4, 4]$	X increment $= .8$
Y interval: $[-.5, 1.5]$	Y increment $= .2$

(b) gray field image

X interval: $[-4, 4]$	X increment $= .8$
Y interval: $[-.5, 1.5]$	Y increment $= .2$

FIGURE 6.32
Comparison of bright field and gray field imaging of object shown in Figure 6.29(a).

Substituting the right side of (11.1) into formula (9.7) and rearranging the integrals, we obtain

$$\psi(\mathbf{X}, t) \approx \frac{e^{ic} e^{i\frac{\pi}{\lambda f}|\mathbf{X}|^2}}{\lambda^2 D f}$$

$$\int A(\mathbf{x})\psi(\mathbf{x}, t) e^{i\frac{\pi}{\lambda D}|\mathbf{x}|^2} \int P(\mathbf{u}) e^{i\frac{\pi}{\lambda D}|\mathbf{u}|^2} e^{-i\frac{2\pi}{\lambda D}\mathbf{u}\cdot\left(\mathbf{x}+\frac{D}{f}\mathbf{X}\right)} \, d\mathbf{u} \, d\mathbf{x}. \quad (11.2)$$

We now show how to simplify the integral

$$\frac{1}{\lambda D} \int P(\mathbf{u}) e^{i\frac{\pi}{\lambda D}|\mathbf{u}|^2} e^{-i\frac{2\pi}{\lambda D}\mathbf{u}\cdot\left(\mathbf{x}+\frac{D}{f}\mathbf{X}\right)} \, d\mathbf{u} \quad (11.3)$$

which is part of (11.2). First, we observe that

$$\left|\mathbf{u} - \left(\mathbf{x}+\frac{D}{f}\mathbf{X}\right)\right|^2 = |\mathbf{u}|^2 - 2\mathbf{u}\cdot\left(\mathbf{x}+\frac{D}{f}\mathbf{X}\right) + \left|\mathbf{x}+\frac{D}{f}\mathbf{X}\right|^2. \quad (11.4)$$

It then follows from (11.4) that

$$e^{i\frac{\pi}{\lambda D}|\mathbf{u}|^2} e^{-i\frac{2\pi}{\lambda D}\mathbf{u}\cdot\left(\mathbf{x}+\frac{D}{f}\mathbf{X}\right)} = e^{i\frac{\pi}{\lambda D}\left|\mathbf{u}-\left(\mathbf{x}+\frac{D}{f}\mathbf{X}\right)\right|^2} e^{\frac{-i\pi}{\lambda D}\left|\mathbf{x}+\frac{D}{f}\mathbf{X}\right|^2}. \quad (11.5)$$

Hence (11.3) can be written as

$$\frac{1}{\lambda D} \int P(\mathbf{u}) e^{i\frac{\pi}{\lambda D}|\mathbf{u}|^2} e^{-i\frac{2\pi}{\lambda D}\mathbf{u}\cdot\left(\mathbf{x}+\frac{D}{f}\mathbf{X}\right)} \, d\mathbf{u}$$

$$= e^{\frac{-i\pi}{\lambda D}\left|\mathbf{x}+\frac{D}{f}\mathbf{X}\right|^2} \left[\frac{1}{\lambda D} \int P(\mathbf{u}) e^{i\frac{\pi}{\lambda D}\left|\mathbf{u}-\left(\mathbf{x}+\frac{D}{f}\mathbf{X}\right)\right|^2} \, d\mathbf{u}\right]. \quad (11.6)$$

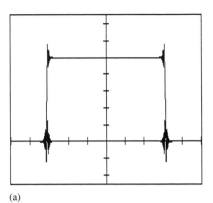

(a)

X interval: $[-64, 64]$ X increment $= 12.8$
Y interval: $[-.5, 1.5]$ Y increment $= .2$

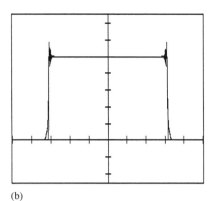

(b)

X interval: $[-64, 64]$ X increment $= 12.8$
Y interval: $[-.5, 1.5]$ Y increment $= .2$

FIGURE 6.33
Comparison of Fresnel diffraction of pupil function from formula (11.7) with that pupil function. (a) Graph of the real and imaginary parts of the pupil function. (b) Graph of magnitude of graph in (a) and the original function.

The integral in brackets is a Fresnel integral, like the one we discussed in Section 2. In fact, as we pointed out in Remark 2.24, it represents a distorted image of the *lens* aperture. For example, suppose one uses a square lens aperture with pupil function

$$P(x, y) = \begin{cases} 1 & \text{if } |x| < 40 \text{ and } |y| < 40 \\ 0 & \text{if } |x| > 40 \text{ or } |y| > 40. \end{cases} \tag{11.7}$$

Then, we show in Figure 6.33(a) the graph of

$$\frac{1}{\lambda D} \int P(\mathbf{u}) e^{\frac{i\pi}{\lambda D} |\mathbf{u} - \mathbf{x}|^2} \, d\mathbf{u}$$

along one axis (using $\lambda D = .5$), and in Figure 6.33(b) compare it to the pupil function in (11.7). Clearly, there is a very close match. We will assume that this matching holds for other pupils as well. Hence, we make the following approximation:

$$\frac{1}{\lambda D} \int P(\mathbf{u}) e^{\frac{i\pi}{\lambda D} \left| \mathbf{u} - \left(\mathbf{x} + \frac{D}{f} \mathbf{X} \right) \right|^2} \, d\mathbf{u} \approx P \left(\mathbf{x} + \frac{D}{f} \mathbf{X} \right). \tag{11.8}$$

Using the right side of (11.8) in (11.6) we have

$$\frac{1}{\lambda D} \int P(\mathbf{u}) e^{\frac{i\pi}{\lambda D} |\mathbf{u}|^2} e^{-i \frac{2\pi}{\lambda D} \mathbf{u} \cdot \left(\mathbf{x} + \frac{D}{f} \mathbf{X} \right)} \, d\mathbf{u}$$

$$\approx e^{\frac{-i\pi}{\lambda D} \left| \mathbf{x} + \frac{D}{f} \mathbf{X} \right|^2} P \left(\mathbf{x} + \frac{D}{f} \mathbf{X} \right). \tag{11.9}$$

Substituting the right side of (11.9) into (11.2) we get

$$\psi(\mathbf{X}, t) \approx \frac{e^{ic}}{\lambda f} e^{\frac{i\pi}{\lambda f}|\mathbf{X}|^2}$$

$$\int A(\mathbf{x})\psi(\mathbf{x}, t) e^{\frac{i\pi}{\lambda D}|\mathbf{x}|^2} e^{\frac{-i\pi}{\lambda D}|\mathbf{x}+\frac{D}{f}\mathbf{X}|^2} P\left(\mathbf{x} + \frac{D}{f}\mathbf{X}\right) d\mathbf{x}. \quad (11.10)$$

Expanding $|\mathbf{x} + D\mathbf{X}/f|^2$ and factoring an exponential involving $|\mathbf{X}|^2$ outside the integral, we can rewrite (11.10) as

$$\psi(\mathbf{X}, t) \approx \frac{e^{ic}}{\lambda f} e^{\frac{i\pi}{\lambda f}\left(1-\frac{D}{f}\right)|\mathbf{X}|^2}$$

$$\int A(\mathbf{x})\psi(\mathbf{x}, t) P\left(\mathbf{x} + \frac{D}{f}\mathbf{X}\right) e^{-i\frac{2\pi}{\lambda f}\mathbf{x}\cdot\mathbf{X}} d\mathbf{x}. \quad (11.11)$$

From (11.11) we obtain, *using the coherency assumption (10.1) as we did in Section 10,*

$$I(\mathbf{X}) \approx \left| \frac{1}{\lambda f} \int A(\mathbf{x}) P\left(\mathbf{x} + \frac{D}{f}\mathbf{X}\right) e^{-i\frac{2\pi}{\lambda f}\mathbf{x}\cdot\mathbf{X}} d\mathbf{x} \right|^2. \quad (11.12)$$

The presence of the pupil factor $P(\mathbf{x} + D\mathbf{X}/f)$ in (11.12) is known as *vignetting*. The object $A(\mathbf{x})$ may not fully appear in the integrand if *the translated pupil* (translated by $-D\mathbf{X}/f$) does not completely contain the object.

One way of minimizing vignetting is to assume that the object described by the function A is small relative to the lens pupil size. Here we are also assuming that the pupil function has the form described in (8.18). Let R_A stand for the *radius of the object A*; in other words, R_A is the smallest positive number for which $A(\mathbf{x}) = 0$ whenever $|\mathbf{x}| > R_A$. Similarly, let R_P denote the radius of the lens aperture. Our *smallness assumption* is that $R_A \ll R_P$, in which case

$$A(\mathbf{x}) P\left(\mathbf{x} + \frac{D}{f}\mathbf{X}\right) = A(\mathbf{x}), \qquad \text{for } |\mathbf{X}| < \tfrac{f}{D}(R_P - R_A) \approx \tfrac{f}{D} R_P. \quad (11.13)$$

Using (11.13), formula (11.12) simplifies to

$$I(\mathbf{X}) \approx \left| \frac{1}{\lambda f} \hat{A}\left(\frac{1}{\lambda f}\mathbf{X}\right) \right|^2, \qquad \text{for } |\mathbf{X}| < \tfrac{f}{D}(R_P - R_A) \approx \tfrac{f}{D} R_P. \quad (11.14)$$

Formula (11.14) shows that the intensity in the back focal plane of the lens is the modulus-squared of the (scaled) Fourier transform of the object function (provided one does not stray too far from the origin). Because of (11.14), when modifications of the transform of the object function A are required (as in gray field imaging or Schlieren imaging, for instance), then *these modifications can be made in the back focal plane of the lens.* For example, in gray field imaging, the partially transmitting material can be placed over the origin of the back focal plane (instead of over the center of the lens aperture).

Modifying the transform of an object in order to change the image is known as *spatial filtering*. For more details, see [Go], [Ii], [Pi], [Mi], and [Me].

REMARK 11.15 If we set $D = 0$, then (11.12) becomes (compare Exercises 6.43 and 6.46)

$$I(\mathbf{X}) \approx \left| \frac{1}{\lambda f} \int A(\mathbf{x}) P(\mathbf{x}) e^{-i\frac{2\pi}{\lambda f}\mathbf{x}\cdot\mathbf{X}} \, d\mathbf{x} \right|^2. \qquad (11.16)$$

Formula (11.16) does not suffer from vignetting. In fact, we only need to assume that $R_A < R_P$ (not *much* less, just less) in order to obtain from (11.16)

$$I(\mathbf{X}) \approx \left| \frac{1}{\lambda f} \hat{A}\left(\frac{1}{\lambda f}\mathbf{X} \right) \right|^2 \qquad (11.17)$$

without the restrictions on $\mathbf{X}$ that are in (11.14). The practice of placing an object within the entrance aperture of a lens ($D = 0$ and $R_A < R_P$) is a very common practice, precisely because of the absence of vignetting effects. (See [Ha-e] for some beautiful examples.) ∎

It is important to note that when the object is placed at a distance of $D = f$ units from the imaging lens, formula (11.11) simplifies to

$$\psi(\mathbf{X}, t) \approx \frac{e^{ic}}{\lambda f} \int A(\mathbf{x})\psi(\mathbf{x}, t) P(\mathbf{x} + \mathbf{X}) e^{-i\frac{2\pi}{\lambda f}\mathbf{x}\cdot\mathbf{X}} \, d\mathbf{x}. \qquad (11.18)$$

If we make the smallness assumption $R_A \ll R_P$ again, then we get

$$\psi(\mathbf{X}, t) \approx \frac{e^{ic}}{\lambda f} \int A(\mathbf{x})\psi(\mathbf{x}, t) e^{-i\frac{2\pi}{\lambda f}\mathbf{x}\cdot\mathbf{X}} \, d\mathbf{x}, \qquad \text{for } |\mathbf{X}| < R_P - R_A \approx R_P. \qquad (11.19)$$

It is also true that when $|\mathbf{X}| > R_P + R_A$, then $A(\mathbf{x})P(\mathbf{x} + \mathbf{X}) = 0$. Hence

$$\psi(\mathbf{X}, t) = 0, \qquad \text{for } |\mathbf{X}| > R_P + R_A \approx R_P. \qquad (11.20)$$

Since $R_P - R_A \approx R_P \approx R_P + R_A$, we combine these last two results by writing

$$\psi(\mathbf{X}, t) \approx \left[\frac{e^{ic}}{\lambda f} \int A(\mathbf{x})\psi(\mathbf{x}, t) e^{-i\frac{2\pi}{\lambda f}\mathbf{x}\cdot\mathbf{X}} \, d\mathbf{x} \right] P(\mathbf{X}). \qquad (11.21)$$

Formula (11.21) shows that, except for the pupil factor $P(\mathbf{X})$, *the light $\psi(\mathbf{X}, t)$ at the back focal plane of the lens is precisely the (scaled) Fourier transform of $A(\mathbf{x})\psi(\mathbf{x}, t)$.* This result is the basis for *two-lens imaging*. A second lens is used to perform a second Fourier transform, thereby obtaining by Fourier inversion an inverted image of the object. Since the basic theory and results are quite similar to what we described in Section 10, we will not discuss two-lens imaging. The interested reader can find good discussions in [Go] and [Ii].

12   Incoherent imaging

The second main type of imaging is imaging under *incoherent illumination*. In Section 1 we gave one definition of incoherent illumination [see (1.13)]. Here, we will adopt the following *model* of incoherency. Let's assume that the cross-correlation function Γ is Gaussian with mean 0 and extremely small variance as a function of $|\mathbf{x} - \mathbf{z}|$. That is, we shall assume that

$$\Gamma(\mathbf{x}, \mathbf{z}) = \frac{1}{\epsilon^2} e^{-\pi|\mathbf{x}-\mathbf{z}|^2/\epsilon^2} \tag{12.1}$$

where ϵ is a positive constant very near to 0. A graph of Γ for $\mathbf{z} = (0,0)$ and $\mathbf{x}$ restricted to the x-axis is shown in Figure 6.34(a). To obtain this graph we took $\epsilon = 0.1$. If this coherency function is integrated against a function of $\mathbf{x}$, like the one shown in Figure 6.34(b), then a convolution is being performed *and this function is closely approximated*, as can be seen from Figure 6.34(c). Generally, we will assume that ϵ is so close to 0 that

$$\int g(\mathbf{z})\Gamma(\mathbf{x}, \mathbf{z}) \, d\mathbf{z} \approx g(\mathbf{x}) \tag{12.2}$$

for all functions g that we examine.

For further discussion of the mathematical details needed to verify (12.2) for any bounded continuous function g, see [Wa, Chapters 6.4 and 6.7].

The assumption that Γ has the form given in (12.1) is mostly for illustrative purposes. The key fact that we need is (12.2), and this result is deducible for many natural sources from the *van Cittert–Zernike Theorem*. See [Go,2] for further discussion of this point.

If we apply (12.2) to the $\mathbf{z}$-integral in the imaging formula (9.13), we find that

$$\int \left[A(\mathbf{z}) e^{\frac{i\pi}{\lambda D}|\mathbf{z}|^2} \hat{P} \left(\frac{\mathbf{z}}{\lambda D} + \frac{\mathbf{X}}{\lambda \Delta} \right) \right]^* \Gamma(\mathbf{x}, \mathbf{z}) \, d\mathbf{z}$$

$$\approx \left[A(\mathbf{x}) e^{\frac{i\pi}{\lambda D}|\mathbf{x}|^2} \hat{P} \left(\frac{\mathbf{x}}{\lambda D} + \frac{\mathbf{X}}{\lambda \Delta} \right) \right]^*. \tag{12.3}$$

Substituting the right side of (12.3) in place of the $\mathbf{z}$-integral in (9.13), we obtain for the intensity $I(\mathbf{X})$ at the observation plane

$$I(\mathbf{x}) \approx \frac{1}{(\lambda^2 D \Delta)^2} \int \left| A(\mathbf{x}) \hat{P} \left(\frac{\mathbf{x}}{\lambda D} + \frac{\mathbf{X}}{\lambda \Delta} \right) \right|^2 d\mathbf{x}. \tag{12.4}$$

Notice that the complex exponential factor does not appear in (12.4), since

$$\left| e^{\frac{i\pi}{\lambda D}|\mathbf{x}|^2} \right|^2 = 1.$$

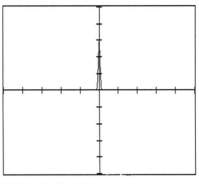

(a) Gaussian function

X interval: $[-4, 4]$	X increment $= .8$
Y interval: $[-17, 17]$	Y increment $= 3.4$

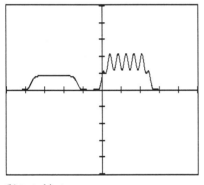

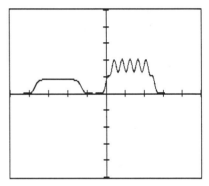

(b) test object

X interval: $[-4, 4]$	X increment $= .8$
Y interval: $[-17, 17]$	Y increment $= 3.4$

(c) convolution

X interval: $[-4, 4]$	X increment $= .8$
Y interval: $[-17, 17]$	Y increment $= 3.4$

FIGURE 6.34
Illustration of formula (12.2).

Unlike coherent illumination, no small object approximation is needed when incoherent illumination is used.

As we did for the coherent case, we substitute $\mathbf{x} = -\mathbf{s}/M$ and $d\mathbf{x} = d\mathbf{s}/M^2$, where $M = \Delta/D$ is the degree of magnification, and we obtain from (12.4)

$$I(\mathbf{x}) \approx \int \left| A\left(\frac{-\mathbf{s}}{M}\right) \right|^2 \left| \frac{1}{(\lambda\Delta)^2} \hat{P}\left[\frac{1}{\lambda\Delta}(\mathbf{X} - \mathbf{s})\right] \right|^2 d\mathbf{s}. \qquad (12.5)$$

The integral in (12.5) is a two-dimensional convolution. As opposed to the coherent case, the function

$$\left| A\left(\frac{-\mathbf{X}}{M}\right) \right|^2 \tag{12.6}$$

is convoluted with

$$\left| \frac{1}{(\lambda\Delta)^2} \hat{P}\left(\frac{1}{\lambda\Delta}\mathbf{X}\right) \right|^2. \tag{12.7}$$

The function in (12.6) is a magnified inverted form of $|A|^2$ that corresponds to the *intensity* of the aperture function (object). The function in (12.7) is the *point spread function (PSF) for incoherent illumination*. Notice that this *PSF* is the modulus-squared of the *PSF* for coherent illumination. The function in (12.7) is the *power spectrum* of the function $P(\lambda\Delta\mathbf{x})$. The operations involved are

$$P(\lambda\Delta\mathbf{x}) \xrightarrow{\ \mathcal{F}\ } \frac{1}{(\lambda\Delta)^2} \hat{P}\left(\frac{1}{\lambda\Delta}\mathbf{X}\right) \xrightarrow{\ |\cdot|^2\ } \left| \frac{1}{(\lambda\Delta)^2} \hat{P}\left(\frac{1}{\lambda\Delta}\mathbf{X}\right) \right|^2. \tag{12.8}$$

When using *FAS* there is a choice for computing a power spectrum automatically; it is part of the *Four T* procedure.

The procedures for computing images under incoherent illumination are quite similar to those described in the previous section, the main difference being that (12.8) is used in place of (10.10) to compute *PSFs*. We will limit ourselves to one example (some others are described in the exercises).

Example 12.9
Suppose that incoherent light having wavelength $\lambda = 5 \times 10^{-4}$ mm illuminates an edge defined by

$$A(x,y) = \begin{cases} 1 & \text{if } x > 0 \\ 0 & \text{if } x < 0 \end{cases}$$

and this edge is imaged by a lens with a square aperture of side length 40 mm and focal length 1 m. Assume that $D = 2$ m and $\Delta = 2$ m. Describe the image that results.   ▯

SOLUTION   Since $|A|^2 = A$, the only difference between this example and Example 10.9 is that the *PSF* is now (*Note*: $\lambda\Delta = 1$)

$$\left| \frac{1}{(\lambda\Delta)^2} \hat{P}\left(\frac{1}{\lambda\Delta}\mathbf{X}\right) \right|^2 = [40\,\text{sinc}\,(40X)]^2\,[40\,\text{sinc}\,(40Y)]^2.$$

In this case, using *Parseval's equality* [see (3.15), Chapter 5]

$$\int_{-\infty}^{\infty} [40\,\text{sinc}\,(40Y)]^2\,dY = \int_{-\infty}^{\infty} [\text{rec}\,(y/40)]^2\,dy$$

$$= \int_{-20}^{20} 1\,dy = 40.$$

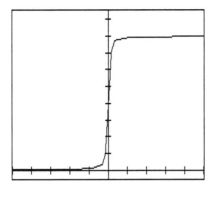

X interval: [−.5, .5]   X increment = .1
Y interval: [−100, 1900] Y increment = 200

FIGURE 6.35
Incoherent imaging of an edge.

Therefore, we will convolute

$$f(X) = \begin{cases} 1 & \text{for } X > 0 \\ 0 & \text{for } X < 0 \end{cases}$$

with $g(X) = 40\,[40\sin c\,(40X)]^2$. The result is shown in Figure 6.35; it was obtained using *FAS* with $(-32, 32)$ as the interval and 4096 points.

It is very interesting to compare this result for incoherent imaging of an edge with the result for coherent imaging from Example 10.9. For incoherent imaging there is no oscillation or Gibbs' effect at the edge's boundary. Furthermore, if the edge boundary is marked by half intensity, then the edge boundary in incoherent imaging equals the edge boundary of the object. For coherent imaging the edge boundary in the image is displaced by a small amount.   ∎

Our remarks at the end of the example above show some of the nice features that an incoherent image of an edge has in comparison to the coherent image of an edge. This one example does not, however, tell the whole story. We will continue to compare these two types of imaging in the exercises. It is also important to note that *processing of images* is more difficult with incoherent imaging (there is no Fourier transform in the focal plane).

A reader who attempts to reproduce the example above will notice that the *form* of the edge occurs as we have shown in Figure 6.35, but the vertical (intensity) scale is much larger in the image than in the original object. This is a *scaling* problem which we show how to repair in Remark 6.72 in the exercises.

References

For more on Fourier optics, see [Go] or [Ii]. For more on image formation and image processing, see [Pi], [Me], [Mi], and [Mi-T]. Coherence is discussed in greater detail in [Bo-W] and [Go,2]. Many beautiful photographs can be found in [Ha-e].

Exercises

Section 1

6.1 Suppose that $\psi(\mathbf{x}, 0, t) = e^{i2\pi\nu t}$. [This is called *plane wave* illumination.] Show that plane wave illumination is coherent.

6.2 Suppose that $\psi(\mathbf{x}, 0, t) = e^{i\frac{2\pi}{\lambda}(\mathbf{a}\cdot\mathbf{x} - ct)}$ where $\mathbf{a}$ is a fixed vector in $\mathbf{R}^2$ and c is the speed of light. [This is also called *plane wave* illumination.] Show that

$$\Gamma(\mathbf{x}, \mathbf{z}) = \left[e^{i\frac{2\pi}{\lambda}\mathbf{a}\cdot\mathbf{x}} \right] \left[e^{i\frac{2\pi}{\lambda}\mathbf{a}\cdot\mathbf{z}} \right]^*.$$

6.3 Suppose that light diffracted from a single point source (as in Example 1.3) illuminates a parallel screen a distance ρ away. Show that, by choosing proper coordinates, the light $\psi(\mathbf{x}, 0, t)$ at that screen is described by

$$\psi(\mathbf{x}, 0, t) = \frac{P(t)e^{i\frac{2\pi}{\lambda}\left[\rho^2 + |\mathbf{x}|^2\right]^{\frac{1}{2}}}}{\lambda\left[\rho^2 + |\mathbf{x}|^2\right]^{\frac{1}{2}}}$$

where $P(t)$ represents the light passing through the point source.

6.4 Show that for the function $\psi(\mathbf{x}, 0, t)$ from Exercise 6.3,

$$\Gamma(\mathbf{x}, \mathbf{z}) = \kappa \left[\frac{e^{i\frac{2\pi}{\lambda}\left[\rho^2 + |\mathbf{x}|^2\right]^{\frac{1}{2}}}}{\lambda\left[\rho^2 + |\mathbf{x}|^2\right]^{\frac{1}{2}}} \right] \left[\frac{e^{i\frac{2\pi}{\lambda}\left[\rho^2 + |\mathbf{z}|^2\right]^{\frac{1}{2}}}}{\lambda\left[\rho^2 + |\mathbf{z}|^2\right]^{\frac{1}{2}}} \right]^*$$

where κ is a positive constant defined by $1/T \int_0^T |P(t)|^2 \, dt$.

6.5 *Generalized Coherence.* A more general definition of coherence than (1.15) is to assume that $\Gamma(\mathbf{x}, \mathbf{z})$ satisfies

    Coherency.

$$\Gamma(\mathbf{x}, \mathbf{z}) = C(\mathbf{x})C(\mathbf{z})^* \tag{6.6}$$

for some function C. Show that:

(a) If Γ satisfies (1.15), then it also satisfies (6.6).

(b) Each of the functions ψ described in Exercises 6.1, 6.2, and 6.3 are coherent in the sense of (6.6).

(c) Given the coherency defined in (6.6), formula (1.20) becomes

$$I(\mathbf{u}, D) = \left| \int_{\mathbf{R}^2} \frac{T(\mathbf{x})}{\lambda|\mathbf{C_x}|} e^{i\frac{2\pi}{\lambda}|\mathbf{C_x}|} \, d\mathbf{x} \right|^2 \tag{6.7}$$

where $T(\mathbf{x}) = A(\mathbf{x})C(\mathbf{x})$.

REMARK 6.8 The function $T(\mathbf{x})$ takes account of the aperture, via the aperture function $A(\mathbf{x})$, *and the nature of the light illuminating the aperture*, via the *coherency factor* $C(\mathbf{x})$. In order to simplify our calculations in subsequent sections, we will work with (1.20) where $A(\mathbf{x})$ is described by (1.19). This situation can be realized approximately if a setup like the one shown in Figure 6.2 is used to generate coherent illumination of the aperture. ∎

6.9 Suppose that $\Gamma(\mathbf{u}, \mathbf{v})$ is defined by

$$\Gamma(\mathbf{u}, \mathbf{v}) = \frac{1}{T} \int_0^T \psi(\mathbf{u}, D, t) \psi^*(\mathbf{v}, D, t)\, dt.$$

Show that if $\psi(\mathbf{u}, D, t)$ is replaced by the integral in (1.17), then $\Gamma(\mathbf{u}, \mathbf{v}) = K(\mathbf{u})K^*(\mathbf{v})$ for some function K. Thus, *light that is coherent over the aperture remains coherent after diffraction.*

Section 2

6.10 Using *FAS*, graph approximations to $I_1(u, D)$ in (2.27) given

$$A_1(x) = \mathrm{rec}\left(\frac{x-2}{2}\right) + \mathrm{rec}\left(\frac{x+2}{2}\right)$$

and $\lambda = 7 \times 10^{-4}$ mm (red light), $D = 100$ mm, 120 mm, 150 mm.

6.11 Sketch diffraction patterns from a rectangular aperture with aperture function

$$A(x, y) = \begin{cases} 1 & \text{if } |x| < 1 \text{ mm and } |y| < 4 \text{ mm} \\ 0 & \text{if } |x| > 1 \text{ mm or } |y| > 4 \text{ mm} \end{cases}$$

when $\lambda = 5 \times 10^{-4}$ mm, $D = 200$ mm, 300 mm, 400 mm.

6.12 Sketch the diffraction patterns resulting from an aperture with aperture function $A(x, y) = A_1(x) \, \mathrm{rec}\,(y)$ where $A_1(x)$ is the function given in Exercise 6.10. Use the same values of λ and D as in Exercise 6.10.

6.13 Sketch the diffraction patterns resulting from an aperture with aperture function $e^{-\pi(x^2+y^2)}$ when $\lambda = 5 \times 10^{-4}$ mm, $D = 100$ mm, 150 mm, 200 mm.

6.14 *Edge Diffraction.* Suppose that $A_1(x)$ in formula (2.27) is given by

$$A_1(x) = \begin{cases} 1 & \text{if } 0 < x < 32 \\ 0 & \text{if } -32 < x < 0. \end{cases}$$

Using *FAS*, plot approximations to $I_1(u, D)$ in formula (2.27). Use 4096 points, interval $[-32, 32]$, $\lambda = 5 \times 10^{-4}$ mm, and $D = 120$ mm. Confirm that $I_1(u, D)$ has the form shown in Figure 6.36.

REMARK The graph in Figure 6.36 is a good approximation to *Fresnel diffraction by an edge.* ∎

6.15 Repeat Exercise 6.14 for $\lambda = 6 \times 10^{-4}$ mm and $\lambda = 7 \times 10^{-4}$ while keeping $D = 120$ mm. Compare your graphs to the result from Exercise 6.14, shown in Figure 6.36. What is the effect of increasing the wavelength?

Section 3

6.16 Show that the two-dimensional Fourier transform defined in (3.9) satisfies the following properties (see Theorem 2.1 in Chapter 5):

(a) *Linearity.* For all complex numbers α and β

$$\alpha A(\mathbf{x}) + \beta B(\mathbf{x}) \overset{\mathcal{F}}{\to} \alpha \hat{A}(\mathbf{u}) + \beta \hat{B}(\mathbf{u}).$$

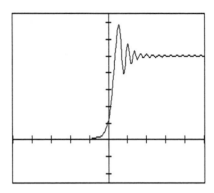

FIGURE 6.36
Fresnel diffraction by an edge.

(b) *Scaling.* If ρ is a positive constant, then

$$A\left(\frac{1}{\rho}\mathbf{x}\right) \xrightarrow{\ \mathcal{F}\ } \rho^2 \hat{A}(\rho\mathbf{u}) \qquad \text{and} \qquad A(\rho\mathbf{x}) \xrightarrow{\ \mathcal{F}\ } \frac{1}{\rho^2}\hat{A}\left(\frac{1}{\rho}\mathbf{u}\right).$$

(c) *Shifting.* For each $\mathbf{c}$ in $\mathbf{R}^2$,

$$A(\mathbf{x}-\mathbf{c}) \xrightarrow{\ \mathcal{F}\ } \hat{A}(\mathbf{u})e^{-i2\pi\mathbf{c}\cdot\mathbf{u}}.$$

(d) *Modulation.* For each $\mathbf{c}$ in $\mathbf{R}^2$,

$$A(\mathbf{x})e^{i2\pi\mathbf{c}\cdot\mathbf{x}} \xrightarrow{\ \mathcal{F}\ } \hat{A}(\mathbf{u}-\mathbf{c}).$$

6.17 Suppose that a rectangular aperture has aperture function

$$A(x,y) = \mathrm{rec}\,(x/a)\,\mathrm{rec}\,(y/b)$$

for $a > 0$ and $b > 0$. Describe the diffraction pattern from such an aperture, labeling all the positions of zero intensity.

6.18 *Vertical slit.* Suppose that a rectangular aperture has the form of a vertical slit. That is, its aperture function is the same as in Exercise 6.17, but $b \gg a$. Show why the diffraction pattern from such a vertical slit has a form like the one shown in Figure 6.37(a). A photograph of such a pattern is shown in Figure 6.37(b) [where b is only about 5 times the size of a].

6.19 Suppose that a sheet of mica is placed over just one of the apertures in Figure 6.11. Assuming that this multiplies the aperture function (for that *single* aperture) by a 90° *phase shift factor* $e^{i\pi/2}$, describe the diffraction pattern.

6.20 Repeat Exercise 6.19 but assume now that the phase shift factor is $e^{i2\pi\tau}$ for some real constant τ.

Section 4

6.21 By changing the x-interval appropriately, estimate the first three zeroes of $\hat{a}(\rho)$ where $a(x)$ is given in (4.19). Also, estimate the first two positive local extrema of $\hat{a}(\rho)$.

6.22 Graph the radial dependence of $I(\mathbf{u}, D)^{1/2}$ for a circular aperture of radius $R > 0$. You should be able to verify Figure 6.38.

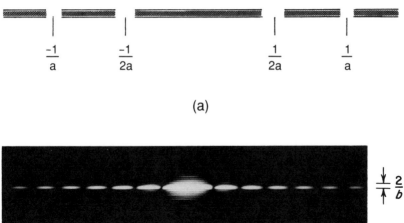

(a)

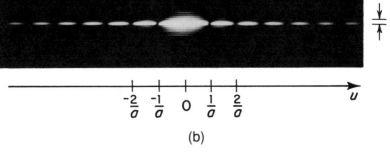

(b)

FIGURE 6.37
**Fraunhofer diffraction by a vertical slit. (a) Graph of the diffraction pattern
(negative image). (b) Photograph of a slit diffraction pattern (reproduced with
permission from [Wa, p. 291]).**

6.23 *Diffraction from Annuli.* Suppose that coherent light of wavelength λ illuminates
an *annular aperture*, with aperture function

$$A(\mathbf{x}) = \begin{cases} 1 & \text{if } R_1 < |\mathbf{x}| < R_2 \\ 0 & \text{if } |\mathbf{x}| < R_1 \text{ or } |\mathbf{x}| > R_2. \end{cases}$$

Graph $\hat{A}(\rho)$ and the radial dependencies of the intensity function $I(\mathbf{u}, D)$ for the
following radii:

(a) $R_1 = 1, R_2 = 2$

(b) $R_1 = 1, R_2 = 1.1$

(c) $R_1 = 1, R_2 = 1.05$ (See Figure 6.39.)

Section 5

6.24 Suppose that the aperture function for Figure 6.40(a) is

$$A(\mathbf{x}) = A_0(\mathbf{x} - \mathbf{c}) + A_0(\mathbf{x} + \mathbf{c})$$

where $\mathbf{c} = (a, b)$ is in $\mathbf{R}^2$, $|\mathbf{c}| = (1/2)\delta$, and A_0 is the aperture function for a
single aperture. Show that

$$I(u, v, D) = I_0(u, v, D) \left[4 \cos^2 \frac{2\pi}{\lambda D} (au + bv) \right]$$

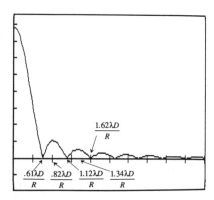

FIGURE 6.38
Radial dependences of the intensity function $I^{1/2}$ for diffraction from a circular aperture of radius R.

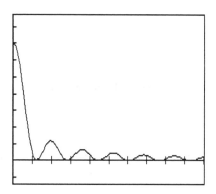

X interval: $[0, 3\lambda D]$ X increment $= .3\lambda D$

FIGURE 6.39
Diffraction by an annulus [see Exercise 6.23(c)]. Intensity as a function of radius.

and that the intensity I is 0 when

$$au + bv = \pm\frac{\lambda D}{4}, \pm\frac{3\lambda D}{4}, \dots, \pm\frac{(2k+1)\lambda D}{4}, \dots.$$

Show that the distance between successive dark fringes is $\lambda D/\delta$.

A diffraction pattern from an aperture like the one in Figure 6.40(a) is shown in Figure 6.40(b).

6.25 Suppose that four circular apertures are arranged at the corners of a parallelogram as shown in Figure 6.41. Describe their Fraunhofer diffraction pattern.

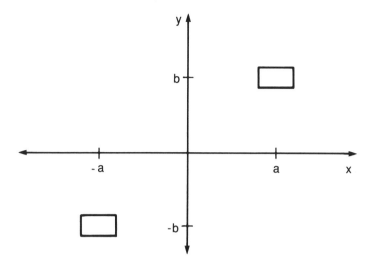

(a) apertures

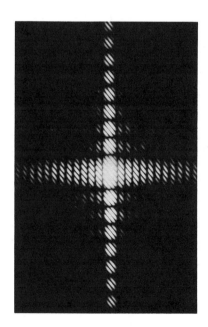

(b) diffraction pattern

FIGURE 6.40
Diffraction from two identical rectangular apertures (reproduced with permission from [Wa, p. 297]).

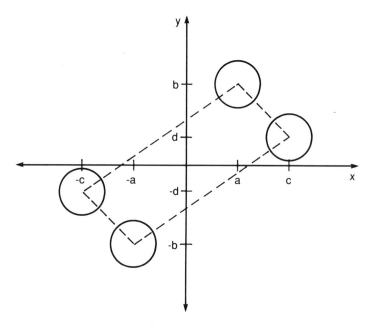

FIGURE 6.41
Four identical circular apertures.

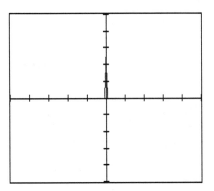

FIGURE 6.42
Unit function for a triangular grating.

Section 6

6.26 Using the procedure described in Example 6.10, plot the instrument functions from 17-slit diffraction gratings, using the following functions as unit functions.
  (a) $\operatorname{rec}(x)(1 - \operatorname{abs}(2x))$ (See Figure 6.42)
  (b) $\operatorname{rec}(x)(.5 + .5\cos(2\pi x))$
  (c) $\operatorname{rec}(2x)$

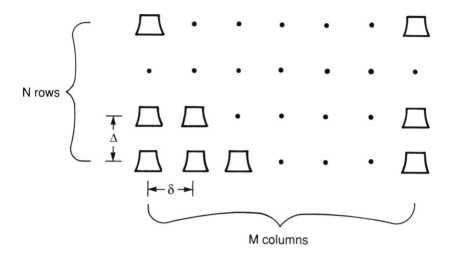

FIGURE 6.43
Rectangular array of MN identical apertures.

For (a), the grating might be called a *triangular grating*; such gratings are used in spectroscopy (usually in reflective gratings).

6.27 Repeat Exercise 6.26, but now use 25-slit gratings.

6.28 Repeat Exercise 6.27, but now use 31-slit gratings, and lower the integer values used to create the comb function from multiples of 100 to multiples of 90.

6.29 Apply the filter function $\exp(-.4x \wedge 2)$ to the diffraction gratings in Exercise 6.26. How does this affect the instrument function?

6.30 *Random Errors.* Modify the integer values in Example 6.10 in the following random way. Instead of entering multiples of 100, enter $100k \pm 10$, where k is an integer from -8 to 8, and you choose $+10$ or -10 depending on whether a coin toss turns up heads or tails. (Enter 4096 for every function value, just as in Example 6.10.)

How does this affect the instrument function? What happens if you use ± 5, or ± 1, instead of ± 10?

6.31 *Random Errors.* Modify the previous exercise by adding $\pm j$ ($j = 0, 1, 2, 3, 4, 5$) to the multiples of 100, the value of j being 1 less than the number obtained from rolling a die and again choosing $+$ or $-$ depending on whether a coin toss shows heads or tails.

6.32 *Random Errors.* Instead of modifying the integer values in Example 6.10, change the function values entered in the following random way. Enter $4096 - 20j$ (where $j = 0, 1, 2, 3, 4, 5$) for the function value, the value of j being 1 less than the number obtained from rolling a die. How does this affect the instrument function?

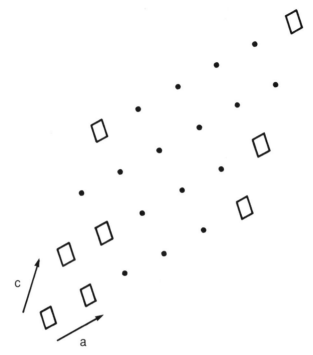

FIGURE 6.44
Parallelogram array of MN **identical apertures.**

6.33 *Array Theorem.* Suppose that MN identical apertures are arranged in N rows and M columns (see Figure 6.43). Show that

$$I(u,v,D) = I_0(u,v,D)\mathcal{S}_M(u)\mathcal{S}_N(v) \qquad (6.34)$$

where

$$\mathcal{S}_M(u) = \frac{\sin^2(\pi M\delta u/\lambda D)}{\sin^2(\pi\delta u/\lambda D)} \ , \qquad \mathcal{S}_N(v) = \frac{\sin^2(\pi N\Delta v/\lambda D)}{\sin^2(\pi\Delta v/\lambda D)} \ .$$

Explain why the diffraction pattern from such an array will consist mainly of high intensity dots located at the points $(u,v) = (m\lambda D/\delta, n\lambda D/\delta)$ where m and n are integers, the intensity at these points being $M^2 N^2 I_0(m\lambda D/\delta, n\lambda D/\delta, D)$.

6.35 *Parallelogram Array.* Suppose MN identical apertures are arranged as shown in Figure 6.44. Show that the aperture function for the array can be written as

$$A(\mathbf{x}) = \sum_{m=0}^{M-1}\sum_{n=0}^{N-1} A_0(\mathbf{x} - m\mathbf{a} - n\mathbf{c})$$

where $\mathbf{a} = (a,b)$ and $\mathbf{c} = (c,d)$ are two vectors in $\mathbf{R}^2$. Show that

$$I(\mathbf{u},D) = I_0(\mathbf{u},D)\mathcal{S}_M^{\mathbf{a}}(\mathbf{u})\mathcal{S}_N^{\mathbf{c}}(\mathbf{v})$$

where

$$\mathcal{S}_M^a(\mathbf{u}) = \frac{\sin^2(\pi M \mathbf{a} \cdot \mathbf{u}/\lambda D)}{\sin^2(\pi \mathbf{a} \cdot \mathbf{u}/\lambda D)}, \qquad \mathcal{S}_N^c(\mathbf{v}) = \frac{\sin^2(\pi N \mathbf{c} \cdot \mathbf{v}/\lambda D)}{\sin^2(\pi \mathbf{c} \cdot \mathbf{v}/\lambda D)}.$$

Explain why the diffraction pattern has a form that consists mainly of high intensity dots located at the points (u, v) which satisfy

$$au + bv = m\lambda D, \qquad cu + dv = n\lambda D \qquad (6.36)$$

where m and n are both integers. Also explain why the intensities at these points are equal to $M^2 N^2$ times the intensity $I_0(u, v, D)$.

Finally, show that the points satisfying both equations in (6.36) *lie on the intersections of lines perpendicular to the vectors* $\mathbf{a}$ *and* $\mathbf{c}$.

REMARK   The results of Exercise 6.35 are important in crystallography, the parallelogram array being a *two-dimensional crystal*. ∎

Section 7

6.37 Using *FAS*, compare the two sides of (7.6), assuming $\lambda D = 1$, $\alpha = .4$, $\beta = .5$, and $N = 100$.

6.38 Using *FAS*, compare the instrument functions from sinusoidal gratings with frequencies of $N = 50$ and $N = 100$, both with $\beta = .5$, $\alpha = .4$, and $\lambda D = 1$.

Section 8

6.39 Suppose that the lens is plano-convex. That is, one side is a plane surface, and the other side is a spherical surface bulging outward. Show that the focal length formula is

$$\frac{1}{f} = (\eta - 1)\frac{1}{R}$$

where R is the radius of curvature of the spherical side.

6.40 Suppose that the lens is a positive meniscus. That is, the lens surface at the entrance plane bulges outward spherically with a radius of curvature of R_1, while the lens surface at the exit plane bulges inward spherically with a radius of curvature of R_2. Show that the focal length formula is

$$\frac{1}{f} = (\eta - 1)\left[\frac{1}{R_1} - \frac{1}{R_2}\right].$$

6.41 Suppose that an aperture is placed directly in front of a lens, as shown in Figure 6.45. Assuming that the lens is double convex with spherical surfaces, derive the following formula:

$$\psi(\mathbf{u}, \Delta + \epsilon, t) \approx \frac{e^{i\delta}}{\lambda\Delta} \int_{\mathbf{R}^2} A(\mathbf{x})\psi(\mathbf{x}, 0, t)P(\mathbf{x})e^{\frac{i\pi}{\lambda\Delta}|\mathbf{u}-\mathbf{x}|^2}e^{\frac{-i\pi}{\lambda f}|\mathbf{x}|^2}\,d\mathbf{x} \qquad (6.42)$$

6.43 Show that, under the assumption of coherent illumination, formula (6.42) leads to

$$I(\mathbf{u}, \Delta + \epsilon) \approx \left|\frac{1}{\lambda\Delta}\int_{\mathbf{R}^2} A(\mathbf{x})P(\mathbf{x})e^{\frac{i\pi}{\lambda\Delta}|\mathbf{u}-\mathbf{x}|^2}e^{\frac{-i\pi}{\lambda f}|\mathbf{x}|^2}\,d\mathbf{x}\right|^2. \qquad (6.44)$$

And that, *if the aperture defined by* $A(\mathbf{x})$ *fits inside the lens aperture*, then [assuming $P(\mathbf{x})$ is described by (8.18)]

$$I(\mathbf{u}, \Delta + \epsilon) \approx \left|\frac{1}{\lambda\Delta}\int_{\mathbf{R}^2} A(\mathbf{x})e^{\frac{i\pi}{\lambda\Delta}|\mathbf{u}-\mathbf{x}|^2}e^{\frac{-i\pi}{\lambda f}|\mathbf{x}|^2}\,d\mathbf{x}\right|^2. \qquad (6.45)$$

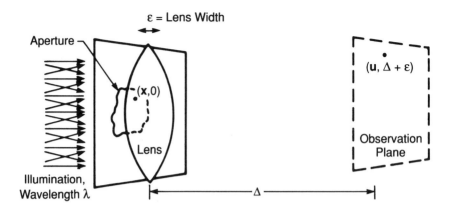

FIGURE 6.45
Lens with an aperture in its entrance pupil.

6.46 Suppose that $\Delta = f$, show that (6.45) becomes

$$I(\mathbf{u}, f + \epsilon) \approx \left| \frac{1}{\lambda f} \hat{A} \left(\frac{\mathbf{u}}{\lambda f} \right) \right|^2. \tag{6.47}$$

Formula (6.47) shows that *the intensity in the back focal plane of the lens is equal to the modulus-squared of the (scaled) Fourier transform of the aperture function.* [Compare with formula (3.10).]

6.48 Show that if $\Delta = f$, then formula (6.42) becomes

$$\psi(\mathbf{u}, f + \epsilon, t) \approx \frac{e^{i\delta}}{\lambda f} e^{\frac{i\pi}{\lambda f} |\mathbf{u}|^2} \int_{\mathbf{R}^2} A(\mathbf{x}) \psi(\mathbf{x}, 0, t) P(\mathbf{x}) e^{-i\frac{2\pi}{\lambda f} \mathbf{u} \cdot \mathbf{x}} \, d\mathbf{x}. \tag{6.49}$$

6.50 Explain why, if a properly selected lens is placed directly behind the observation plane in Figure 6.45 (so that its entrance plane coincides with the observation plane), then (6.49) can be replaced by

$$\psi(\mathbf{u}, f + \epsilon', t) \approx \frac{e^{i\delta'}}{\lambda f} \mathcal{P}(\mathbf{u}) \int_{\mathbf{R}^2} A(\mathbf{x}) \psi(\mathbf{x}, 0, t) P(\mathbf{x}) e^{-i\frac{2\pi}{\lambda f} \mathbf{u} \cdot \mathbf{x}} \, d\mathbf{x} \tag{6.51}$$

where $\mathcal{P}(\mathbf{u})$ is the pupil function for this second lens. (ϵ' and δ' are positive constants due to the thickness of the second lens.)

Section 9

6.52 Suppose that

$$\frac{1}{D} + \frac{1}{\Delta} - \frac{1}{f} = \pm\delta$$

where δ is a small positive constant. (This is known as *defocus.*) Show that the pupil function $P(\mathbf{u})$ in (9.11) must be replaced by $P(\mathbf{u})e^{\pm i\pi\delta|\mathbf{u}|^2/\lambda}$.

6.53 Suppose fourth power terms are included in the approximations in (8.9). (This would be more necessary the larger the lens aperture.) Show that the pupil function in (9.11) must then be replaced by $P(\mathbf{u})e^{i\pi\sigma|\mathbf{u}|^4}$ where σ is a constant. (This is known as *spherical aberration*; the constant σ is the *spherical aberration constant*.)

6.54 Describe what shape of lens surfaces would be needed to insure that no spherical aberration occurs (of the type described in the previous exercise).

Section 10

6.55 Repeat Example 10.9, but now use the following wavelengths: $\lambda = 6 \times 10^{-4}$ mm, 7×10^{-4} mm, and 4×10^{-4} mm. What is the effect of varying the wavelength on the edge image?

6.56 Suppose that we have a one-dimensional pupil function $P(\lambda \Delta x) = \text{rec}(x/40)$. Using an initial interval of $[-512, 512]$ and 2048 points, Fourier transform this pupil function to get a *PSF* for coherently imaging the following object:

$$A(x) = \exp(-(x/.8) \wedge 20) - \text{rec}(100x), \qquad -1 \leq x \leq 1.$$

You should observe a pronounced effect in the image due to the spreading out of the point-like part of the object, $-\text{rec}(100x)$, which represents a dust particle absorbing light.

6.57 Repeat Exercise 6.56, but now use the object function

$$A(x) = \exp(-(x/.8) \wedge 20) - \text{rec}(100(x - .04)) - \text{rec}(100(x + .04)).$$

6.58 Suppose that instead of an edge, the object in Example 10.9 is changed to

$$A(x,y) = \begin{cases} 0 & \text{if } x \leq -2 \\ \frac{1}{2}(x+2) & \text{if } -2 \leq x \leq 0 \\ 1 & \text{if } x \geq 0. \end{cases}$$

Keeping all other parameter values the same as in that example, graph the resulting image. Compare your results with those in Example 10.9. What happens if the image detector (film, or charge coupled device) describes $I(X,Y)^{1/2}$ instead of $I(X,Y)$?

6.59 Suppose that a *Gaussian filter* is placed over the lens aperture in Example 10.9, so that the pupil function is now

$$P(x,y) = e^{-.1(x^2+y^2)} = e^{-.1x^2}e^{-.1y^2}. \tag{6.60}$$

Keeping all other parameter values the same as in Example 10.9, graph the image of an edge. Compare your results with those in Example 10.9. In which case is the imaged edge closer to the true edge? Is there a bright band along the image edge when this Gaussian pupil is used? What happens with other Gaussian pupils (wider or narrower ones)?

REMARK 6.61 Using a pupil function like (6.60) is described in detail in [Mi-T]. This would be an *excellent* paper for the reader to study after finishing this chapter. ∎

6.62 Repeat Example 10.22, but now use

$$A(x) = \exp(-(x/3) \wedge 10)(1 + \beta \cos(12\pi x)) \tag{6.63}$$

for $\beta = .04, .06, .08, .1, .12, .14,$ and $.16$. Estimate the improvement in contrast over the bright field image for each value of β. At what value of β do false details begin to appear?

6.64 Repeat Exercise 6.62, but change the gray field pupil to

$$P_{\text{GF}}(\lambda \Delta x) = \begin{cases} 0 & \text{if } 40 < |x| \\ 10 & \text{if } 3 < |x| < 40 \\ 2 & \text{if } |x| < 3. \end{cases}$$

Describe the effects of raising the value of P_{GF} for $|x| < 3$.

6.65 The *Schlieren* method of imaging consists of placing an opaque material over a half-plane immediately behind the lens aperture. This exercise describes a one-dimensional model of Schlieren imaging. Assume that the pupil function is

$$P(\lambda \Delta x) = \begin{cases} 1 & \text{if } 0 < x < 40 \\ 0 & \text{if } x < 0 \text{ or } x > 40 \end{cases}$$

and transform this function over $[-128, 128]$ using 2048 points to create a *complex PSF*. Compare Schlieren imaging with gray field imaging of the function $A(x)$ in (10.18).

Section 11

6.66 Explain why, assuming that the object is small relative to the lens aperture, the approximation in (11.8) need only hold moderately close to the origin (in particular, the spikes near the edges of the graph in Figure 6.33(a) can then be ignored).

6.67 Why is there no Fourier transforming property (at the back focal plane, or any plane) when incoherent illumination is used?

Section 12

6.68 Graph the *PSF* of the square lens aperture from Example 10.13, assuming all parameter values are the same, but under incoherent illumination. What advantages does the *PSF* for incoherent illumination have over the coherent one? What would the image of several tiny point-like objects (like dust particles) look like for these two types of illumination?

6.69 Given that

$$P(\lambda \Delta x) = \begin{cases} 1 & \text{if } |x| < 10 \\ 0 & \text{if } |x| > 10 \end{cases} \tag{6.70}$$

compare the one-dimensional images of the object

$$\exp(-(x/3) \wedge 10)(1 + \cos(22\pi x))$$

under coherent and incoherent illumination. *Note*: To obtain the *PSFs* you should compute the Fourier transform and the power spectrum of $P(\lambda \Delta x)$ over $[-128, 128]$ using 2048 points.

6.71 Repeat Exercise 6.69 but use for the object

$$\exp(-(x/3) \wedge 10)(1 + \cos(k\pi x))$$

where $k = 17, 19, 21, 23$, and 25. What advantages or disadvantages do you see in using coherent or incoherent illumination, especially in regard to *relative contrast* [see (10.21)] of the oscillations near 0? There is some advantage for coherent illumination when $k = 17$ and 19; what is it?

REMARK 6.72   In studying incoherent imaging it is a standard practice to divide the *PSF* by the value of its integral over $\mathbf{R}^2$. *This forces its integral over $\mathbf{R}^2$ to be 1.* This new function is called the *normalized PSF*. By Parseval's equality, the integral of the *PSF* over $\mathbf{R}^2$ is equal to the square of the 2-Norm of $P(\lambda \Delta \mathbf{x})$. Therefore, the normalized *PSF* can be found by computing the power spectrum of

$$P(\lambda \Delta \mathbf{x}) / \|P(\lambda \Delta \mathbf{x})\|_2. \tag{6.73}$$

The function in (6.72) is called the *normalized pupil function*.

When using *FAS* it is easy to do the operation in (6.73). You just have to create $P(\lambda \Delta x)$, then find its 2-Norm (using *Norm* from the *Graph* choice on the display menu, and entering 2 for the power of the norm), and then go back and create $P(\lambda\Delta x)$ *divided by this 2-Norm.* ▮

6.74 Show that the integral of the normalized *PSF* over $\mathbf{R}^2$ is 1. Also, show that the power spectrum of the function in (6.73) is indeed the same as the normalized *PSF*.

6.75 Consider the image of an edge discussed in Example 12.9. Show that the normalized *PSF along the X-direction* is given by $40\,\mathrm{sinc}^2(40X)$. Using this result, obtain the image of the edge along the X-direction by convolution of the edge function with $40\,\mathrm{sinc}^2(40X)$. (Use the edge function given in Example 12.9, and convolute over $[-32, 32]$ using 4096 points.) Your result should be a close approximation to the original edge.

6.76 Repeat Exercise 6.71, but now use the normalized pupil function for incoherent imaging (follow the instructions at the end of Remark 6.72). Your comparison of coherent and incoherent imaging should now be done *directly* without appeal to the relative contrast formula.

6.77 Compute the inverse Fourier transform of the coherent *PSF* and the normalized incoherent *PSF* from the previous exercise. Explain the results of that exercise in terms of these inverse tranforms. [*Hint*: Use the *convolution theorem*; the images are the Fourier transforms of the products of the object transforms times the inverse transforms of the *PSFs*.]

REMARK   The inverse Fourier transform of the coherent *PSF* is called the *coherent transfer function*, and the inverse Fourier transform of the normalized incoherent *PSF* is called the *optical transfer function*. For further discussion, see [Go, Chapter 6]. ▮

6.78 *Blurring.* This exercise describes a one-dimensional model of image blurring. For the normalized *PSF*, use the function $4\,\mathrm{rec}\,(4x)$. (This models a *blurred point*.) Using *FAS*, convolute this normalized *PSF* with the following object intensity functions over $[-1, 1]$ using 1024 points.
   (a) $\exp(-(x/.8) \wedge 10)(1 + .5\cos(4\pi x))$
   (b) $\exp(-(x/.8) \wedge 10)(1 + .5\cos(8\pi x))$
   (c) $\exp(-(x/.8) \wedge 10)(1 + .5\cos(12\pi x))$
   (d) $\exp(-(x/.8) \wedge 10)(1 + .5\cos(16\pi x))$
   (e) $\exp(-(x/.8) \wedge 10)(1 + .5\cos(20\pi x))$
The resulting images are blurred. Notice, in particular, that in the image for (c) the oscillations are *out of phase* with the original object's oscillations by about $90°$. Explain this effect, as well as the *absence* of oscillations in images (b) and (d), by comparing the optical transfer function with the transforms of the objects (a)–(e).

A

User's Manual for FAS

In this appendix, we describe all the aspects of using Fourier Analysis Software.

Components

Fourier Analysis Software (*FAS*) can be found on the disk accompanying this book under the file name FA.EXE. At the DOS prompt, entering FA or FA.EXE will load the program. If you have a PC operating under DOS 2.1 or higher, and you have CGA video (or EGA video, or VGA video), then *FAS* should run properly. It will run best with 400K of RAM.

FAS consists of the following eight components, listed on the menu line at the bottom of the screen (see Figure A.1):

1. *Four S*: Computes Fourier series and filtered Fourier series.

2. *Sine S*: Computes Fourier sine series and filtered Fourier sine series.

3. *Cos S*: Computes Fourier cosine series and filtered Fourier cosine series.

4. *Four T*: Computes Fourier transforms and power spectra. Also, functions can be filtered before transforming.

5. *Sine T*: Computes Fourier sine transforms and allows you to filter before transforming.

6. *Cos T*: Computes Fourier cosine transforms and allows you to filter before transforming.

7. *Conv*: Computes convolutions and autocorrelations.

8. *Graphs*: Draws graphs of functions without doing any Fourier analysis. Also, one can create a *GraphBook*, which is a series of previously saved screen images that can be viewed like a slideshow.

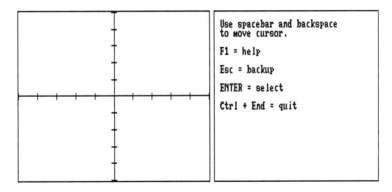

Choose a type of Fourier analysis.

Four. S.   Sine S.   Cos. S.   Four. T.   Sine T.   Cos. T.   Conv.   Graphs

FIGURE A.1
Initial menu.

Initial setup

For each of the first seven choices above, there is a common sequence of steps that are performed initially. These first three steps are

1. *Choice of interval type,* $(-L, L)$ or $(0, L)$. After choosing which interval type you want, you are then asked for the value of L. You may enter a positive real number (double precision), or a number of the form #π where # represents a real number. This format allows for multiples of π to be used. To enter the symbol π, hold down the *Ctrl key* and press *p*. [*Note*: Low resolution CGA monitors will not produce a readable character for π. *FAS* will, however, still process the value correctly. You can also enter pi as a substitute for π.]

2. *Choice of number of points.* After choosing an interval, you are presented with a menu of the form

$$128 \quad 256 \quad 512 \quad 1024 \quad 2048 \quad 4096$$

from which you can select the number of points you want to work with. Choosing 128, 256, or 512 gives very short processing time in exchange for low resolution, while choosing 1024, 2048, or 4096 gives higher resolution in exchange for longer processing time. The choice of 1024 gives a useful mean between the two.

3. *Choice of function.* After choosing the number of points, you are presented with the following function menu:

User   Piece   Point   Data

Note: You may backup from each menu to the previous menu by pressing the *Esc* key. If you are at a step requiring input of data, you can still backup by pressing the *Esc* key.

Functions

There are four choices listed on the function menu (see above). Here is a description of each of these choices.

USER

This choice allows you to compute functions of x defined by formulas involving elementary functions. The prompt

$$f(x) =$$

followed by a blinking cursor, will appear. You may then type in up to three lines of text for your formula. The functions allowed are

sin( )	[sine function]
cos( )	[cosine function]
tan( )	[tangent function]
exp( )	[exponential function]
sqr( )	[positive square root function]
abs( )	[absolute value function]
log( )	[natural logarithm function]
atn( )	[arctangent function]
sinc( )	[sinc function]
sinh( )	[hyperbolic sine function]
cosh( )	[hyperbolic cosine function]
tanh( )	[hyperbolic tangent function]
rec( )	[rectangle function]
gri( )	[greatest integer function]

The arctangent function is, of course, also known as the inverse tangent function. The sinc and rec functions are defined by

$$\operatorname{sinc}(x) = \begin{cases} 1 & \text{for } x = 0 \\ \frac{\sin \pi x}{\pi x} & \text{for } x \neq 0, \end{cases} \quad \operatorname{rec}(x) = \begin{cases} 1 & \text{for } |x| < 0.5 \\ 0.5 & \text{for } |x| = 0.5 \\ 0 & \text{for } |x| > 0.5. \end{cases}$$

You may combine these functions using composition and the five binary operations: ∧ (raise to a power), / (division), ∗ (multiplication), − (subtraction), and + (addition). Multiplication may also be done using *juxtaposition*. The order of operations is the familiar one from elementary algebra: raising to a power is first, then division, multiplication, subtraction, and addition is last.

For example, the following are all valid formulas:

$$2\sin(3x) - 4\cos(5x)$$

$$\exp[\sin(3x)] - 4\cos(3\pi x)\text{atn}(4x)$$

$$3\exp\{-(2x - 1)(3x - 4)\}$$

$$x - x \wedge 3/3! + x \wedge 5/5! - x \wedge 7/7!$$

Notice that brackets, [ ], or curly brackets, { }, are allowed instead of parentheses. And the factorial symbol ! is allowed, too. The factorial symbol will only give the correct values if non-negative integers are used before it. As explained above, π is entered by holding down the *Ctrl* key and pressing *p*. [*Note:* Low resolution CGA monitors will not produce a readable character for π. In this case, you can enter "*pi*" as a substitute for π.]

It is clear from these examples that formulas can be entered in a fairly natural way.

The editor used for editing formulas will respond to the *Home* key, the *End* key, and the arrow keys. *Home* or *End* will take you to the beginning or end of a line, respectively. The arrow keys allow you to move about the formula while typing. If you press the *Del* key, the letter you are on will be deleted. Pressing the *Ins* key allows you to toggle between typing over previous letters (normal mode) and inserting new letters (insert mode).

If a formula has been previously entered, then that formula will appear again after the "$f(x) =$" prompt. This allows for simple modifications for creating related formulas. To clear the formula completely, press the *F9* key.

When you have finished entering a formula, press the *Enter* key and the program will graph the function described by that formula.

If the formula does not have the proper syntax, e.g. too many left or right parentheses, a beeping sound will result and the following error message will appear:

Invalid formula. Press any key to continue.

When you press a key, your previous formula will appear and you may correct it. Similarly, if the formula is correct but an overflow occurs during computation or the domain is too large (which can happen with sqr and log), then the program will beep and the following error meggage will appear:

Run Time Error. Press any key to continue.

Again, after you press a key, you may modify your function to handle the problem.

To recall a formula that was previously saved (see *SAVE*), press the *F3* key. A list of formula file names will appear in the help box, and you can select a formula.

To back out of the *User* procedure, press the *Esc* key and you will return to the function menu.

PIECE

This choice allows you to graph piecewise defined functions. These functions are described by defining a separate formula over each of a finite number of subintervals of the interval that you chose at the start. For example, suppose you chose the interval $[0, 8]$ and 1024 points and you want to graph the following function:

$$f(x) = \begin{cases} 2x & \text{for } 0 < x < 2 \\ \sin \pi x & \text{for } 2 < x < 5 \\ -\exp(-(x-6)) & \text{for } 5 < x < 7 \\ -1 & \text{for } 7 < x < 8. \end{cases}$$

Then, after choosing the interval $[0, 8]$ and 1024 points and selecting *Piece* from the function menu, you should respond as follows (entered data is underlined):

The interval is $[0, 8]$.

Enter the number of subintervals: 4

The first subinterval is $[0, x]$. x? 2

$f(x) = 2x$

FAS will produce the graph shown in Figure A.2. After pressing *Enter* to proceed to the next step, you should respond as follows:

The next subinterval is $[2, x]$. x? 5

$f(x) = \sin(\pi x)$

FAS will produce the graph shown in Figure A.3. After pressing *Enter* to proceed to the next step, you should respond as follows:

The next subinterval is $[5, x]$. x? 7

$f(x) = -\exp(-(x-6))$

FAS will produce the graph shown in Figure A.4. Pressing *Enter* to proceed to the last step, you should respond as follows:

$f(x) = -1$

FAS produces the graph shown in Figure A.5, which is the completed graph.

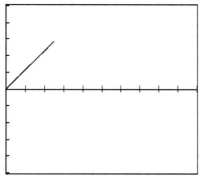

Graph partially done, 1 interval out of four
X interval: $[0, 8]$ X increment = .8
Y interval: $[-7, 7]$ Y increment = 1.4

FIGURE A.2
Construction of piecewise function.

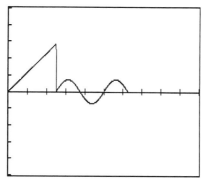

Graph partially done, 2 intervals out of four
X interval: $[0, 8]$ X increment = .8
Y interval: $[-7, 7]$ Y increment = 1.4

FIGURE A.3
Construction of piecewise function.

It is very easy, for instance, to graph step functions using the *Piece* procedure. You just have to successively enter constants for the functions on each subinterval.

If you enter the wrong formula at some point, and wish to backup, you just press the *Esc* key. The backup, however, will often be to the very beginning of the *Piece* procedure.

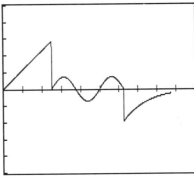

Graph partially done, 3 intervals out of four
X interval: [0, 8]   X increment = .8
Y interval: [−7, 7]   Y increment = 1.4

FIGURE A.4
Construction of piecewise function.

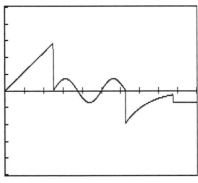

Piecewise function
X interval: [0, 8]   X increment = .8
Y interval: [−7, 7]   Y increment = 1.4

FIGURE A.5
Construction of piecewise function.

POINT

This choice allows you to compute discrete delta functions. This is done by assigning values to specific integer indices. For example, suppose you chose the interval (−16, 16) and 1024 points. You will be asked how many integers you want values assigned to and then what values to assign. If you respond as

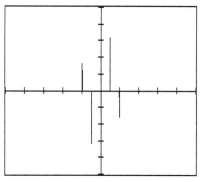

Point function
X interval: $[-16, 16]$ X increment $= 3.2$
Y interval: $[-1602, 1602]$ Y increment $= 320.4$

FIGURE A.6
Point function.

follows (the underlined numbers are entered values):

How many integers do you want values assigned to? <u>4</u>
Enter integer 1 out of 4: <u>50</u>   Function value? <u>1024</u>
Enter integer 2 out of 4: <u>100</u>   Function value? <u>−512</u>
Enter integer 3 out of 4: <u>−50</u>   Function value? <u>−1024</u>
Enter integer 4 out of 4: <u>−100</u>   Function value? <u>512</u>

then the graph will appear as in Figure A.6. The spikes appear at the x-values:

$$50(32/1024) = 1.5625$$
$$100(32/1024) = 3.125$$
$$-50(32/1024) = -1.5625$$
$$-100(32/1024) = -3.125$$

The scale factor $32/1024$ is equal to the length of the interval, 32, divided by the number of points, 1024. You should enter integers between $\pm 1/2$(number of points), in this case ± 512.

If you chose the interval $[0, 16]$, then negative integers are *not* allowed. If you respond as follows:

How many integers do you want values assigned to? <u>3</u>
Enter integer 1 out of 3: <u>50</u>   Function value? <u>512</u>
Enter integer 2 out of 3: <u>100</u>   Function value? <u>−1024</u>
Enter integer 3 out of 3: <u>300</u>   Function value? <u>1024</u>

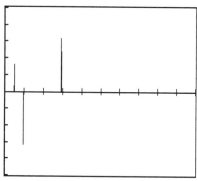

Point function
X interval: $[0, 16]$ X increment $= 1.6$
Y interval: $[-1602, 1602]$ Y increment $= 320.4$

FIGURE A.7
Point function.

then the graph will appear as in Figure A.7. The spikes appear at the x-values:

$$50(16/1024) = .78125$$

$$100(16/1024) = 1.5625$$

$$300(16/1024) = 4.6875$$

The scale factor $16/1024$ is equal to the length of the interval, 16, divided by the number of points, 1024. You should enter integers between 0 and the number of points.

Note: You may backup by pressing the *Esc* key for any of the numerical data entries.

DATA

This choice allows you to load numerical data files that were previously saved (using *SAVE* on the display menu). When you select this choice, a list of previously saved numerical data file names will appear in the Help box (see Figure A.8). Using the spacebar and backspace, or the up and down arrow keys, you can place the cursor on the file you want and then press *Enter*. (If there are more file names than the Help box can hold, then you will scroll through the list of file names once you reach the end of the first list.)

Once you select it, either a file will load and a graph will appear on the screen or, if the number of points or interval are inconsistent with present values, an error message will appear at the bottom of the Help box.

The files listed in the Help box were retrieved using the path listed at the

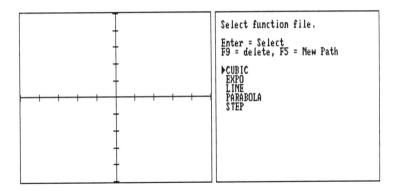

A:\FA

FIGURE A.8
List of function files.

bottom left corner of the screen (see Figure A.8). To change the path, you go through the following steps. If you press the *F5* key, then the following prompt will appear (followed by a blinking cursor):

$$* >$$

where $*$ represents the letter of the drive you are in. You then may enter either a new drive using the format

$$* :$$

where $*$ represents the letter of the new drive, or change directories using the format

$$cd \; pathname$$

where *pathname* stands for any valid DOS path designation. For example, if you enter

$$cd \; \backslash data$$

then you will change to the directory DATA on the current drive. If you enter

$$cd \; a : \backslash data$$

then the default directory on drive A will be changed to DATA. If A is the current drive, then you will change to this directory. If A is not the current drive, then you will *not* change over to drive A. To do that, you have to enter

$$a:$$

Since the default directory on drive A is DATA you will now change to drive A and directory DATA. In other words, the files that are listed now will be obtained using the path A : \DATA (and this path will appear at the bottom left corner of the screen).

You may also load external data files if they are in the following format. They must have a name with format *.FDT where * represents a valid eight-character DOS file name. These external files must be ASCII sequential files. The first data value must be the number of points (either 128, 256, 512, 1024, 2048, or 4096). The next entry must be the left endpoint of the interval; it must be either 0 or the negative of the right endpoint of the interval. The entry after that must be the right endpoint of the interval; it must be positive. The remaining entries are the function data values, either 129, 257, 513, 1025, 2049, or 4097 values. Notice that each entry in this list is *one* more than the corresponding entry in the list of numbers of points. This has to do with the way *FAS* processes the other function choices, and this choice is forced to be consistent with that method. All these numerical values must be in IEEE double-precision format. *Note*: Many devices for obtaining data collect some power of 2 number of data values. There is then no unique way to assign the last data entry. (One possible way is to give it the value of the first entry. This insures that transforms are calculated correctly. For Fourier series, however, a better method would be to assign to it some average of the last few data values.)

This completes our discussion of the initial setup of function data. We will now discuss each of the eight choices available on the initial menu (see Figure A.1).

FOUR S

Once you have completed the initial setup, then this choice allows you to compute Fourier series partial sums for your function data. The first question you are asked is

$$\text{How many harmonics do you want?} \qquad (A.1)$$

The number of harmonics is the number of complex exponential terms (counting positively and negatively indexed harmonics together, and excluding the constant term). For example, the function $f(x) = x$ has the following complex Fourier series over the interval $[0, 2\pi]$:

$$\pi + \sum_{k=-\infty}^{\infty}{}' \frac{i}{k} e^{ikx}$$

(where the prime on the summation sign means the $k = 0$ term is omitted). If you choose to graph, say, 15 harmonics from this Fourier series, then *FAS* will compute (via an FFT algorithm) the partial sum

$$\pi + \sum_{k=1}^{15}\left(\frac{i}{k}e^{ikx} - \frac{i}{k}e^{-ikx}\right) \qquad (A.2)$$

If you press f instead of entering a number of harmonics, then *FAS* will graph the original function.

After you choose the number of harmonics you want (except if you press f), then you are asked:

$$Do\ you\ want\ to\ apply\ a\ filter?\ (Press\ y\ or\ n) \qquad (A.3)$$

If you press n, then the partial sum in (A.2) is graphed. If you press y, then a list of filters is given. After selecting a filter, the sum in (A.2) is modified by multiplying by constants $\{F_k\}$ as follows:

$$F_0\pi + \sum_{k=1}^{15}F_k\left(\frac{i}{k}e^{ikx} - \frac{i}{k}e^{-ikx}\right). \qquad (A.4)$$

The nature of the constants $\{F_k\}$ depends on which filter you choose. Some information on this is given in the Help file; just press *F1* when the cursor is on the particular filter you want information about. Further discussion of these filters can be found in Chapter 4. After the graph is displayed, if you press *Enter* while the menu cursor is on *NEXT*, you will be sent to the next part of the *Four S* routine.

After graphing a partial sum, you are asked the following question:

$$Do\ another\ partial\ sum?\ (Press\ y\ or\ n) \qquad (A.5)$$

If you press y, then you are sent back to the question about the number of harmonics and the procedure begins all over again. If you press n, then you are asked:

$$Do\ a\ series\ for\ another\ function?\ (Press\ y\ or\ n) \qquad (A.6)$$

If you press y, you are sent back to the function menu, where you can create a new functon and enter the *Four S* procedure again. If you press n, you are presented with the following message:

$$Press\ e\ to\ end\ the\ program,\ Enter\ to\ continue \qquad (A.7)$$

If you press e, *FAS* terminates and you are sent back to DOS. If you press *Enter*, you are sent back to the initial menu.

For each of the questions in the *Four S* procedure, if you press the *Esc* key, you are sent back to the previous question [or to the Display menu if you press *Esc* at question (A.5)].

SINE S

The structure of the *Sine S* procedure is essentially the same as the *Four S* procedure. The only difference is that instead of computing a Fourier series you are computing a Fourier sine series. For example, if you choose to compute 15 harmonics of the Fourier sine series for $f(x) = x - \pi$ over the interval $[0, 2\pi]$, then *FAS* will graph

$$\sum_{n=1}^{15} \frac{-2}{n} \sin nx. \qquad (A.8)$$

Other than what is being graphed [the sine series in (A.8) instead of the Fourier series in (A.2)], the questions you are asked and how *FAS* responds to your answers is *identical* to what is described above for the *Four S* procedure.

COS S

The structure of the *Cos S* procedure is essentially the same as the *Four S* procedure. The only difference is that instead of computing a Fourier series you are computing a Fourier cosine series. For example, if you choose to compute 15 harmonics of the Fourier cosine series for $f(x) = x^2/4$ over the interval $[0, 2\pi]$, then *FAS* will graph

$$\frac{\pi^2}{3} + \sum_{n=1}^{15} \frac{1}{n^2} \cos nx. \qquad (A.9)$$

Other than what is being graphed [the cosine series in (A.9) instead of the Fourier series in (A.2)], the questions you are asked and how *FAS* responds to your answers is identical to what is described above for the *Four S* procedure.

FOUR T

This choice is unique in that you may work with complex-valued functions if you wish. First, you are presented with the following menu:

$$\textit{Real} \qquad \textit{Complex} \qquad (A.10)$$

If you want to transform a real-valued function, select *Real*. If you want to transform a complex-valued function, select *Complex*.

If you select *Real*, the initial setup is the same as described above: you must choose an interval type and a number of points and then enter your function data. When you choose *Complex*, however, you must enter two functions. These functions will be the real and imaginary parts of the complex-valued function that you are transforming. You will be asked to supply the real part as the first function and the imaginary part as the second function. Once you have finished entering the function data, *FAS* will begin approximating Fourier transforms.

First, you are presented with the following menu of two choices:

$$\text{Four Tr} \qquad \text{Power Spectrum} \qquad (A.11)$$

If you select *Four Tr*, you will be asked the following:

$$\text{Positive transform exponent } (y = pos., \ n = neg.)? \ (Press \ y \ or \ n) \qquad (A.12)$$

If you press *y*, then *FAS* will approximate the following Fourier transform:

$$\tilde{f}(x) = \int_{-\infty}^{\infty} f(s)e^{i2\pi xs} \, ds, \qquad (A.13)$$

while if you press *n*, *FAS* will approximate the following Fourier transform:

$$\hat{f}(x) = \int_{-\infty}^{\infty} f(s)e^{-i2\pi xs} \, ds. \qquad (A.14)$$

FAS will then ask you the following question:

$$\text{Do you want to apply a filter? } (Press \ y \ or \ n) \qquad (A.15)$$

If you press *y*, then a menu of filter choices will appear. If you choose a filter, then *FAS* will approximate the filtered transform

$$\int_{-\infty}^{\infty} F(s)f(s)e^{\pm i2\pi xs} \, ds$$

where the sign in the exponent depends on how you answered question (A.12). The purposes of the filtering process are described in Chapter 5.

FAS will first display the real part of the transform (filtered or unfiltered). If you choose *NEXT* from the display menu, then you will be asked the following question:

$$\text{Leave previous graph(s) on the screen? } (Press \ y \ or \ n) \qquad (A.16)$$

If you press *n*, then *FAS* will graph the imaginary part of the transform. If you press *y*, *FAS* will graph the real and imaginary part of the transform.

After finishing the display portion of the *Four Tr* procedure, you will be asked the following question:

$$\text{Do another transform? } (Press \ y \ or \ n) \qquad (A.17)$$

If you press *y*, then you are returned to the menu in (A.11) and the *Four Tr* procedure begins again *using the same function you initially chose to transform.*

You can perform a transform with a different filter (or without a filter) or you can choose to do a *Power Spectrum.* If you press *n*, then you are asked the following question:

> *Do you want to transform another function? (Press y or n)*   (*A*.18)

If you press *y*, then you will be returned to menu (A.10). Pressing *n* will bring up the end of program message (A.7). Press *e* to end the program (and return to DOS) or *Enter* to return to the initial menu.

Press the *Esc* key to back up from each stage to the previous one.

If you select *Power Spectrum* from menu (A.11), then *FAS* will approximate the modulus-squared of the transform $|\hat{f}(x)|^2$. If you chose *Real* in (A.10), then it does not matter whether formula (A.13) or (A.14) is used to compute $|\hat{f}(x)|$; the result for $|\hat{f}(x)|^2$ is the same. Consequently, you are not asked question (A.12) about what sign exponent you want (*FAS* just chooses positive automatically). If you chose *Complex*, however, it does matter whether formula (A.13) or (A.14) is used, so you will be asked question (A.12).

The structure of the *Power Spectrum* portion of the *FOUR T* procedure is the same as described above for the *Four Tr* subprocedure. Therefore, we will end our discussion of the *FOUR T* procedure here.

SINE T

The structure of the *SINE T* procedure is almost the same as the *Four Tr* procedure described above for real-valued functions. The only difference is that with the *SINE T* procedure *FAS* approximates the following sine transform:

$$\hat{f}_S(x) = 2 \int_0^\infty f(s) \sin(2\pi x s)\, ds. \qquad (A.19)$$

Since there is no exponent in (A.19), you are not asked question (A.12) about the sign of the exponent, as in the *Four Tr* procedure. Other than that, the questions you are asked are the same as in the *Four Tr* procedure.

COS T

The structure of the *COS T* procedure is almost the same as the *Four Tr* procedure described above for real-valued functions. The only difference is that with the *COS T* procedure *FAS* approximates the following cosine transform:

$$\hat{f}_C(x) = 2 \int_0^\infty f(s) \cos(2\pi x s)\, ds. \qquad (A.20)$$

Since there is no exponent in (A.20), you are not asked question (A.12) about the sign of the exponent, as in the *Four Tr* procedure. Other than that, the questions you are asked are the same as in the *Four Tr* procedure.

CONV

This choice computes either convolution or autocorrelation. After initial setup, choosing interval type and number of points, you are presented with the following menu:

$$\text{convolution} \qquad \text{autocorrelation} \qquad (A.21)$$

If you choose convolution, the following menu appears:

> *Choose first function.*
> *User   Piece   Point   Data* $(A.22)$

from which you then proceed to create your first function; call it $f(x)$. After selecting *NEXT* from the display menu, *FAS* asks you to choose a second function; call it $g(x)$. The following menu appears:

> *Choose second function.*
> *User   Piece   Point   Data* $(A.23)$

After this second function is graphed and you have selected *NEXT* from the display menu, *FAS* approximates the convolution of f and g. There are several kinds of convolution. You are first asked the following question:

$$\text{Divide by interval length? (Press y or n)} \qquad (A.24)$$

If you press *y* and the interval you chose was $[-L, L]$, then *FAS* will approximate the integral

$$\frac{1}{2L} \int_{-L}^{L} f(s)g(x - s)\, ds \qquad (A.25)$$

which is the convolution of f and g over the finite interval $[-L, L]$. If the interval you chose was $[0, L]$, then *FAS* will approximate the integral

$$\frac{1}{L} \int_0^{L} f(s)g(x - s)\, ds \qquad (A.26)$$

which is the convolution of f and g over the interval $[0, L]$. See Chapters 4 and 5 for further details about the theory of convolution.

On the other hand, if you press n in response to question (A.24), then, when the interval you chose was $[-L, L]$, *FAS* will compute

$$\int_{-L}^{L} f(s)g(x-s)\,ds. \qquad (A.27)$$

For certain kinds of functions, f and g, the integral in (A.27) will be a good approximation to the integral

$$\int_{-\infty}^{\infty} f(s)g(x-s)\,ds \qquad (A.28)$$

which is the convolution of f and g over $(-\infty, \infty)$. If the interval you chose was $[0, L]$, then *FAS* will compute

$$\int_{0}^{L} f(s)g(x-s)\,ds. \qquad (A.29)$$

For certain kinds of functions, f and g, the integral in (A.29) will be a good approximation to

$$\int_{0}^{x} f(s)g(x-s)\,ds \qquad (A.30)$$

which is the convolution of f and g over $[0, \infty)$.

If you choose autocorrelation from the menu in (A.21), you will only be asked to create one function. The following menu appears:

> *Choose a function.*
> *User   Piece   Point   Data* $\qquad (A.31)$

After you have graphed this function and selected *NEXT* from the display menu, you will be asked question (A.24). If you press n and you chose the interval $[-L, L]$, then the integral

$$\int_{-L}^{L} f(s)f(s-x)\,ds \qquad (A.32)$$

will be computed. If you chose the interval $[0, L]$, then the integral

$$\int_{0}^{L} f(s)f(s-x)\,ds \qquad (A.33)$$

will be computed. If you press y in response to question (A.24), then the integral in (A.32) will be divided by $2L$ while the integral in (A.33) will be divided by L. These integrals are approximations to autocorrelations over finite and infinite intervals.

When the convolution and autocorrelation procedures are finished, you are then asked the following question:

$$\text{Do another convolution operation? (Press y or n)} \qquad (A.34)$$

If you press y, then you are sent back to the menu in (A.21) and the *CONV* procedure begins again. If you press n, then the end of the program message (A.7) appears. If you press e, you will return to DOS; if you press *Enter* you will return to the initial menu.

GRAPHS

The last procedure available on the initial menu is the *GRAPHS* procedure. This procedure begins by displaying the following menu:

$$\text{Create Graph} \quad \text{Create GraphBook} \quad \text{View GraphBook} \qquad (A.35)$$

Here is a description of each of these choices.

1) *Create Graph*. This procedure allows you to overlay graphs without doing any Fourier analysis. It is useful for overlaying, for comparison purposes, graphs from separate procedures. For example, you might wish to compare different convolutions, or Fourier transforms of different functions, or certain Fourier series with Fourier transforms, etc. This procedure can also be used as a graphing program for other purposes, such as comparing functions with their power series partial sums.

After initial setup, the procedure will graph the first function you choose to graph. When you want to graph functions created in the other procedures of *FAS*, you must first save them at the time they are created and then retrieve them using the *Data* choice on the function menu.

After you have graphed a function and selected *NEXT* from the display menu, you are asked the following question:

$$\text{Graph another function? (Press y or n)} \qquad (A.36)$$

If you press y, you are sent to the initial function menu. This gives you the opportunity to overlay graphs for comparison purposes. If you press n, the end of program message (A.7) appears. Pressing e returns you to DOS; pressing *Enter* will return you to the initial menu.

2) *Create GraphBook*. This procedure allows you to create a series of previously saved screen images that can be viewed like a slideshow. When you are at the display menu, you can save a screen image (see the section entitled *SAVE*). These screen images are what the present procedure will link together.

The procedure begins by displaying the prompt

$$* >$$

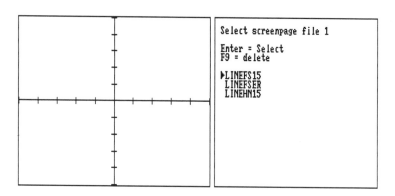

FIGURE A.9
List of screen files

where ∗ is the letter of the drive you are in. You can then change the drive or directory you want to be in (or remain in the same directory by pressing *Enter*). The procedures involved are described above in the *Data* procedure. After this step, the following message appears:

Enter the name of your GraphBook: (*A*.37)

followed by a blinking cursor. The name you should give is any valid eight-character DOS file name. The extension .GBK is automatically attached to your *GraphBook* name and the result is the DOS file name for your *GraphBook*.

After this, the screen files that are available are listed for selection in the Help Box (see Figure A.9). Use the spacebar and backspace (or the up and down arrow keys) to place the cursor on your choice, then press *Enter* to select the screen image file. The screen image you saved in this file will appear on the screen (the delay involved is minimal if you are using a hard disk; retrieving the file from a floppy disk takes a little longer). You are then asked the following question:

Include this screen in your GraphBook? (Press y or n)   (*A*.38)

If you press *y*, then the screen is included in your *GraphBook*. It is also removed from the list of available screen files. It is easy, however, to put it back if you wish to do so (this will be explained in the next section).

After this [or if you pressed *n* in response to question (A.38)], you are asked the following question:

Are you finished? (Press y or n)   (*A*.39)

If you press *n*, then you are sent to the list of available screen files, and the

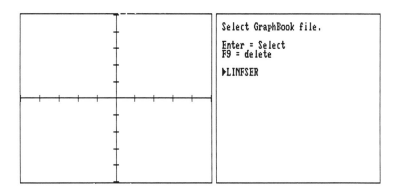

FIGURE A.10
List of GraphBook files

procedure begins again. If you press *y*, then your *GraphBook* is saved under the file name you created above.

3) *View GraphBook*. This procedure allows you to view a *GraphBook* created by the previous procedure. The procedure begins by displaying the prompt

$$* >$$

(followed by a blinking cursor) where $*$ is the letter of the drive you are in. You can then change the drive or directory you want to be in (or press *Enter* to remain in the same directory). The procedures involved are described above in the *Data* procedure. After this step you must select which *GraphBook* you want to view. The *GraphBook* file names appear in the Help Box (see Figure A.10). Use the spacebar and backspace (or the up and down arrow keys) to place the cursor on your choice, then press *Enter* to select that *GraphBook*. The first screen image in the *GraphBook* will then appear. This will happen quickly if you are using a hard disk; it will take a little longer if you are retrieving the file from a floppy disk. To view successive screen images just press the *PgDn* key to go forward and the *PgUp* key to go backward.

When a screen image is displayed you will see the full display menu (see the section entitled DISPLAY below). You can use this menu to make modifications to the screen image, such as changing the x or y intervals, adding a note, modifying the note that appears on the screen, etc. A modified screen image can be saved as part of the *GraphBook*. First, you select *SAVE* from the display menu and the following menu appears:

$$auto \quad screen \quad function \quad Function\ (ASCII) \qquad (A.40)$$

By selecting *auto*, a modified screen image is automatically saved as part of the *GraphBook* that you are viewing.

If you select *function* from menu (A.40), you can save one of the graphs from the screen image. This graph will then be accessible from the *Data* choice on the function menu. If you select *screen*, you can store the screen image in a screen file. This screen file will then be accessible from the *Create GraphBook* choice on the *GRAPHS* menu (A.35). (This is how screen images can be restored to the list of screen files, after they are removed when a *GraphBook* is created).

The choice *Function (ASCII)* allows you to save one of the graphs from the screen image into an ASCII text file. This ASCII file is readable by other software (including many graphics programs). Graphs saved in this format are still accessible from the *Data* choice on the function menu.

When you have finished viewing all the screen images in a *GraphBook*, the following question appears:

$$\text{Are you finished? (Press y or n)} \qquad (A.41)$$

If you want to add more screen images to the *GraphBook*, press *n*. You will then be asked the following question:

$$\text{Add more screen pages? (Press y or n)} \qquad (A.42)$$

Pressing *y* will allow you to choose again from the list of available screen files. If you answer question (A.41) by pressing *y* or (A.42) by pressing *n*, you are sent back to menu (A.35), the beginning of the *GRAPHS* procedure.

DISPLAY

All of the procedures described above eventually utilize the DISPLAY procedure for displaying graphs. When a graph is displayed on the screen, the following menu appears:

$$\text{NEXT \quad X-int \quad Y-int \quad Graphs \quad Cap \quad PrSc \quad SAVE \quad Note} \qquad (A.43)$$

We will now describe each of the eight procedures listed on this menu.

NEXT

If you make this choice, it means that you are ready to proceed to the next step in one of the procedures of *FAS*. If you are unsure what will happen next, then you can, of course, consult this manual. You could also just select *NEXT* to see what happens, since almost every procedure can be backed up *one* step to the DISPLAY procedure.

X-int

Select this choice in order to change the x-interval. You will be presented with the following message:

Enter X-interval (left endpt., right endpt.): $(A.44)$

followed by a blinking cursor. You then type in an interval in the form of two real numbers separated by a comma (you need not include parentheses or brackets around the numbers). The first number will be the new left endpoint and the second number will be the new right endpoint of the x-interval. If the right endpoint is not greater than the left endpoint, *FAS* will beep and you will be asked to redo your input. This will also happen if the new interval does not lie within the original x-interval that you chose during the initial setup.

Pressing the *Esc* key will return you to the Display menu (A.43).

Y-int

Select this choice in order to change the y-interval. You will be presented with the message

Enter Y-interval (left endpt., right endpt.): $(A.45)$

followed by a blinking cursor. You then type in an interval in the form of two real numbers separated by a comma (you need not include parentheses or brackets around the numbers). The first number will be the new left endpoint and the second number will be the new right endpoint of the y-interval. If the right endpoint is not greater than the left endpoint, *FAS* will beep and you will be asked to redo your input.

Pressing the *Esc* key will return you to the Display menu (A.43).

Graphs

This procedure allows you to do some algebraic operations on graphs. The following menu will appear:

Norm   Norm Difference   Combine   Sum Squares   Multiply   Delete

$(A.46)$

Here is a description of each choice.

1) *Norm*. With this choice you can approximate the p-Norm or the Sup-Norm of a graph on the screen. The p-Norm of a function $f(x)$ over an interval $[a, b]$ is defined to be

$$\left[\int_a^b |f(x)|^p \, dx \right]^{\frac{1}{p}} \qquad (A.47a)$$

while the Sup-Norm is defined by

$$\sup_{a \le x \le b} |f(x)| \qquad (A.47b)$$

where sup stands for *supremum* and means (essentially) the maximum of $|f(x)|$ over the interval $[a, b]$. You are asked the following question:

$$\textit{Power norm?} \qquad (A.48)$$

followed by a blinking cursor. If you enter a real number greater than or equal to 1, that real number is used for the power p in computing the p-Norm in (A.47a). If you enter a number less than 1, then you will be told to redo your input. By entering the word *sup* in response to question (A.48), you will tell *FAS* to use (A.47b) to compute a Sup-Norm.

The Norms of all the graphs on the screen are approximated and the results are displayed on the caption line below the graphs.

Note: These norms are computed for $[a, b]$ equal to the present x-interval. If this interval is not the original x-interval, then you will have to change back to that original one in order to approximate norms over it.

2) *Norm Difference*. With this choice you can approximate either the p-Norm difference or the Sup-Norm difference between two graphs on the screen. Depending on your answer to question (A.48), either formula (A.47a) or (A.47b) is used for $f(x) = h(x) - g(x)$, where $h(x)$ and $g(x)$ are the two functions whose norm difference is being calculated. If there are just two graphs on the screen, then the norm difference between those two graphs is computed and the result is displayed in the caption below the graphs. If there are more than two graphs, you are presented with the following request for input:

$$\textit{Numbers of graphs (separate numbers by a comma):} \qquad (A.49)$$

followed by a blinking cursor. You may then enter the numbers for which you want the norm difference calculated (1,2 for graph numbers 1 and 2, etc.). The norm difference is then approximated and the result is displayed in the caption line below the graphs.

Note: These Norm Differences are computed for $[a, b]$ equal to the present x-interval. If this interval is not the original x-interval, then you will have to change back to that original one in order to approximate Norm differences over it.

3) *Combine*. This choice allows you to create a new graph that is a linear combination of the graphs displayed on the screen. For example, if you have graphed the functions

Graph 1: $\cos x$

Graph 2: x

Graph 3: $\sin x$

then you may graph the function

$$2\cos x - 3\sin x$$

by choosing *Combine* and responding as follows (entered data is underlined):

Enter the coefficient for graph 1: <u>2</u>

Enter the coefficient for graph 2: <u>0</u>

Enter the coefficient for graph 3: <u>−3</u>

Pressing the *Esc* key at any point will take you back one step.

Note: If you are in the Four S, Sine S, or Cos S procedures, then the new graph generated by the Combine procedure becomes the function for which new partial sums are calculated.

4) *Sum Squares*. This choice allows you to create a new graph that is the sum of the squares of the values for two graphs displayed on the screen. This is used for computing intensities from real and imaginary parts of Fourier transforms or for creating Fresnel diffraction patterns.

First, you are asked the following question:

$$\textit{Normalize (take square roots)? (Press y or n)} \qquad (A.50)$$

If you press *n*, the sum of squares of two graphs will be computed. If you press *y*, the *square root* of the sum of squares of the two graphs will be computed.

If there are only two graphs on the screen, the above calculations are performed and you are asked the following question:

$$\textit{Leave the previous graph(s) on the screen? (Press y or n)} \qquad (A.51)$$

If you press *n*, only the new graph (sum of squares, or square root of the sum of squares) will be displayed. If you press *y*, the new graph will be displayed as well as the old graphs.

Note: If you are in the Four S, Sine S, or Cos S procedures, this new graph becomes the function for which new partial sums are calculated.

5) *Multiply*. This choice allows you to multiply any two graphs (including a graph times itself). If only one graph is displayed, then that graph will be multiplied times itself when you select *Multiply*. If more than one graph is displayed, then you will be asked to name the two graphs that you wish to multiply. (If you list the same number twice, then you will cause that graph to be multiplied by itself.)

6) *Delete*. This choice allows you to delete one of the graphs displayed on the screen. (If only one graph is displayed, this choice does not appear.) When you select this choice, the following request for input appears:

$$\textit{Enter the number of the graph to be deleted:}$$

followed by a blinking cursor. After entering the number of the graph you want deleted, that graph will be removed from the screen. *Be careful using this choice!* The data from the deleted graph will be lost unless you have previously saved it.

Cap

Each time a graph is displayed, a caption is automatically supplied with that new graph. This choice allows you to modify the caption or insert a new one. The following request will appear:

$$\textit{Caption:} \hspace{4cm} (A.52)$$

followed by the present caption. You may then modify that caption, or, by pressing the *F9* key, erase it and type in a new caption. After you press *Enter*, the caption you have created will appear in the caption line. If you press the *Esc* key before you have entered the new caption, the old caption will be retained and you will be back at the display menu (A.43).

PrSc

This choice allows you to feed a screen image to a printer to make a printed copy. To utilize this procedure, you must have loaded a program for graphics dumping *before* loading FA.EXE. For dot matrix printers, the DOS command file GRAPHICS.COM will serve this purpose; for laser printers, there are similar programs. It saves a lot of trouble if you write a batch file that loads this graphics program followed by FA.EXE. Consult your DOS manual for instructions on how to create batch files.

After you select *PrSc*, if you have a monochrome video, the display menu is removed from the screen. If you then press the *Print Screen* button (or *Shift + PrSc* on some PC's), the screen image will be fed to the printer *provided* you have loaded the graphics program prior to loading FA.EXE. After printing, press any key to continue with *FAS*.

If you have color video, then the following menu appears:

$$\textit{Black \& White} \hspace{1cm} \textit{Color} \hspace{3cm} (A.53)$$

If you have an ordinary black and white printer, then you should select *Black & White*. The screen converts to a monochrome image. If you then press the *Print Screen* button (or *Shift + PrSc*) the screen image will be fed to the printer. After printing, press any key to continue with *FAS*. If you have a color printer that can print the colors on the screen, you may want to choose the *Color* choice on menu (A.53). Then the display menu will vanish and you can print the screen as we described above.

SAVE

This procedure allows you to save graphs or screen images. Normally, the following menu appears:

$$\textit{function} \hspace{0.5cm} \textit{screen} \hspace{0.5cm} \textit{formula} \hspace{0.5cm} \textit{Function (ASCII)} \hspace{2cm} (A.54)$$

The other type of menu is (A.40), which we described above in the *View Graph-Book* procedure.

If you select *function* from menu (A.54) or (A.40), you are asked the following question:

$$\text{\textit{Change caption? (Press y or n)}} \qquad (A.55)$$

If you press *y*, then the prompt (A.52) appears followed by the present caption. After making your changes, you press *Enter* and the new caption will be saved with the graph that you save (however, the caption displayed on the screen will not change). If you press *n* in response to question (A.55), then the program will proceed to the next step. If more than one graph is displayed, then you will be asked which graph you want saved. The following message will appear:

$$\text{\textit{Enter the number of the graph you want saved:}} \qquad (A.56)$$

followed by a blinking cursor. After entering the number of the graph that you want saved, the following request for input appears:

$$\text{\textit{File Name:}} \qquad (A.57)$$

followed by a blinking cursor. You may enter any valid eight-character DOS file name. The extension .FDT is automatically appended to your file name [if you are doing a (Fourier) series filter, the extension is .FFF; if you are doing a function filter (within a transform procedure), the extension is .TFF] and the result is the function data file name. This function data file will be accessible if you select *DATA* from the function menu. If the extension is .FDT this will be *Data* on the initial function menu. If the extension is .FFF or .TFF this will be *Data* on the function menu that you obtain after selecting *User* from one of the filter menus.

If you select *screen* from menu (A.54) or (A.40), the following request for input appears:

$$\text{\textit{File name:}} \qquad (A.58)$$

followed by a blinking cursor. You may enter any valid eight-character DOS file name. The extension .FNB is automatically appended to your file name and the result is the screen file name. This screen file will be accessible if you select *Create GraphBook* from the *GRAPHS* procedure.

If you select *formula* from menu (A.54), you are asked the following question:

$$\text{\textit{Check formula? (Press y or n)}} \qquad (A.59)$$

If you want to check to make sure that you are saving the correct formula, press *y*. The formula that is presently stored in memory will then appear. If it is the one you want to save, press *Enter*. If you want to change the formula, you may create a new one and then press *Enter*. This new formula will be the one saved and it becomes the new formula stored in memory. The last step of this procedure is the saving of the formula; the following request for input appears:

$$\textit{File Name:} \qquad\qquad\qquad (A.60)$$

followed by a blinking cursor. You may enter any valid eight-character DOS file name. The extension .UFM is automatically appended to the name you select and the result is the formula file name. The resulting formula will be accessible whenever you are creating a formula by pressing the *F3* key to access the list of formula files.

If you select *Function (ASCII)*, you are asked the same questions, and file names are assigned in the same way, as if you had chosen *function*. The only difference is that the data for this choice is saved in an ASCII text format. This ASCII file is readable by other software (including many graphics programs). The data can also be read by FAS using the *Data* command, as described above.

NOTE

This procedure allows you to create a note on the right side of the screen, which can be used to describe the various graphs displayed. You can type in a note using the letters on the keyboard and using the following special keys to do some simple editing:

Home = return to left margin

End = move to end of line

PgDn = new page (up to 20 pages are allowed)

PgUp = back to previous page

Del = delete character

Ins = toggle between *typeover* and *insert* modes

Up Arrow = move cursor up one line

Down Arrow = move cursor down one line

Left Arrow = move cursor left one space

Right Arrow = move cursor right one space

Enter = start a new line

Esc = end recording of note

If you choose to write a note while one is already displayed, the following menu appears:

$$\textit{Modify} \qquad \textit{Create} \qquad\qquad\qquad (A.61)$$

If you choose *Modify* from menu (A.61), then you can modify the note that is presently displayed. If you choose *Create*, then the present note is erased (it is *lost* unless you have saved it using the *screen* choice from the *SAVE* procedure). You are then free to create a new note using the procedure described above.

Since the *PgDn* key can be used to create new pages in a note (up to a maximum of 20 pages), this same key is used when you return to the display menu to view these pages. You may also use *PgUp* to view a previous page.

HELP

There is a limited amount of on-line help available. If you are at a menu, then you can obtain some information about the choice marked by the cursor by pressing the *F1* key. It is important to keep the file FA.HLP in the same directory as FA.EXE in order for the help to be accessible. If FA.HLP is not in the same directory as FA.EXE then the following message will appear in the Help box:

Help file not found.

Press any key to continue

To obtain more information on a particular topic, you should consult either this user's manual or Chapters 1–5.

SPECIAL CHARACTERS (π and |)

There are two special characters that can be entered easily from the keyboard. To enter π, you just hold down the *Ctrl* key and press *p*. To enter |, which is used for indicating absolute values in captions and notes, hold down the *Ctrl* key and press the key with the backslash character \.

It is important to remember that | is not allowed in formulas (the function abs ( ) is used for computing absolute values in formulas). If you use this character in a formula, then *FAS* will beep and the following error message will appear:

| *or* \ *not allowed in formulas. Press any key to continue*   (*A*.62)

After pressing a key, your formula will reappear and you can correct it.

Note: On low resolution CGA monitors, the symbols π and | will not be visible (a garbage character appears). For using π in formulas, you can get around this problem by entering pi instead. *FAS* will recognize the two letter combination "pi" as a substitute for π.

B

Some Computer Programs

In this appendix we include listings of some computer programs for algorithms discussed in Chapter 3. These programs are:

1. SUB BITREV   This procedure performs bit reversal of complex function data.

2. SUB SINES   This procedure generates the sines necessary for performing the FFTs in FTBFLY, REALFFT, and InvRFFT.

3. SUB FTBFLY   This procedure performs the butterflies needed for performing a complex FFT with weight $e^{i2\pi/N}$. In order for it to perform properly, BITREV must first be called in order to permute your data into bit reversed form. Also, FTBFLY uses the sines generated by the procedure SINES, so SINES must also be called before FTBFLY.

4. SUB REALFFT   This procedure performs an FFT, with weight $e^{i2\pi/N}$, on real function data. Just like FTBFLY, it requires that SINES be called first in order to generate the necessary sines.

5. SUB InvRFFT   This procedure performs the inverse of an FFT obtained from real function data (using the procedure REALFFT). Just like FTBFLY, it requires that SINES be called first in order to generate the necessary sines.

6. SUB COSTRAN   This procedure performs a fast cosine transform of real function data. It requires that SINES be called first in order to generate the necessary sines.

7. SUB SINETRAN   This procedure performs a fast sine transform of real function data. It requires that SINES be called first in order to generate the necessary sines.

These programs are written in the QuickBASIC programming language (a product of Microsoft Corporation). Some of this language's nice features are that there are no line numbers, variables are local to the procedures, and the programs are meant to be *compiled*. Indeed, these procedures behave pretty much like procedures in a structured language like PASCAL. Unfortunately, the author has not made an effort to write these programs in a highly structured format. One unintended benefit of this, however, is that the reader might feel the need to write his or her own programs for FFTs (there is no better way to learn about FFT programming!).

When these programs were compiled and a 1024-point complex FFT was computed on a PC, operating at 16 MHz with a 10-MHz coprocessor, the execution time was about 1.9 seconds. For a 1024-point real FFT, the execution time was about 1.2 seconds.

For the reader's convenience, these programs are also available in an ASCII file (readable by standard programming editors) labeled FA.BAS on the disk accompanying this book.

Procedure declarations

Here are the procedure declarations. These declaration statements belong at the very beginning of any program that utilizes the procedures. (See *Sample Programs* below.)

```
DECLARE SUB BITREV (F#( ), G#( ), R%)
DECLARE SUB SINES (N%, R%, S#( ), Z#)
DECLARE SUB FTBFLY (F1S#( ), F2S#( ), SA#( ), N%, C9%)
DECLARE SUB InvRFFT (N%, R%, F#( ), G#( ), SA#( ))
DECLARE SUB REALFFT (N%, F1#( ), F2#( ), S#( ), R%)
DECLARE SUB SINETRAN (N%, F#( ), FH#( ), SA#( ), R%, ZF#)
DECLARE SUB COSTRAN (N%, F#( ), FH#( ), SA#( ), R%, ZF#)
```

Procedures

Note: In each of these procedures, all of the arrays are assumed to have been declared in such a way that their indices have initial value 0. This is the convention, for example, in the REDIM statement in each of the sample programs listed below.

---

1 The procedure BITREV

This procedure performs bit reversal permutations of the double precision arrays F( ) and G( ). The number of points in each of these arrays is $M = 2^R$. The

positive integer R is passed to the procedure when it is called, as are the arrays
F( ) and G( ).

The first part of this procedure calculates bit reversal numbers for the integer
array J( ), using the method of Buneman. The second part of the procedure (after
the statement label BR) calculates the indices for swapping and performs the
swaps, according to the second algorithm described in Section 2 of Chapter 3.
To increase the speed of the program, the equations described in formulas (2.13)
and (2.17) of that section have been programmed so that successive additions
are performed instead of multiplications.

```
DEFINT A–D, H–N, P–R, Y
DEFDBL E–G, O, S–X, Z
SUB BITREV (F( ), G( ), R)
M = 2 ∧ R: N2 = M / 2: N = M: M7 = 0: R2 = R / 2: N1 = N: Y = 0: C = 1
IF R MOD 2 = 1 THEN N1 = N2: R2 = (R − 1) / 2: C = 0
N9 = SQR(N1): IF R MOD 2 = 1 THEN N8 = N9 + N9 ELSE N8 = N9
' $STATIC
DIM J(64)
J(0) = 0: J(1) = 1: L9 = 2
FOR I9 = 2 TO R2
FOR J9 = 0 TO L9 − 1
J(J9) = J(J9) + J(J9): J(J9 + L9) = J(J9) + 1
NEXT J9: L9 = L9 + L9
NEXT I9
BR:
FOR L = 1 TO N9 − 1
M7 = M7 + N8: N6 = J(L) + Y: T = G(M7): G(M7) = G(N6): G(N6) = T
T = F(M7): F(M7) = F(N6): F(N6) = T
FOR K = 1 TO L − 1
M6 = M7 + J(K): N6 = N6 + N8
T = G(M6): G(M6) = G(N6): G(N6) = T
T = F(M6): F(M6) = F(N6): F(N6) = T
NEXT K
NEXT L
Y = N9: M7 = N9: C = C + 1: IF C < 2 THEN GOTO BR
END SUB
```

2   The procedure SINES

This procedure calculates the values of the sines needed for FFTs and stores
them in the double precision array S( ). The integer R has the value of the

power of 2 used for FFTs, and the integer N is the number of points $(N = 2^R)$. The program uses Buneman's recursion procedure for computing sines for FFTs (see section 4 of Chapter 3). The double precision variable Z retains the last value of Z in that recursion procedure; it is needed if COSTRAN or SINETRAN is called.

```
DEFINT A–D, H–N, P–R, Y
DEFDBL E–G, O, S–X, Z
SUB SINES (N, R, S( ), Z)
Z = 0: N2 = N / 2: H = N2 / 2: S(0) = 0: S(H) = 1: D = 1
FOR J = 0 TO R − 3
Z = SQR(2 + Z): H9 = H: H = H / 2: L = H
FOR K = 1 TO D
S(L) = (S(L + H) + S(L − H)) / Z
L = H9 + L
NEXT K: D = D + D
NEXT J
END SUB
```

3   The procedure FTBFLY

This procedure computes a complex FFT using weight $e^{i2\pi/N}$. The double precision arrays F1S( ) and F2S( ) are the real and imaginary parts of the complex function data. The double precision array SA( ) holds the sines needed for doing FFTs. The integer N is the number of points. The integer C9 is the step size for walking through angles (retrieving their sines). For a complex FFT, C9 should be given the value 1 (see the sample programs below). For a real FFT, C9 is given the value 2 (see the calling of FTBFLY in the procedure REALFFT).

The first part of the program (up to statement label BY) executes the first stage of the FFT, which consists of $(1/2)N (= N2)$ 2-point DFTs. Those DFTs only involve additions and subtractions, so in order to save time they are programmed separately from the other DFTs in the FFT algorithm.

The second part of the program (from statement BY to the end) performs the higher order DFTs in the remaining $\log_2 N - 1$ stages of the FFT. The first C-loop contains the butterflies which only involve multiplication by i or $-i$. *In terms of real and imaginary parts*, these are not really multiplications. Therefore, for greater speed, they are programmed separately. The rest of the program performs the butterflies for which Buneman's algorithm, described in Section 3 of Chapter 4, is needed. The real and imaginary parts of the FFT are returned in the arrays F1S( ) and F2S( ), respectively.

Note: There is a simple algorithm for inverting this FFT; we described it in Exercise 3.14. (See Sample Program 2 below.)

```
DEFINT A–D, H–N, P–R, Y
DEFDBL E–G, O, S–X, Z
SUB FTBFLY (F1S( ), F2S( ), SA( ), N, C9)
N2 = N / 2: L = 0: M2 = (C9 * N) / 2: M4 = M2 / 2
FOR K = 0 TO N2 − 1
T1 = F1S(L) + F1S(L + 1): T2 = F2S(L) + F2S(L + 1)
F1S(L + 1) = F1S(L) − F1S(L + 1): F2S(L + 1) = F2S(L) − F2S(L + 1)
F1S(L) = T1: F2S(L) = T2: L = L + 2
NEXT K
Q = C9 * N2: D = 2: Q2 = N2
BY: Q2 = Q2 / 2: Q = Q / 2: D2 = D: D = D2 + D2
A1 = 0: L = 0: A = A1: D4 = D2 / 2
FOR C = 1 TO Q2
B1 = A + D2
T = F1S(B1): S = F2S(B1): F1S(B1) = F1S(A) − T: F2S(B1) = F2S(A) − S
F1S(A) = F1S(A) + T: F2S(A) = F2S(A) + S
AV = A + D4: BV = B1 + D4
T = F1S(BV): S = F2S(BV): F1S(BV) = F1S(AV) + S: F2S(BV) = F2S(AV) − T
F1S(AV) = F1S(AV) − S: F2S(AV) = F2S(AV) + T: A = A + D
NEXT C
L = L + Q: A1 = A1 + 1: A = A1
FOR K = 2 TO D4
SL = SA(L): TL = (1 − SA(M4 − L)) / SL
FOR C = 1 TO Q2
B1 = A + D2: V = F2S(B1) + TL * F1S(B1): T3 = F1S(B1) − V * SL
T4 = T3 * TL + V: F1S(B1) = F1S(A) − T3: F2S(B1) = F2S(A) − T4
F1S(A) = F1S(A) + T3: F2S(A) = F2S(A) + T4: AV = A + D4
BV = B1 + D4: V = F1S(BV) − TL * F2S(BV): T3 = −F2S(BV) − V * SL
T4 = T3 * TL + V: F1S(BV) = F1S(AV) − T3: F2S(BV) = F2S(AV) − T4
F1S(AV) = F1S(AV) + T3: F2S(AV) = F2S(AV) + T4: A = A + D
NEXT C: L = L + Q: A1 = A1 + 1: A = A1
NEXT K
IF D < N THEN GOTO BY
END SUB
```

4 The procedure REALFFT

This procedure produces the FFT of the real data contained in the double precision array F1( ). The double precison array S( ) contains the sines created by the SINES procedure. And, the integer R is the power of 2.

The first part of the program creates a new set of data by putting the odd indexed values of F1( ) into the first half of F2( ) and putting the even-indexed values of F1( ) into the first half of F1( ). Then, a $\frac{1}{2}$N-point FFT is performed by calling BITREV and FTBFLY (note that C9 is fed the value 2).

The remainder of the program performs the butterflies needed for the last stage of the real FFT algorithm and completes the FFT calculation according to equations (6.3)–(6.5) of Chapter 3. The real and imaginary parts of the FFT are returned in the arrays F1( ) and F2( ), respectively.

```
DEFINT A-D, H-N, P-R, Y
DEFDBL E-G, O, S-X, Z
SUB REALFFT (N, F1( ), F2( ), S( ), R)
N2 = N / 2: K = 0: N4 = N2 / 2
FOR I = 0 TO N - 1 STEP 2
F1(K) = F1(I): F2(K) = F1(I + 1): K = K + 1
NEXT I
CALL BITREV(F1( ), F2( ), R-1)
CALL FTBFLY(F1( ), F2( ), S( ), N2, 2)
FOR L = 1 TO N4 - 1
SL = S(L): SLQ = S(N4 - L): TL = (1 - SLQ) / SL
T1 = F1(L) - F1(N2 - L): T2 = F2(L) + F2(N2 - L)
V = T2 + TL * T1: T3 = T1 - V * SL: T4 = T3 * TL + V
F1(N - L) = (F1(L) + F1(N2 - L) + T4) / 2
F2(N - L) = (-F2(L) + F2(N2 - L) + T3) / 2
LJ = N4 + L
SL = S(N2 - LJ): TL = SL / (1 - S(LJ - N4))
T1 = F1(LJ) - F1(N2 - LJ): T2 = F2(LJ) + F2(N2 - LJ)
V = T2 + TL * T1: T3 = T1 - V * SL: T4 = T3 * TL + V
F1(N - LJ) = (F1(LJ) + F1(N2 - LJ) + T4) / 2
F2(N - LJ) = (-F2(LJ) + F2(N2 - LJ) + T3) / 2
NEXT L
F1(N - N4) = F1(N4): F2(N - N4) = -F2(N4)
FOR L = 1 TO N2 - 1
F1(L) = F1(N - L): F2(L) = -F2(N - L)
NEXT L  F1(N2) = F1(0) - F2(0): F1(0) = F1(0) + F2(0): F2(0) = 0: F2(N2) = 0
END SUB
```

---

5   The procedure InvRFFT

This procedure inverts a real FFT (obtained from the procedure REALFFT). The integer N is the number of points. The integer R is the power of 2, the double precision arrays F( ) and G( ) are the real and imaginary parts of the complex function data. And, the double precision array SA( ) contains the sines generated by the SINES procedure.

The first part of the program (up to the calling of BITREV and FTBFLY) involves the creation of the auxiliary arrays for doing Fast Sine and Fast Cosine Transforms (see Sections 9 and 7 of Chapter 3). Also, the starting values for the recursions in those algorithms are computed in this part of the procedure.

The second part of the procedure involves the simultaneous calculation of the Fast Sine and Cosine Transforms (of order $(1/2)N$). Both involve doing an FFT on the auxiliary arrays (which are *real*). Therefore, these two real arrays are FFT'd simultaneously. This is performed by calling BITREV and FTBFLY. The K-loop that follows performs the separation of the two FFTs (as described in Section 5 of Chapter 3). The remainder of the program implements the two recursion formulas for Fast Sine and Cosine Transforms, and then reconstructs the inverse FFT from them (as described in Section 9 of Chapter 3).

Note: The output of this procedure should be thought of as the array F( ). It will be the same (except for rounding errors) as the real array F1( ) which was initially input for the procedure REALFFT. There is no guarantee, however, that the output G( ) will be identically zero.

```
DEFINT A–D, H–N, P–R, Y
DEFDBL E–G, O, S–X, Z
SUB InvRFFT (N, R, F( ), G( ), SA( ))
N2 = N / 2: N4 = N / 4: SPV2 = F(N2): SPV1 = −F(0)
J = 1: SPV3 = 0: SPV4 = 0
FOR K = 1 TO N2 − 1
J = −J: SPV3 = SPV3 + J * F(K): SPV4 = SPV4 + F(K)
NEXT K
SPV3 = SPV3 + SPV3: SPV4 = SPV4 + SPV4
SPV3 = F(0) + F(N2) + SPV3: SPV4 = SPV4 + SPV2 − SPV1
F1 = F(0): G(0) = 0: G1 = 0
FOR K = 1 TO N4 − 1
F1 = F1 + SA(K) * (F(N4 − K) − F(N4 + K))
NEXT K
FOR K = 1 TO N4 − 1
T = SA(K)
S = (F(K) + F(N2 − K)) / 2: U = F(K) − F(N2 − K)
F(K) = S − T * U: F(N2 − K) = S + T * U
S1 = (G(K) − G(N2 − K)) / 2: U1 = T * (G(K) + G(N2 − K))
G(K) = S1 + U1: G1 = G1 + U1: G(N2 − K) = U1 − S1
NEXT K
U = G(N4): G1 = G1 + U: G(N4) = U + U
CALL BITREV(F( ), G( ), R − 1)
CALL FTBFLY(F( ), G( ), SA( ), N2, 2)
FOR K = 1 TO N4 − 1
F(N2 + K) = G(K) − G(N2 − K): F(N − K) = G(N2 − K) + G(K)
NEXT K
G(1) = F1 + F1
FOR K = 1 TO N4 − 1
G(K + K) = F(N2 − K) + F(K): G(K + K + 1) = G(K + K − 1) + F(N2 + K)
```

```
F(N2 + K) = F(N2 - K) - F(K)
NEXT K
F(1) = G1 + G1
FOR K = 1 TO N4 - 1
F(K + K) = F(N2 + K)
F(K + K + 1) = F(K + K - 1) + F(N - K)
NEXT K
F(0) = SPV4
FOR K = 1 TO N2 - 1
SPV2 = -SPV2: U = SPV1 + SPV2 + G(K)
F(N - K) = U - F(K): F(K) = U + F(K)
NEXT K
F(N2) = SPV3
END SUB
```

6 The procedure COSTRAN

This procedure computes a Fast Cosine Transform of the real data contained in the double precision array F( ). The other double precision array FH( ) is auxiliary to the procedure (it should have all values equal to 0 when it is input to this procedure). The double precision array SA( ) holds the sines generated by the SINES procedure. The integer R is the power of 2. The double precision variable ZF is to be given the value of Z obtained from the SINES procedure.

The first part of the program forms the auxiliary array and computes a starting value for the recursion used in the Fast Cosine Transform algorithm described in Section 7 of Chapter 3. Then, a real FFT is done on this auxiliary array. After REALFFT is called, the program finishes by recursively computing the Fast Cosine Transform. The values of this Fast Cosine Transform are then contained in the array F( ).

```
DEFINT A-D, H-N, P-R, Y
DEFDBL E-G, O, S-X, Z
SUB COSTRAN (N, F( ), FH( ), SA( ), R, ZF)
N2 = N / 2: F1 = F(0): V = 1 / SQR(2 + ZF)
FOR K = 1 TO N2 - 1
IF K MOD 2 = 1 THEN
T = (SA((K + 1) / 2) + SA((K - 1) / 2)) * (-V) ELSE
T = SA(K / 2)
END IF
F1 = F1 + T * (F(N2 - K) - F(N2 + K))
NEXT K
FOR K = 1 TO N2 - 1
IF K MOD 2 = 1 THEN
T = (SA((K + 1) / 2) + SA((K - 1) / 2)) * V
```

```
ELSE
T = SA(K / 2)
END IF
S = (F(K) + F(N − K)) / 2: U = F(K) − F(N − K)
F(K) = S − T * U: F(N − K) = S + T * U
NEXT K
CALL REALFFT(N, F( ), FH( ), SA( ), R)
F(N − 1) = F1
FOR K = N2 − 1 TO 1 STEP −1
F(K + K) = F(K): F(K + K − 1) = F(K + K + 1) − FH(K)
NEXT K
END SUB
```

---

7   The procedure SINETRAN

This procedure computes a Fast Sine Transform of the real data contained in the double precision array F( ). The other double precision array FH( ) is auxiliary to the procedure (it should have all values equal to 0 when it is input to this procedure). The double precision array SA( ) holds the sines generated by the SINES procedure. The integer R is the power of 2. The double precision variable ZF is to be given the value of Z obtained from the SINES procedure.

The first part of the program forms the auxiliary array and computes a starting value for the recursion used in the Fast Sine Transform algorithm described in Section 7 of Chapter 3. Then, a real FFT is done on this auxiliary array. After REALFFT is called, the program finishes by recursively computing the Fast Sine Transform. The values of this Fast Sine Transform are then contained in the array F( ).

```
DEFINT A–D, H–N, P–R, Y
DEFDBL E–G, O, S–X, Z
SUB SINETRAN (N, F( ), FH( ), SA( ), R, ZF)
F(0) = 0: N2 = N / 2: V = 1 / SQR(2 + ZF): J = −1
FOR K = 1 TO N2 − 1
IF K MOD 2 = 1 THEN
T = (SA((K + 1) / 2) + SA((K − 1) / 2)) * V
ELSE
T = SA(K / 2)
END IF
S = (F(K) − F(N − K)) / 2: U = T * (F(K) + F(N − K)) J = −J
F(K) = S + U: F1 = F1 + J * U: F(N − K) = U − S
NEXT K
U = F(N2): F1 = F1 − U: F(N2) = U + U
```

```
CALL REALFFT(N, F( ), FH( ), SA( ), R)
F(N - 1) = F1: F(0) = 0: F(N) = 0
FOR K = N2 - 1 TO 1 STEP -1
F(K + K - 1) = F(K + K + 1) - F(K): F(K + K) = FH(K)
NEXT K
END SUB
```

Sample programs

Here are two sample programs that utilize the procedures described above. These programs are not included in the file FA.BAS.

Sample Program 1

This program checks that the complex FFT and the real FFT of a sequence of real data produce the same results. The author used it to obtain the FFT execution times given in the introduction to this appendix.

```
DEFINT A–D, H–N, P–R, Y
DEFDBL E–G, O, S–X, Z
DECLARE SUB BITREV (F#( ), G#( ), R%)
DECLARE SUB SINES (N%, R%, S#( ), Z#)
DECLARE SUB FTBFLY (F1S#( ), F2S#( ), SA#( ), N%, C9%)
DECLARE SUB REALFFT (N%, F1#( ), F2#( ), S#( ), R%)
CLS
INPUT "Enter an integer from 1 to 12: ", IR
CLS
M = 2 ∧ IR
REDIM FR(M), FI(M), SN(M / 4 + 1)
FOR J = 0 TO M - 1
FR(J) = EXP(-J): FI(J) = 0
NEXT J
T = TIMER
CALL SINES(M, IR, SN( ), ZI)
CALL BITREV(FR( ), FI( ), IR)
CALL FTBFLY(FR( ), FI( ), SN( ), M, 1)
T = TIMER - T
PRINT "Complex FFT, "; M; "Points. Elapsed time "; T; "seconds"
PRINT
IF M < 16 THEN
PRINT "Index", "Real Part", "Imaginary Part"
FOR J = 0 TO M - 1
PRINT J, CSNG(FR(J)), CSNG(FI(J))
```

```
NEXT J
END IF
REDIM FR(M), FI(M), SN(M / 4 + 1)
FOR J = 0 TO M - 1
FR(J) = EXP(-J): FI(J) = 0
NEXT J
T = TIMER
CALL SINES(M, IR, SN( ), ZI)
CALL REALFFT(M, FR( ), FI( ), SN( ), IR)
T = TIMER - T
PRINT "Real FFT, "; M; "Points. Elapsed time "; T; "seconds"
IF M < 16 THEN
PRINT
PRINT "Index", "Real Part", "Imaginary Part"
FOR J = 0 TO M - 1
PRINT J, CSNG(FR(J)), CSNG(FI(J))
NEXT J
END IF
END
```

Sample Program 2

This program shows how to invert a complex FFT, without generating any new sines. It implements the method described in Exercise 3.14.

```
DEFINT A-D, H-N, P-R, Y
DEFDBL E-G, O, S-X, Z
DECLARE SUB BITREV (F#( ), G#( ), R%)
DECLARE SUB SINES (N%, R%, S#( ), Z#)
DECLARE SUB FTBFLY (F1S#( ), F2S#( ), SA#( ), N%, C9%)
CLS
INPUT "Enter an integer from 1 to 12: ", IR
CLS
M = 2 ^ IR
REDIM FR(M), FI(M), SN(M / 4 + 1)
FOR J = 0 TO M - 1
FR(J) = J: FI(J) = -J
NEXT J
CALL SINES(M, IR, SN( ), ZI)
CALL BITREV(FR( ), FI( ), IR)
CALL FTBFLY(FR( ), FI( ), SN( ), M, 1)
PRINT "Complex FFT, "; M; "Points"
PRINT
PRINT "Index", "Real Part", "Imaginary Part"
FOR J = 0 TO M - 1
PRINT J, CSNG(FR(J)), CSNG(FI(J))
NEXT J
```

```
FOR J = 0 TO M − 1
FI(J) = −FI(J)
NEXT J
CALL BITREV(FR( ), FI( ), IR)
CALL FTBFLY(FR( ), FI( ), SN( ), M, 1)
FOR J = 0 TO M − 1
FR(J) = FR(J) / M: FI(J) = −FI(J) / M
NEXT J
PRINT
PRINT "Inverse Complex FFT"; M; "Points"
PRINT
PRINT "Index", "Real Part", "Imaginary Part"
FOR J = 0 TO M − 1
PRINT J, CSNG(FR(J)), CSNG(FI(J))
NEXT J
END
```

Bibliography

[Ba]   Bartle, R. W., *The Elements of Real Analysis*. Wiley, New York, 1965.

[Bo-W]   Born, M. and E. Wolf., *Principles of Optics*. Pergamon, Oxford, 1965.

[Br]   Bracewell, R. N., *The Fast Hartley Transform*. Oxford University Press, Oxford, 1986.

[Br,2]   Bracewell, R. N., *The Fourier Transform and its Applications*. McGraw-Hill, New York, 1978.

[Bri]   Brigham, E. O., *The Fast Fourier Transform*. Prentice-Hall, Englewood Cliffs, New Jersey, 1974.

[Ch-B]   Churchill, R. V. and J. W. Brown, *Fourier Series and Boundary Value Problems*. McGraw-Hill, New York, 1978.

[Da]   Davis, H. F., *Fourier Series and Orthogonal Functions*. Allyn and Bacon, Boston, 1963.

[Da-H]   Davis, P. J. and R. Hersh, *The Mathematical Experience*. Houghton-Mifflin, 1982.

[El-R]   Elliot, D. F. and K. R. Rao, *Fast Fourier Transforms: Algorithms, Analysis, and Applications*. Academic Press, New York, 1982.

[Fe]   Feynman, R. P., *QED, The Strange Theory of Light and Matter*. Princeton University Press, Princeton, NJ, 1985.

[Fo]   Fourier, J., *The Analytical Theory of Heat*, translated by A. Freeman. Dover, New York, 1955.

[Go]   Goodman, J. W., *Introduction to Fourier Optics*. McGraw-Hill, New York, 1968.

[Go,2]   Goodman, J. W., *Statistical Optics*. Wiley, New York, 1985.

[Ha]   Hamming, R. W., *Digital Filters*. Prentice-Hall, Englewood Cliffs, New Jersey, 1977.

[Ha-e]   Harburn, G., C. A. Taylor, and T. R. Welberry, *Atlas of Optical Transforms*. Cornell University Press, Ithaca, New York, 1975.

[Ii]   Iizuka, K., *Engineering Optics*. Springer-Verlag, New York, 1985.

[Je]   Jerri, A., "The Shannon Sampling Theorem — Its Various Extensions and Applications: A Tutorial Review." *Proceedings of the IEEE* **65**(11): 1565 (1977).

[Ka]   Katznelson, Y., *An Introduction to Harmonic Analysis*. Wiley, New York, 1968.

[Me]   Meyer-Arendt, J. R., "Microscopy as a Spatial Filtering Process," *Advances in Optical and Electron Microscopy*, Vol. 8, Academic Press, London, 1982.

[Mi]   Misell, D. L., "The Phase Problem in Electron Microscopy," *Advances in Optical and Electron Microscopy*, Vol. 7, Academic Press, London, 1978.

[Mi-T]   Mills, J. P. and B. J. Thompson, "Effect of Aberrations and Apodizations on the Performance of Coherent Optical Systems (I and II)." *Journal of the Optical Society of America, A* **3**: 694 (1986).

[Mo]   Monforte, J., "The Digital Reproduction of Sound." *Scientific American* **251**(6): 78 (1986).

[Nu]   Nussbaumer, H. J., *Fast Fourier Transform and Convolution Algorithms*. Springer-Verlag, New York, 1982.

[Op-S]   Oppenheim, A. V. and R. W. Schaffer, *Digital Signal Processing*. Prentice-Hall, Englewood Cliffs, New Jersey, 1975.

[Pi]   Pincus, H. J., "Optical Diffraction Analysis in Microscopy." *Advances in Optical and Electron Microscopy*, Vol. 7, Academic Press, London, 1978.

[Pr-e]   Press, W. H., B. P. Flannery, S. A. Teukolsky, and W. T. Vetterling, *Numerical Recipes*. Cambridge University Press, Cambridge, 1986. [*Note:* There are editions of this book with computer programs in *Fortran*, *Pascal*, or *C*.]

[Ra]   Rao, K. R. (Ed.), *Discrete Fourier Transforms and their Applications*. Van Nostrand Reinhold, New York, 1985.

[Ra-G]   Rabiner, L. R. and B. Gold, *Digital Signal Processing*. Prentice-Hall, Englewood Cliffs, New Jersey, 1975.

[Ra-R]   Rabiner, L. R. and C. M. Radar (Eds.), *Digital Signal Processing*. IEEE Press, New York, 1972.

[Ru]   Rudin, W., *Real and Complex Analysis*. McGraw-Hill, New York, 1974.

[Ru,2]   Rudin, W., *Principles of Mathematical Analysis*. McGraw-Hill, New York, 1964.

[Sn]   Sneddon, I. N., *Fourier Transforms*. McGraw-Hill, New York, 1951.

[St]   Stearns, S. D., *Digital Signal Processing*. Hayden Book Co., New York, 1975.

[Str] Strang, G., *Introduction to Applied Mathematics*. Wellesley-Cambridge Press, Wellesley, Mass., 1986.

[St-W] Stein, E. and G. Weiss, *Fourier Analysis on Euclidean Spaces*. Princeton University Press, Princeton, NJ, 1971.

[To] Tolstov, G. P., *Fourier Series*. Prentice-Hall, Englewood Cliffs, New Jersey, 1962.

[Wa] Walker, J. S., *Fourier Analysis*. Oxford University Press, Oxford, 1988.

[Wa,2] Walker, J. S., "A New Bit Reversal Algorithm." *IEEE Transactions in Acoustics, Speech, and Signal Processing* **38**(8): 1472 (1990).

[We] Weinberger, H. F., *A First Course in Partial Differential Equations*. Wiley, New York, 1965.

[Zy] Zygmund, A., *Trigonometric Series*. Cambridge University Press, Cambridge, 1968.

Index